Spices, Plantation Crops, Medicinal and Aromatic Plants: A Handbook

About the Author

S.K. Tyagi is working as subject Matter Specialist (Horticulture) at RVSKVV, Krishi Vigyan Kendra, Khargone (Madhya Pradesh). He has qualified National Eligibility Test (NET) twice i.e. in the year 2001 and 2006 respectively, conducted by Agriculture Scientists Recruitment Board, New Delhi. Author has more than ten years of experience in the field of research, farm management and agriculture extension. 08 research papers, 20 abstracts, 06 books, 02 training manuals, 10 success stories, 09 extension bulletins, 10 folders and 130 popular articles goes to his credit, which have been published in various reputed journal and magazines. His contribution in the subject was recognized the Best Writer Award-2009 bestowed by Bhoomi Nirman National News Paper, Bhopal (Madhya Pradesh). The Book Authored by him would go a long way in serving the interest of students

Spices, Plantation Crops, Medicinal and Aromatic Plants: A Handbook

S.K. Tyagi
SMS, Horticulture
KVK, Khargone, Madhya Pradesh

NEW INDIA PUBLISHING AGENCY
New Delhi – 110 034

NEW INDIA PUBLISHING AGENCY
101, Vikas Surya Plaza, CU Block, LSC Market
Pitam Pura, New Delhi 110 034, India
Phone: + 91 (11)27 34 17 17 Fax: + 91(11) 27 34 16 16
Email: info@nipabooks.com
Web: www.nipabooks.com

Feedback at feedbacks@nipabooks.com

ISBN: 978-81-930144-8-6

Typeset and Designed by New Delhi, India

Printed and bound in India by Replika Press Pvt. Ltd.

Preface

Precise database pertinent to related subject is an essential pre-requisite to get through competitive examinations. The competitive examinations have become procedure of recruitment and admission for higher studies such as JRF, SRF, NET, Ph.D., ARS, and other competitive examinations. All these examinations are mostly based on objective type questions and students always look for study material that is ready to use and easy to grasp. Thus, one has to prepare him / her very strongly for these competitive exams and for the success, need to read authentic and authorised reading materials. It was long felt need of the aspirants that there should be a vast collection of recent information and data related to Spices, Plantation Crops, Medicinal and Aromatic Plants in a single book. Keeping this in mind the manuscript has been prepared. Humble efforts have been made here to compile basic and applied knowledge of these crops, which will be of immense value and help for all those who want to pursue career in this field.

This book is a single compendium which deals with all aspects and facts of Spices, Plantation Crops, Medicinal and Aromatic Plants which may meet the requirements of all those who are pursuing for JRF, SRF, NET, Ph.D., ARS, and other competitive examinations.

The book has been compiled on the basis of available information for guidance and not for legal purposes. Readers are also advised to consult the latest recommendation for pesticides/herbicides before their use and for the latest botanical and zoological names site www.sp2000.org may be consulted.

However, I have put my best efforts in preparing this book, but if any error or whatsoever has been skipped out, I will by heart welcome your suggestions with the hope that this book will be extremely useful to the students.

S.K. Tyagi

Contents

Preface .... *v*

1. Spices and Condiments

1. Importance of Spices and Condiments 1
2. Technical Information on Spices and Condiments 7
3. Production Techniques of Spices and Condiments 11
4. Recent Initiatives in Crop Improvement 87
5. Recent Initiatives in Crop Production 103
6. Recent Initiatives in Crop Protection 117
7. Recent Initiatives in Post Harvest Technology 135
8. Diseases of Major Spices and Condiments 139
9. Insect Pests of Major Spices and Condiments 151

2. Plantation Crops

10. Importance of Plantation Crops 159
11. Technical Information on Plantation Crops 165
12. Production Techniques of Plantation Crops 167
13. Recent Initiatives in Crop Improvement 213

14. Recent Initiatives in Crop Production 231

15. Recent Initiatives in Crop Protection 251

16. Recent Initiatives in Post Harvest Technology 273

17. Diseases of Major Plantation Crops 277

18. Insect Pests of Major Plantation Crops 291

3. Medicinal and Aromatic Plants

19. Importance of Medicinal and Aromatic Plants 315

20. Technical Information on Medicinal Plants 323

21. Production Techniques of Medicinal Plants 329

22. Technical Information on Aromatic Plants 371

23. Production Techniques of Aromatic Plants 375

24. Recent Initiative in Crop Improvement 399

25. Recent Initiative in Crop Production 401

26. Recent Initiatives in Crop Protection 409

27. Diseases of Major Medicinal and Aromatic Plants 413

28. Insect Pests of Major Medicinal and Aromatic Plants 421

Glossary 427

Appendices 433

Appendix-I: *List of Spices under the Purview of the Spices Board* .. 433

Appendix-II: *Item-Wise Export of Spices from India* 435

Appendix-III: *Major Item/Country-Wise Export of Spices from India* 436

Appendix -IV: *List of 178 Medicinal Plant Species in High Volume Trade* 450

Appendix-V: *ICAR Institutions, Deemed Universities, National Research Centres, National Bureaux, Directorate/Project Directorates* 460

Appendix-VI: *List of Authorities/Agencies/Institutions/ Functionaries* 464

Appendix-VII: *International Agricultural Organizations* 465

Appendix-VIII: *List of Agricultural Universities* 467

Appendix-IX: *List of Various Boards/ organizations* 470

Appendix-X: *List of National and International Horticulture's Journals and Magazines* 472

Appendix-XI: *List of Acts and Rules* 473

Appendix-XII: *Agricultural revolutions in India* 476

Appendix-XIII: *Popular Name of Horticultural Crops* 477

Appendix XIV: *List of People Considered Father of a Scientific Field* 478

Appendix XV: *List of Pesticides / Pesticides Formulations Banned in India* 481

Appendix XVI: *Conversion Chart* 484

References 485

List of Horticulture Books 487

1. Spices and Condiments

1

Importance of Spices and Condiments

- The term Spice and Condiments applies to natural plant or vegetable products or mixtures in whole or ground form, which are used for imparting flavour, aroma and piquancy to the food items.
- Under the act of Parliament, a total of 52 spices are brought under the purview of Spices Board.
- However 109 spices are notified in the ISO list.
- India is the largest producer, consumer and exporter of spices and spice products in the world and produces more than 50 spices.
- The spices production in India is of the order of 5.74 million MT from an area of about 3.07 million ha.
- Chilli is the major spice crop occupying about 25.8 percent of area under cultivation and contributing about 22.7 percent of total spices production in the country.
- Garlic accounts for 21.9% of production and 8.1% of area Turmeric accounts for 16.9% of production and 6.3% of area.
- Rajasthan occupies major area under spices owing to seed spices cultivation in the state followed by Gujarat, Andhra Pradesh, Madhya Pradesh and Karnataka.
- Andhra Pradesh, which is the foremost state producing chilli and turmeric in the country, ranks first in terms of production.

- Indian spices exports have been able to record strident gains in both volume and value spices exports have registered substantial growth during the last five years, registering an compound annual average growth rate of 21% in value and 12% in volume and India commands a formidable position in the World Spice Trade.
- During 2013-14, a total of 8,17,250 tons of spices and spice products valued Rs.13735.39 crore (US$2267.67 Million) has been exported from the country as against 7,26,613 tons valued Rs.12112.76 crore (US$ 2212.13 Million) in 2012-13 registering an increase of 12% in volume and 13% in rupee terms and 3% in dollar terms of value.
- The total export of Spices during 2013-14 has exceeded the target in terms of both quantity and value. Compared to the target of 6,24,700 tons valued Rs.10394.00 crore (US$1825 million) for the financial year 2013-14, the achievement is 131% in terms of quantity and 132 % in rupee and 124% dollar terms of value.
- Spices and aromatic plants are mainly used for imparting flavour, aroma, pungency and for seasoning the food.
- Many of them are also used in medicines and in perfumery.
- India is considered as the magical land of spices.
- Black pepper, cardamom, ginger, turmeric, capsicum, cinnamon, clove, nutmeg, tamarind, and vanilla constitute the major spices.
- Seed spices like coriander, cumin, fennel, fenugreek, dill, caraway, anise and herbal spices like saffron, lavender, thyme, oregano, celery, anise, sage and basil are also important.
- India is the native home of many important spices like black pepper, cardamom, tamarind, curry leaf and to certain extent ginger, turmeric, garcinia and cinnamon where a good variability exists.
- In fact there is no state in India that does not grow spices, which in turn play an important role for the lives of the people and for their own economic sustainability.
- From the Indian sub-continent these spices spread over to most of the tropical part of the world.
- Because of the varying climates - from tropical to sub-tropical to temperate- almost all spices grow splendidly in India.
- In reality almost all the states and union territories of India grow one or the other spices.

- Spices and condiments are a major commercial crop in India, and earn a major part of foreign exchange annually.

Spices Board India

- Spices Board (Ministry of Commerce and Industry, Government of India) is the flagship organization for the development and worldwide promotion of Indian spices.
- The Board is an international link between the Indian exporters and the importers abroad.
- The Board has been spearheading activities for excellence of Indian spices, involving every segent of the industry.
- The Board has made quality and hygiene the corner stones for its development and promotional strategies.

Area, production and productivity of major spices (2012-13)

S. No.	Crop	Area in 000' ha	% share of total area	Production in 000' MT	% share of total production	Productivity MT/ha
1.	Chillies	794	25.8	1304	22.7	1.6
2.	Garlic	248	8.1	1259	21.9	5.1
3.	Turmeric	194	6.3	971	16.9	5.0
4.	Ginger	136	4.4	683	11.9	5.0
5.	Coriander	543	17.7	524	9.1	1.0
Total including others		3075.9		5743.5		1.9

Source: Indian Horticulture Database 2013

State-wise area, production & productivity of spices (2012-13)

S. No.	State	Area in 000' ha	% share of total area	Production in 000' MT	% share of total Production	Productivity MT/ha
1.	Andhra Pradesh	312.6	10.2	1187.7	20.7	3.8
2.	Gujarat	551.7	17.9	882.1	15.4	1.6
3.	Rajasthan	720.6	23.4	860.9	15.0	1.2
4.	Karnataka	213.8	7.0	369.9	6.4	1.7
5.	Madhya Pradesh	299.9	9.7	461.2	8.0	1.5
6.	Total including others	3075.9		5743.5		1.9

Source: Indian Horticulture Database 2013

Crop wise area, production and productivity of spices (2012-13)

S. No.	Crop	India (2012-13)					Leading country in the world in production	India's rank in the world in production
		Area in 000' ha	Production in 000' MT	PDY M T/ha	Leading state in area	Leading state in production		
1.	Chilli	794	1304	1.6	Andhra Pradesh	Andhra Pradesh	India	I
2.	Garlic	248	1259	5.1	Rajasthan	Gujarat	China	II
3.	Turmeric	194	971	5.0	Andhra Pradesh	Andhra Pradesh		
4.	Ginger	136	683	5.0	Karnataka	Odisha	India	I
5.	Coriander	543	524	1.0	Rajasthan	Rajasthan		
6.	Cumin	594	394	0.7	Rajasthan	Gujarat		
7.	Tamarind	58	202	3.5	Tamil Nadu	Karnataka		
8.	Fennel	100	143	1.4	Gujarat	Gujarat		
9.	Fenugreek	93	113	1.2	Rajasthan	Rajasthan		
10.	Pepper	125	53	o.4	Kerala	Kerala	Vietnam	III
11.	Ajwan	35	27	0.8	Rajasthan	Rajasthan		
12.	Cardamom (small)	92	18	0.2	Kerala	Kerala		
13.	Tejpat/cinnamon	3	5	1.7	Meghalaya	Meghalaya		
14.	Cloves	2	1	0.5	Kerala	Tamil Nadu	Indonesia	
15.	Nutmeg	17.49	12.62		Kerala	Kerala	Guatemala	III
16.	Saffron / Vanilla	7.10	1.07		Jammu & Kashmir	Jammu & Kashmir		
Total including others		3075.9	5743.5	1.9				

Source: Indian Horticulture Database 2013 for World Data, FAO Website, December 2014

□□□□

2

Technical Information on Spices and Condiments

Common name, botanical name, family, centre of origin, chromosome number, type of inflorescence, type of fruit, pollination system, propagation, commercial part, phytochemical constituents of major spices and condiments

S. No	Common name	Botanical name	Family	Centre of origin	Chromosome number	Type of inflorescence	Type of fruit	Pollination system	Propagation	Commercial part	Main Phytochemical constituents
1.	Garlic	*Allium sativum* L.	Alliaceae	Central Asia	16	Umbel		Cross pollination	Cloves	Bulbs	Allicin (diallyl disulphide oxide)
2.	Cumin	*Cuminum cyminum* Linn.	Apiaceae	Mediterranean region	2n=14	Compound umbel	Schizocarp	Cross-pollination	Seeds	Seed	Cuminol (Cumaldehyde)
3.	Fennel	*Foeniculum vulgare* Mill.	Apiaceae	Southern Europe and the Mediterranean area	2n=22	Umbel		Cross-pollination	Seeds	Seed	Anethole and d-fenchone
4.	Dill	*Anethum graveolens* L.	Apiaceae	Southern Europe and Western Asia	2n=22	Umbel	Schizocarps	Cross-pollination	Seeds	leaf and fruit	Carvone
5.	Celery	*Apium graveolens* L.	Apiaceae	Mediterranean region	2n=22	Umbel	Schizocarp	Cross-pollination	Seeds	Leaf and seed	d-limonene
6.	Bishop's Weed (Ajwan)	*Trachyspermum ammi* (L.) Sprague	Apiaceae	Egypyt	2n=18	Compound umbel		Cross-pollination	Seeds	Seed	Thymol
7.	Coriander	*Coriandrum sativum* L.	Apiaceae	Mediterranean region	2n=22	Compound umbel	Schizocarp	Highly cross-pollination	Seeds	Leaf and seed	Dextro linalol
8.	Fenugreek	*Trigonella foenum graecum* L.	Fabaceae	South eastern Europe and west Asia	2n=16	Racemose	Pod	Self-pollination	Seeds	Seed	Diosgenin
9.	Tamarind	*Tamarindus indica* Linn.	Fabaceae	Tropical Africa	2n=24	Terminal or axillary racemes	Pod	Cross-pollination	Seeds / Softwood grafting	Pod	Tartaric acid
10.	Saffron	*Crocus sativus* L.	Iridaceae	Mediterranean area, Asia Minor, and Iran	x=8, 2n=3x=24			Hand-pollination	Corms	Stiga	Picrocrocin
11.	Cinnamon	*Cinnamomum verum* Bercht. and Presl	Lauraceae	Srilanka	2n=24	Paniculate cymose	One-seeded berry	Cross-pollination	Rooted cuttings, air layering and seedlings	Bark, leaf	Cinnamic aldehyde

(Contd.)

S. No	Common name	Botanical name	Family	Centre of origin	Chromosome number	Type of inflorescence	Type of fruit	Pollination system	Propagation	Commercial part	Main Phytochemical constituents
12.	Nutmeg	*Myristica fragrans* Houtt.	Myrsticaceae	Indonesia	2n=42	Umbellate cymes	One seeded drupe	Cross-pollination	Epicotyl grafting	Seed	Terbein
13.	Clove	*Syzigium aromaticum* (L.) Merrill & Perry	Myrtaceae	Indonesia	2n=22	Terminal cymes		Self-pollination	Seeds	Unopened flower bud	Eugenol
14.	Allspice	*Pimenta dioica* (Linn.) Merill	Myrtaceae	West Indies	2n = 22	Racemose cymes	Berry	Cross-pollination	Seeds and Air layering	Fruit and seed	Eugenol, 4-hydroxy-3-methoxy cinnamaldehyde, 3,4-dimethoxycinnam aldehyde, vanillin and 3-(4-hydroxy-3-methoxypheny) pro-pane-1,2 diol
15.	Vanilla	*Vanilla planifolia* Andrews	Orchidaceae	Brazil	2n=32	Raceme	Capsule	Cross-pollination	Stem cuttings	Pod	Vanillin
16.	Black pepper	*Piper nigrum L.*	Pipereceae	Western Ghats of India	x=13, 2n=52	Pendent spike	Drupe	cross-pollination	Shoot cuttings	Dried mature berry	Piperine
17.	Curry leaf	*Murraya koenigii* Spreng.	Rutaceae	India, Sri Lanka and other south Asian countries	2n=18	Terminal cyme		Self-pollination	Seeds	Leaf	Pinene, sabinene, caryophyllene, cadinol and cadinene
18.	Chilli	*Capsicum annum* L.	Solanaceae	South America	2n=24	Solitary axillary	Berry	Often cross pollinated crops	Seeds	Green as well as ripe and dried pod (fruit)	Capsaicin, capsanthin
19.	Cardamom (small)	*Elettaria cardamomum* (Maton.)	Zingiberaceae	India	x=12, 2n=48	Panicle	Capsule	Cross-pollination	Suckers	Fruit (Capsule)	Cineole
20.	Ginger	*Zingiber officinale* Rosc.	Zingiberaceae	South East Asia	x=11, 2n=22	Spike	Capsule		Rhizomes	Rhizome	Gingerol, shogoal and zingerone
21.	Turmeric	*Curcuma longa* L.	Zingiberaceae	South East Asia	x=21, 2n=63	Spike	Capsule	Cross-pollination	Rhizomes	Rhizome	Curcuminoids, curcumin

3

Production Techniques of Spices and Condiments

Garlic

Botanical name	*Allium sativum* L.
Family	Alliaceae
Origin	Central Asia
Chromosome number	2n=2x=16
Type of inflorescence	Umbel
Pollination system	Cross- pollination
Commercial part	Bulbs
Main Phytochemical constituent	Allicin (diallyl disulphide oxide)

- Garlic belongs to family *Alliaceae*, genus- *Allium* and spp. *sativum* L.
- The chromosome number is 2n=2x=16.
- It is one of the most important bulb crops grown throughout India.
- It is used as a spice or condiment throughout India.
- The subterranean reserve structure derived from a leaf is the used plant part.
- It has a strong and characteristic odour, which is markedly different in fresh and fried state.
- The pungency of fresh garlic vanishes after cooking or frying.

- The name garlic comes from the **Anglo Saxon garleac** in which gar, a spear, refers to the pointed leaves and leac is a plant.
- The origin of garlic is in **Central Asia**, but cultivated all over the world, in Spain, France, Egypt, Bulgaria, Hungary, USA, Mexico and Brazil.
- It is a perennial of the **Alliaceae** family.
- It grows to a height of about 60 cm. It has short, flat upright leaves of 15-30 cm.
- The tall single flower stem bears spherical head of pale pink or greenish-white blooms, often mixed with tiny bulbils.
- The subterranean white-skinned bulb or corn is subdivided into numerous 'cloves'.
- It is a **sexually sterile diploid**, unknown in the wild.
- It contains a wealth of sulphur compounds; most important for the taste is **allicin** (diallyl disulphide oxide), which is produced enzymatically from **alliin** (S-2-propenyl-L-cysteine sulfoxide) if cells are damaged; its biological function is to repel herbivorous animals.
- **Allicin** is deactivated to diallyl disulphide; therefore, minced garlic changes its aroma if not used immediately.
- In the essential oil from steam distillation, diallyl disulphide (60%) is found besides diallyl trisulphide (20%), diallyl sulfide, ajoene and minor amounts of other di- and polysulphides.
- It is used as a flavouring, vegetable and medicinal herb that has accumulated superstitions over the centuries. Some cuisines like salads and sauces are fond of raw garlic.
- It aids digestion and prevent flatulence. It is considered to be beneficial in the treatment of diabetes.
- Pharmaceutical preparations of garlic are manufactured throughout Europe, some of them standardized to allicin, one of its proven active constituents.

Production techniques

Climate	Garlic can be grown at elevation of 1200–2000 m from the mean sea level. It requires short days, cool (12–18°C) moist period during vegetative growth.
Soil	A well drained friable soil rich in organic matter with a pH range of 6 – 7 is ideal. Extreme acidic soils as well as heavy soils are not suited for this crop.
Propagation	Garlic is propagated vegetatively by cloves.
Seed rate	500 kg cloves of 8-10mm dia are required to plant one hectare.
Season	October – November.
Transplanting	Cloves are planted at 15 x 10 cm spacing.
Manuring	During last ploughing incorporate 50 t/ha of FYM; Apply *Azospirillum* 2 kg and *Phosphobacteria* 2 kg/ha, 40:75:75 kg/ha NPK, 50 MgSO4 and 1 ton Neem cake as basal and N 35 kg/ha at 45 days after planting.
Controlling rubberisation	Avoid excessive application of N to minimise rubberisation. N should not. be applied in the form of Urea. Use Ammonium sulphate. Spray 1500 ppm of CCC on 30th day of planting. Reduce the frequency of irrigation to minimise rubberisation. Apply 0.2 % Boron 0.1 % Sodium molybdate on 30th, 60th and 90th days after planting.
Irrigation	Irrigate before and after planting; life irrigation is given on third day of planting and at weekly intervals of time.
Weed control	Weeding should be done on 30th and 60th after planting.
Harvesting	Turning of tops to yellowish or brownish colour and starts drying is the ideal harvest index for harvesting garlic.
Yield	The average yield is 10-20 tonnes/ha depending upon the variety.
Storage	Cold storage of garlic is possible at 0.5°-2.2°C temperature and 60-70% relative humidity.

Improved varieties

Variety/Hybrid	Salient features
NHRDF, Nasik	
Agrifound white (G-41)	Bulbs are compact, silvery and white with creamy flesh. A bigger elongated clove with 20-25 in number and diameter of bulb is 3.5 to 4.5 cm. Variety is susceptible to purple blotch and stempthylium blight. Total soluble solid is 41%, dry matter 42.78% and pyruvic acid content is 27 micro mole/g. Good for storage and average yield is 130 q/ha.
Yamuna Safed (G-1)	Bulbs are compact, silvery white skin with creamy flesh, diameter 4.0-4.5 cm. Sickle shaped cloves with 25 – 30 in number and diameter of cloves 0.8 to 1.00 cm. Variety is tolerant for purple blotch, stemphylium blight and thrips. Total soluble solid, dry matter and pyruvic acid content is 38% and 39.5 % and 29 micro mole/g respectively. Good for storage and average yield 150-175 q/ha.

Contd.

Variety/Hybrid	Salient features
Yamuna Safed-2 (G-50)	Bulbs are compact, attractive, white creamy flesh and diameter 3.5-4.0 cm. Number of cloves 35-40, diameter of cloves 0.75-1.4 cm and clove size index 1.75-2.5 cm2. Total soluble solid, dry matter and pyruvic acid content is 38-40%, 40-41% and 26 micro mole/g respectively. Average yield is 150-200 q/ha.
Yamuna Safed-3 (G-282)	Bulbs are creamy white and bigger sized and 4.5-6 cm in diameter. Diameter of clove 1.2-1.5 cm and cloves/bulb is 15-16. Total soluble solid, dry matter and pyruvic acid content is 38-42%, 39-43% and 25 micro mole/g respectively. Medium in storage and average yield is 175-200 q/ha/ Variety is suitable for export.
Yamuna Safed-4 (G-323)	Bulbs are silvery white and bigger sized 3.5-4.0 cm in diameter. Diameter of clove 1.2-1.25 cm, clove size index 2.50-3.00 cm2 and number of cloves 20-25/ bulb. Total soluble solid and dry matter content is 40-42% and 41-42% respectively.
Yamuna Safed-5 (G-189)	Dark green straight erect leaves, sturdy plant growth, erect shape wide 8-10 leaves/plant. Bulbs compact, attractive creamy white colour, creamy flesh, big size bulbs (4.5-5.0 cm in diameter) and cloves, oblong shape, number of cloves per bulbs 22-30. Variety is table purpose pungent with good aroma. Total soluble solids and dry matter content is 39% and 41 % respectively. Good storer and better adaptability and tolerant to stemphylium blight, purple blotch and thrips under field condition. Average yield is 150 – 160 q/ha.
Agrifound Parvati (G-313)	Bulbs are of bigger in size 5.0-6.5 cm in diameter, creamy white colour with pinkish tinge, diameter of cloves 1.5-1.8 cm, 10-16 cloves/bulb. Total soluble solid, dry matter and pyruvic acid content is 35-36%, 37-38% and 22-23 micro moles/g respectively. Medium in storage and average yield is 175-225 q/ha.
Agrifound Parvati-2 (G-408)	Bulbs are of bigger in size 5.0-6.0 cm in diameter, creamy white colour, diameter of cloves 1.5-1.6 cm, 1 -14 cloves/bulb. Total soluble solid, dry matter and pyruvic acid content is 36-37%, 37-38% and 22 – 23 micro moles/g respectively. Average yield is 175 – 200 q/ha.

DOGR, Rajguru Nagar, Pune

Bhima Omkar	Average yield of the variety is 107 q/ha. Gujarat, Haryana, Rajasthan and Delhi.
Bhima Purple	The bulbs are purple in colour with TSS 33.6% and allicin @ 2.9 mg/g (fresh weight basis) and 9.6 mg/g (dry weight basis). The average yield is 68 q/ha.

G.B.U.A.&T, Pant Nagar

Pant Lohit	It is moderately resistant to purple blotch. Bulbs are compact and bold weighing 18-20g. There are 15-18 cloves per bulb having light purple colour. It is a high yielder with the bulb yield of 125-130 q/ha.

Contd.

Variety/Hybrid	Salient features
HAU, Hisar	
HG-17	High yielding, good storage life, bulb weight is better than over varieties and has least incidence of purple blotch disease. Average yield 120-125 q/ha.
TANU, Coimbatore	
Ooty.1	Bigger bulbs weighing 30-40 g each. Average yield 17 t/ha.

Cumin

Botanical name	*Cuminum cyminum* Linn.
Family	Apiaceae
Origin	Mediterranean region
Chromosome number	2n=14
Type of inflorescence	Compound umbel
Pollination system	Cross- pollination
Type of fruit	Schizocarp
Commercial part	Seed
Main Phytochemical constituent	Cuminol (Cumaldehyde)

- Cumin (*Cuminum cyminum*) is a herbaccous annual plant of the apiaceae family, widely cultivated in the Mediterranean region of Europe and in India.
- Primary cultivation of cumin is in Europe, Asia, the Middle East, and North Africa with India and Iran as the largest cumin exporters.
- The valued portion of the plant is the dried fruit called cumin seed, which is esteemed as a condiment.
- It is the dried, white fruit with greyish brown colour of a small slender annual herb.
- The surface of the fruit has 5 primary ridges, alternatively has 4 less distinct secondary ridges bearing numerous short hairs.
- The plant is 15 to 50 cm high. The aromatic seed like fruit is elongated, ovoid, 3 to 6 mm long, slightly bitter and has a warm flavour.
- The inflorescence is a compound umbel with white or pinkish flowers.
- The flowers are small and either pink or white coloured.
- The fruit is a schizocarp, 4-5 mm long, containing two mericarp with a single seed.

- It is a cross pollinated crop, and bees often help in pollination.
- The chromosome number of cumin is 2n=14.
- The odour and flavour of cumin is derived largely from the essential oil, which contains cumaldehyde or cuminic aldehyde as the main constituent.
- Other ingredients of the oil are dihydrocuminaldehyde, dl-pinene, d—pinene, para-cymene, dipentene, and cuminyl alcohol.
- The dried seed of cumin has 2.5 to 5 percent essential oil on a dry weight basis and is obtained by steam distillation.
- Cumin is indigenous to Northern Egypt, Syria, the Mediterranean region, Iran and India.
- Cumin seed have an aromatic odour and bitter taste.
- It is used as a condiment, and is an ingredient in curry powders, seasonings of breads, cakes and cheese.
- In medicine, it is used as a stimulant, carminative, stomachic and astringent.
- Cumin seed oil is used in perfumery and for flavouring liqueurs and cordials.
- Nutritional value of cumin seeds is as follows: 17.7% protein, 23.8% fat, 35.5% carbohydrate and 7.7% minerals.

Production techniques

Climate	Cool and dry climate. The plant, which needs mild temperatures during a three to four month growing season, is intolerant of long periods of dry heat. Plant the seeds in a sunny location after the soil has become warm in spring.
Soil	Well-drained, loamy soils that are rich in organic matter are best for cumin cultivation. For commercial cultivation of cumin, a field in which cumin crop has not been taken up at least during last 3 years should be selected.
Varieties	S-404, MC-43, Gujarat Cumin-1(GC-1), GC-2, GC-3, RS-1, UC-198, RZ-19, etc.
Propagation	Seed propagation is commercially practiced.
Seed rate	12-15kg/ha.
Sowing	Sowing is done from 1st week of November to 1st week of December by broadcasting or in rows drilled at 30 cm.
Manuring	10–15 tons of farmyard manure/ha is added at the time of land preparation. Afterwards, a dose of 20kg P2O5/ha should be applied at the time of sowing, 30 kg N/ha may be applied as top P2O5 dressing either in single dose 30 days after sowing or in 2 equal splits.
Irrigation	A light irrigation is done soon after sowing and thereafter second irrigation should be applied 8–10 days after first irrigation. Depending upon the soil type and climatic conditions the subsequent irrigations may be given at 15–25 days intervals.

Weed control	Weed is a severe problem in cumin cultivation. Weeding at 30 and 60 days after sowing is necessary. Thinning should also be done during first hoeing and weeding to remove the excess plants. Chemical weed control by the application of herbicides may also be practiced. Application of pre-emergent Terbutryn or Oxcadiazone @ 0.5–1.0kg/ha or pre-plant Fluchloralin or pre-emergent Penimethalin @ 1.0kg/ha is very effective.
Harvesting	Field is cleaned and wilt affected plants are uprooted before harvesting. Harvesting is done by cutting the plants with sickle. The plants are stacked on clean threshing floor for sun-drying. After drying, seeds are separated by light beating with sticks by winnowing.
Yield	7-8q/ha.

Improved varieties

Variety	Salient features
RZ 19	A tall variety of cumin with erect stems, pink flowers and bold pubescent grains; tolerant to wilt as well as blight; matures in 120–140 days with an average yield of 5.6 q/ha.
RZ 209	An erect-growing variety of cumin with pink flowers and bold, grey, pubescent grains, resistant to wilt and blight diseases; matures in 140–150 days with an average yield of 6.5 q/ha.
GC 1	An erect-growing variety of cumin with pink flowers and bold, linear, oblong, ash brown colour grains; tolerant to wilt disease; matures in 105–110 days with an average yield of 7.0 q/ha.

Fennel

Botanical name	*Foeniculum vulgare* Mill.
Family	Apiaceae
Origin	Southern Europe and the Mediterranean area
Chromosome number	2n=22
Type of inflorescence	Umbel
Sex expression	Protoandrous
Pollination system	Cross- pollination
Commercial part	Seed
Main Phytochemical constituent	Anethole and d-fenchone

- ***Foeniculum vulgare* Mill.** is a biennial medicinal and aromatic plant belonging to the family **Apiaceae** (Umbelliferaceae). It is a hardy, perennial–umbelliferous herb with yellow flowers and feathery leaves.
- Fennel is an erect growing perennial herb native to **southern Europe and the Mediterranean area**, commonly grown for flavouring purposes.
- It is a stout, glabrous biennial or perennial, aromatic herb, with leaves dissected into numerous filiform segents.

- The flowers are in large, flat, terminal umbels, with from thirteen to twenty rays, and destitute both of general and partial involucres.
- The corolla consists of five petals, which, as well as the stamens, are golden yellow. The fruit is ovate, and of a dark colour, especially in the grooves.
- The fruit is a dry seed 4–10 mm long.
- The mature seeds are the dried fruits of this herb which are used commercially, and the young tender shoots and leaves of the sweet fennel are used in foods in European countries.
- The name comes from the Latin *foenum,* a variety of fragrant hay, reflecting the plant's odour.
- It is a greyish-green, strong-smelling herbaceous perennial, with slim stems, bearing soft lacy, dark green leaves with thread-like lobes and swollen bases.
- Creeping rootstock gradually extends plant into a sparse clump. Reaching a height of 1.5 meters, the plant has small mustard-yellow flowers on a compound umbel.
- The ripe fruit (seed) is small, oblong, cylindrical, 6.8mm long, straight or slightly curved, greenish yellow, deeply furrowed, 5 ridged and having agreeable aroma.
- It contains 1-3% of a volatile oil which is composed of about 50-60% anethole and about 20% d-fenchone.
- Both the seeds and the oil distilled from them are used for flavouring.
- As a medicinal plant, fennel seed has been used as an antispasmodic, carminative, diuretic, expectorant, laxative, stimulant, and stomachic.

Production techniques

Climate	It comes up well in fairly mild with frost free climate. The dry and cold weather favours high seed production. Prolonged cloudy weather at the time of flowering is conducive to diseases and pests.
Soil	Rich in organic matter with well drained loamy soil or black or sandy soil
Varieties	CO 1, UF 32, PF 35 and Gujarat Fennel 1.
	Mutant varieties: Mutant-1,2,3.
	Mass selection: UF-32.
Seed rate	Direct sowing: 10-12 kg/ha, Transplanting: 3-4 kg/ha.

Time of sowing	Mid September to Mid October.
Transplanting	5- 6 weeks old seedlings are transplanted at a spacing of 60 x 30 cm.
Manuring	Basal: Apply 25:10 kg N and P/ha.
	Top dressing: Apply 25 kg N/ ha at the time of flowering.
Intercultural operations	Thinning: In direct sown crop thin 4 – 5 weeks old seedlings to 25 – 30 cm. spacing. The plants are earthed up after 3 months.
Irrigation	First irrigation immediately after sowing, second on the third day and subsequent irrigations at 7-10 days intervals.
Weed control	Pre-emergence application of Pendimethalin @ 1 kg a.i./ha.
Harvesting	The crop matures in 7-8 months. The umbels are harvested 4-5 times at 10 – 15 days intervals and dried in sun for 4-5 days and seeds are thrashed.
Yield	9-10-q/ha.

Dill

Botanical name	*Anethum graveolens* L.
Family	Apiaceae (Umbelliferae)
Origin	Southern Europe and Western Asia
Chromosome number	2n=22
Type of inflorescence	Umbel
Pollination system	Cross- pollination
Type of fruit	Schizocarps
Commercial part	Young leaves and the fully developed green fruit are used for flavouring purposes
Main Phytochemical constituent	Carvone

- Dill (*Anethum graveolens*) is a fennel like annual or biennial herb of the parsley family.
- The genus name *Anethum* is derived from Greek word aneeson or aneeton, which means strong smelling.
- *Anethum graveolens* L. is the sole species of the genus *Anethum*, though classified by some botanists in the related genus *Peucedanum* as *Peucedanum graveolens* (L.).
- Native to Southern Europe and Western Asia, dill grows wild in Spain, Portugal and Italy. It is now cultivated in India, Germany, Rumania, and England and to some extent in North and South America as a commercial crop.
- It is an annual herb of parsley family, 45-75 cm in height, with finely feathered blue-green fern-like leaves and hollow stems.

- It produces small open umbels of creamy-yellow flowers in summer followed by dark brown seeds.
- The yellow flower develops into umbels.
- The seeds are not true seeds. They are the halves of very small, dry fruits called schizocarps.
- Dill fruits are oval, compressed, winged about one-tenth inch wide, with three longitudinal ridges on the back and three dark lines or oil cells (vittae) between them and two on the flat surface. The taste of the fruits somewhat resembles caraway. The seeds are smaller, flatter and lighter than caraway and have a pleasant aromatic odour.
- The fruit, or seed, is broadly oval in shape, about 0.14 inch (3.5 mm) long, with three longitudinal dorsal ridges and two wing-like lateral ridges.
- Dill seed contains 2-5% volatile oil. Its main constituent is carvone and the other components are d-limonene and phellandrene. A recent study also found eugenol and vanillin present in the seed.
- *Anethum* is used as an ingredient in gripe water, given to relieve colic pain in babies and flatulence in young children
- The whole plant is aromatic. The young leaves and the fully developed green fruit are used for flavouring purposes.

Production techniques

Climate	It requires warm to hot summers with huge sunshine levels; even partial shade will reduce the yield substantially. The plant quickly runs into seeds in dry weather.
Soil	Dill prefers rich well-drained, loose soil and full sun. It tolerates a pH in the range 5.3 to 7.8.
Propagation	Propagation through seeds.
Seed rate	Seed rate for line sown and broadcast crops are 5-6kg/ha and 10kh/ha respectively.
Sowing	Usually seeds are drilled in 1-2vm deep in lines at 45cm spacing, followed by light planking.
Manuring	30kg of N, and 40kh of P_2O_5 are given at land preparation and 30kg of N/ha is drilled in rows at flower bud initiation stage.
Intercultural operations	It requires 2-3 weedings and 6 light irrigations.
Harvesting	The seed crop is harvested at full ripe, grey-green maturity stage during early April.
Yield	The seed crop yields 10-14q/ha of grain yield.

Celery

Botanical name	*Apium graveolens* L.
Family	Apiaceae (Umbelliferae)
Origin	Mediterranean region
Chromosome number	2n=22
Type of inflorescence	Umbel
Pollination system	Cross- pollination
Type of fruit	Schizocarp
Commercial part	Leaves and seeds
Main Phytochemical constituent	d-limonene

- The celery plant (*Apium graveolens)* belonging to family Apiaceae is a hardy biennial, occasionally annual, widely cultivated for its fleshy leafstalk used as a vegetable and seeds which yield essential oil.
- It is a Mediterranean herb, brought into cultivation in France and England during sixteenth century. Later, it spread to many parts of temperate Europe and USA. The crop was introduced in India around 1940 from France for its seed crop.
- It is a herbaceous plant, usually 60 to 120 cm. high with white flowers borne in compound umbels.
- The inflorescence is a compound, many-flowered umbel, sessile or short-peduncled, and terminal or opposite the leaves.
- The umbellules are 6-25 flowered. They are hermaphrodite flowers, 5 merous, and white to greenish-white.
- The fruit is schizocarp, splitting into 2 mericarps with each up to 1.5mm long and with 5 light-coloured ribs.
- Root is succulent, well developed with numerous lateral roots.
- Stems are branched, angular, jointed and light green in colour.
- Leaves are oblong, 7-18 cm. long, pinnate or trifoliate. It throws up a flowering head in later part of autumn producing a mass of fruits.
- The fruit is made up of two united umbels, each containing a seed.
- Celery seeds are very small, about 1-2 mm. in length, oval and greenish brown in colour.
- The celery seed has a pleasingly crisp texture and subtle flavour but a slightly bitter taste.

- Commercially, celery is available as seeds, vegetable, seed oil and oleoresin.
- The seed contains 2.2% essential oil (2.2 – 3.0 %). The oil is pale yellow in colour, contains d-limonene (60%), ß-selinene (10-12 %), sedanoic acid anhydride (0.5 %) and sedanolide (2.5-3.0 %).
- The later two components contribute to its characteristic odour.
- The leaves are rich source of minerals like Ca, P, Fe, vitamin A and vitamin C for which it is consumed in salad.
- Celery sees are used as spice and condiment as a food flavorant.
- The seed oil is used in the flavouring of different kind of food-canned soups, meats, sausages and particularly in the flavouring of the popular celery salts and culinary sauces.
- Celery leaves and stalks are used as salad and sauces, vegetables juices, stew soups etc.
- Celery seeds are used as a stimulant and tonic in asthma and for liver diseases.
- As a domestic medicine, seeds are used as nervine, sedative and tonic.
- The celery seed oil is used as an anti-spasmodic and nerve stimulant. The oil has been successfully employed in rheumatoid arthritis and probably acts as an intestinal antiseptic.
- The oil is also used as an important ingredient in novel perfumes.

Production techniques

Climate	It is a winter (Rabi) crop, needs warm days and cool nights, low humidity and plenty of sun-shine. Generally, cool temperature at sowing and warm dry weather at seed ripening is ideal. A week long cold spell during mid seasons can induce bolting and lower seed yield.
Soil	The crop thrives well on silt loam to loamy soil (pH 6 – 8.5) with moderate organic matter. It is not suited to clayey soil and cannot withstand water-logging.
Varieties	Ajmer Celery-1 (ACel-1), EC-99249-1 and PRL-85-1.
Propagation	Through seeds
Sowing	The crop can be grown from seeds, by raising the seedlings in nursery-beds. The seed can be sown directly in field during October at 2 cm deep in rows when temperature is 15–20°C. Seedlings emerge in 2 weeks and rate of germination is around 80%.
Seed rate	1 kg/ha.

Transplanting	Transplant 60 days old seedlings when 12–15 cm tall with 7 – 8 cm root growth in the field to ensure a good crop stand; this enables it to escape early season frost and low temperature.
Spacing	The spacing in field is 30 x 30or 40 x 25 cm to ensure a high rate of plant population.
Manuring	The crop is given 20 to 30 t/ha of FYM at land preparation and 100-150 kg of N/ha (in 3 splits), P_2O_5 and K_2O. One-third of N, together with P_2O_5 and K_2O at 40 to 60 kg/ha is basically applied.
Irrigation	The crop is given irrigation at fortnightly intervals in early part and weekly nearing seed maturity. Irrigation at CPE (cumulative pan evaporation) 25/50 is optimum for high seed yield (13q/ ha).
Weeding	Weed control is desirable and at least 2 weeding cum hoeing are given. Use of Basalin (2 – 2.5 l./ha) as preplanting application controls early weed growth and reduces cost on weeding.
Harvesting	The crop is ready for harvesting when umbels turn light to dark brown in colour during later part in the month of March. The fruits mature by early April. The crops are harvested when 80% of the umbels turn light brown in colour. The plants are cut at ground level in early morning hrs by sickle to avoid seed shedding.
PHT	The crop is stalked and allowed to dry over threshing floor for a few days. Thereafter it is threshed, winnowed and graded with the help of sieve. The seed is further dried and stored in cool dry places.
Yield	The yield is 10-13 q./ha. of dry seed.
Drying	After the harvested crop is dried in the sun for a considerable period, the seeds are obtained by threshing and then cleaned by winnowing. Prolonged exposure to sun during the drying period may result in loss of oil.
Storage	The seed can be stored for 1-3 years without much loss in oil content and oil quality.

Bishop's Weed (Ajwan)

Botanical name	*Trachyspermum ammi* (L.) Sprague
Family	Apiaceae
Origin	Egypyt
Chromosome number	2n=18
Type of inflorescence	Compound umbel
Pollination system	Cross-pollination
Commercial part	Seed
Main Phytochemical constituent	Thymol

- *Trachyspermum ammi* is a native of Egypt and is cultivated in Iraq, Iran, Afghanistan, Pakistan, and India.
- It belonging to family Apiaceae is a highly valued medicinally important seed spice.
- It is a cross-pollinated crop and has a somatic chromosome number of 2n=18. The flowers are self-fertile but cross-pollination occurs through insects.
- It is an erect, glabrous or minutely pubescent, branched annual.
- The seeds contain 2–4.4% brown coloured oil known as ajwain oil. The main component of this oil is thymol, which is used in the treatment of gastro-intestinal ailments, lack of appetite and bronchial problems.
- The stems are striate; the leaves are rather distant, 2-3-pinnately divided, the segents linear.
- The flowers occur in terminal or seemingly-lateral pedunculate, compound umbels, white and small; the fruits are ovoid, muricate, aromatic cremocarps, greyish brown; the mericarps, which are the components of the fruit, are compressed, with distinct ridges and tubercular surface, 1-seeded.
- In India, the major Ajwain producing states are Rajasthan and Gujarat, where Rajasthan produces about 90% of India's total production.
- The essential oil (2.5 to 5% in the dried fruits) is dominated by thymol (35 to 60%); furthermore, a-pinene, b-cymene, limonene and e-terpinene have been found.
- Usage of ajowan is almost confined to Central Asia and Northern India. Ajowan is particularly popular in savoury Indian recipes like savoury pastries, snacks and breads.
- It is much used as a medicinal plant is ayurvedic medicine for its antispasmodic, stimulant, tonic and carminative properties.
- The seeds are used to ease asthma and indigestion. It is also widely used to treat diarrhoea and flatulence. In the West, thymol is used in medicines against cough and throat irritation.

Production techniques

Climate	*Trachyspermum ammi* requires a warm and long frost free growing season. It requires warm weather during seed development. The crop has moderate tolerance to drought and possesses wider climatic adaptability.
Soil	Ajowan grows on all kinds of soil but does well on loamy to sandy loam and even in black soils.
varieties	Ajmer Ajwain-1, Ajmer Ajwain-2 (AA-1,2)
Propagation	Through seeds.
Sowing	The seeds are sown, broadcasted or drilled in rows 45 cm apart during September-October at a depth of 1 cm. The plant to plant spacing should be maintained at 20-30 cm.
Seed rate	4kg/ha.
Manuring	FYM 10-15 tonnes, 80-100 kg N, 30-50 kg P_2O_5, 30-50 kg K_2O /ha.
Harvesting	The plants flower in about two months and the fruits become ready for harvesting when then flower heads turn brown. They are pulled out, dried on mats and the fruits are separated by rubbing by hands or feet. The harvesting is usually done from February to May.
Yield	The yield is 4-6 q/ha under a rain-fed farming system and 12-20 q/ha under irrigated conditions.

Coriander

Botanical name	*Coriandrum sativum* L.
Family	Apiaceae (previously classified under Umbelliferae)
Origin	Mediterranean region
Chromosome number	2n=22
Type of inflorescence	Compound umbel
Sex expression	Adromonoecious
Pollination system	Highly cross pollinated crop
Type of fruit	Schizocarp
Commercial part	Leaf and seed
Main Phytochemical constituent	Dextro linalol

- Coriander (*Coriandrum sativum*) is an annual plant of the Apiaceae family, native to the eastern Mediterranean region and is found in many other parts of the world.
- It is valued for the dry ripe fruits, called coriander seeds and also the fresh green leaves called cilantro.
- The herb is produced in Morocco, Romania, Mexico, Argentina, and the People's Republic of China, Bangladesh, Bulgaria, Canada, Egypt, India, Indonesia, Nigeria, Poland, Syria, the United States, the USSR, and Yugoslavia.

- It is one of the oldest recorded spice, mentioned in ancient Sanskrit texts and in Exodus.
- Seeds have been found in the tombs of the Pharaohs. The name originated from koris, the Greek word for a bed bug, so given because of the similarity between the smell of coriander leaves and the offending bug.
- It is a rigid, strong-smelling annual with pronounced taproot, and slender branching stems up to 60 cm reaching a height of 1 meter, the adromonoecious plant flowers in July and August.
- The plant is a thin stemmed, small, bushy herb, 25 to 50 cm in height with many branches and umbels.
- Leaves are alternate, compound. The whole plant has a pleasant aroma.
- Inflorescence is a compound umbel comprises 5 smaller umbels.
- The germination is epigeal and the plant has a tap root.
- Flowering starts with the primary umbel. In every umbel the peripheral umbellets, and in every umbellet the peripheral flowers are the first ones to flower. These flowers are protandrous.
- The central flowers of the umbellets are staminiferous or sometimes sterile.
- It has an inferior ovary and the five calyx teeth surrounding the stylopodium are still visible in the ripe fruit.
- The five calyx teeth are of different length, as are the petals in peripherally situated flowers.
- The fruits are globular or ovate, with a diameter of up to 6 mm. Usually, the schizocarp does not spontaneously split into two mericarps.
- Fruit has delicate fragrance; seeds are pale white to light brown in colour.
- There are two distinct morphological types, one erect and tall with a comparatively stronger main shoot and the other bushy with a relatively weaker main shoot and longer spreading branches.
- The seed capsules are round red-brown which are aromatic when ripe.
- It is an important spice crop having a prime position in flavouring food.
- The pleasing flavour of the coriander fruit is not thoroughly developed until it is completely dry.

- The essential oil content of the weight of ripe and dried fruits of coriander varies between 0.03 and 2.6%, and the content of fatty oil varies between 9.9 and 27.7%.
- The oil contains d-linalool (also known as coriandrol), camphor, pinenes, camphene, sabinene, myrcene, terpinenes, limonene, and other constituents.
- Coriander leaf or cilantro contains about 4% volatiles, on a wet leaf basis, primarily 2-decenal and 2-dodecenal.
- The fruits (seeds) are widely used as condiments with or without roasting in the preparation of curry powders, sausages and seasonings.
- In medicines it is used as a carminative, refrigerant, diuretic, and aphrodisiac. In household medicines, it is used against seasonal fever, stomach disorders, and nausea.
- Coriander oil and oleoresins are primarily used in seasonings for sausages and other meat products.

Production techniques

Climate	Cool and comparatively dry, frost free climate.
Soil and	Well drained black cotton soil and red loamy soil.
Varieties	CO 1, CO 2, CO 3, CO (CR) 4, GAU 1, UD 1, UD 2, UD 20, UD 21, GC-1, GC-2, RCr-20, Sindhu, Sadhana. RCr 41: Resistant to stem gall and wilt and moderately tolerant to powdery mildew. RD-44 (Rajendra Swati): Resistant to stem gall and moderately resistant to aphids and weevil.
Seed rate	12-15 kg/ha (irrigated crop), 25 – 30 kg/ha (rainfed crop).
Time of sowing	June-July and October-November.
Sowing	Sow the split seeds at a spacing of 20 x 15 cm. The seeds will germinate in about 8-15 days.
Manuring	Basal: Apply 10: 40: 20 kg of NPK/ ha for rainfed and irrigated crops. Top dressing: Top dressing may be done at 10 kg N/ha 30 days after sowing for the irrigated crop only.
Intercultural operations	Thinning is done 30 days after sowing by keeping 2 plants per hill.
Irrigation	Depending upon the climatic conditions, moisture retaining capacity of soil and variety used 4-5 irrigations.
Weed control	Pre-plant Fluchloralin @ 0.75kg/ha, pre-emergence Oxyfluorfen @0.15 kg/ha or Pendamithalin @ 1kg/ha are effective herbicides ml in 500 lit of water per ha.
Growth regulator	Spray CCC @ 250 ppm one month after sowing for inducing drought tolerance in rainfed crops.

Harvesting	The plants are pulled just when the fruits are fully ripe but green and start drying. The plants are dried and thrashed with sticks, winnowed and cleaned. For leaf, pull out the plants when they are 30-40 days old.
Yield	Seed 12-15q/ha.

Fenugreek

Botanical name	*Trigonella foenum graecum* L.
Family	Fabaceae (Leguminosae)
Origin	South-eastern Europe and west Asia
Chromosome number	2n=16
Type of inflorescence	Racemose
Pollination system	Self-pollination
Type of fruit	Pod
Commercial part	Seed
Main Phytochemical constituents	Diosgenin

- Fenugreek is a slender annual herb of the pea family (Fabaceae).
- The Latin species name *foenum graecum* means Greek hay, referring to both the intensive hay fragrance of dried fenugreek herb.
- It is a native of South Eastern Europe and West Asia, now cultivated in India, Argentina, Egypt and Mediterranean countries (Southern France, Morocco and Lebanon).
- Two type of flowers –cleistogamous (closed type) and aneictgamous (open type)- have been reported for fenugreek (Petropoulos 2002).
- Cleistogamous flowers ensure a high level of self-pollination. However, the rate of cross-pollination in *T. foenum-graecum* has ranged from 2.1 to 7.0% (Choudhary 2003).
- The inflorescence is racemose and flowers are bracteates, pedicellate, complete, zygomorphic, bisexual and hypogynous.
- The fruits are pods or legumes narrow and oblong, resembling a sickle.
- This robust herb has light green leaves, is 30-60 cm tall and produces slender, beaked pods, 10-15 cm long, each pod contains 10-20 small hard yellowish brown seeds, which are smooth and oblong, about 3mm long, each grooved across one corner, giving them a hooked appearance.
- The slender pods are up to 6 inches (15 cm) long, curved and beaked, and contain ten to twenty yellow-brown seeds—flat rhomboids characterized by a deep furrow, less than 0.2 inch (1 cm) long.

- The seeds are about 0.5 cm in diameter and are irregularly shaped, very hard, and tan or mustard coloured.
- In India it is grown extensively in Rajasthan, Gujarat, Madhya Pradesh, Uttar Pradesh, Maharashtra and Punjab.
- It is used both as a food and food additive as well as in medicines. Fresh tender pods, leaves and shoots are eaten as curried vegetable.
- The volatile oil content of fenugreek is very small (less than 0.02%). It also contains fixed oils at about 5%-7%. They contain the alkaloids trigonelline and choline and a yellow colouring matter. It is rich in proteins, minerals and vitamins.
- It contains diosgenin, a compound used as a starting material for sex hormones in the pharmaceutical industry.
- The seeds are used in colic flatulence, dysentery, diarrhoea, dyspepsia, chronic cough and enlargement of liver and spleen, rickets, gout and diabetes.
- It is also used as a carminative, tonic, and aphrodisiac. Fenugreek oil is used in the manufacture of hair tonics.

Production techniques

Climate	Cool and comparatively dry, frost free climate.
Soil	A rich well drained loamy soil is best suited.
Varieties	Co 1, Pusa Early Bunching, Lam selection 1, Rajendra Kranti, Hissar Sonali, RMT1and CO 2.
Seed rate	25kg/ha.
Time of sowing	October-November.
Sowing	Seeds are sown at a spacing of 20 X 15 cm.
Manuring	Basal: Apply 30:25:40 kg of N, P, K /haTop dressing: Apply 20 kg of N at 30 days after sowing.
Intercultural operations	Plants are thinned at 20-25 days after sowing and the thinned seedlings are used as greens. One pinching at a height of about 4" will encourage branching
Irrigation	Give first irrigation immediately after sowing, second on the third day and subsequently at 7-10 days intervals.
Weed control	Spray pre-emergence herbicide Fluchloralin 700 ml in 500 lit of water per ha.
Harvesting	20-25 days for greens, 90-100 days for grains.
Yield	10-11q/ha.

Tamarind

Botanical name	*Tamarindus indica* Linn.
Family	Fabaceae (Leguminosae)
Origin	Tropical Africa
Chromosome number	2n=24
Type of inflorescence	Terminal or axillary racemes
Pollination system	Cross- pollination
Pollinators	Honey bees
Type of fruit	Pod
Commercial part	Pods
Main Phytochemical constituent	Tartaric acid

- Tamarindus (*Tamarindus indica*) is an evergreen tree, of the pea family native to tropical Africa.
- It is widely cultivated in other regions as an ornamental and for its edible fruit.
- It is semi-evergreen, tropical tree that grows to about 24 m tall and has long drooping branches with alternate, pinnately compound (feather-formed) leaves; the leaflets are about 2 cm (0.75 inch) long.
- The inflorescences are terminal or axillary racemes, 5-12 cm long, with 10-15 flowers with four sepals (five, two of which are merged together), pink at the beginning, then white, retroflexed after opening, three obovate upper petals, 1-1.5 cm long, of a yellowish colour with purple red venations, and 2 lower ones reduced to scles.
- The yellow flowers, about 2.5 cm across, with a red stripe are borne in small clusters.
- The fruits are usually between 5 to 14 cm in length and approximately 2 cm wide. It is an indehiscent legume, with a hard, brown shell called as pods.
- The dark brown fruit is a plump pod 7.5-24 cm long that does not split open.
- It contains 1 to 12 large, flat seeds embedded in a soft, brownish pulp.
- This pulp has a high tartaric acid content that imparts for its sourness.
- The ripe fruit of tamarind tree is used as a condiment.
- In India, it is chiefly grown in Madhya Pradesh, Andhra Pradesh, Tamil Nadu and Karnataka.

- Tamarind pulp is used in numerous culinary preparations. It is also a raw material for the preparation of wine like beverages.
- The tamarind kernel powder is found to be extensively used for its sizing properties, in textile, confectionary, cosmetics and pharmaceutical industries.
- The testa is used in dyeing and tanning industry. The tender leaves and flowers are used as vegetables.
- In medicine, it is used as appetizing, laxative, healing and anti-helmintic. It is also used against fluorosis.

Production techniques

Climate	The absolute maximum temperature varies from 36-47.5° C and the absolute minimum temperature varies from 0-17.5° C. Rainfall requirement – 750-1900 mm. Altitude – up to 100 m above MSL.
Soil and	Grown on variety of soils ranging from poor degraded, eroded, gravely, saline and alkaline soils. Productivity is higher in red loamy, deep well drained soils.
Varieties	PKM 1, Urigam, Hasanur and Tumkur, Prathisthan, DTS 1, Yogeshwari.
Propagation	Through seeds / Softwood grafting.
Planting season	June-November.
Spacing	Plant at a spacing of 10 m x 10 m.
Manuring	Apply 200:150:250 g of NPK per tree per year along with 25 kg of FYM and 2 kg of Neem cake.
Intercultural operations	Remove the rootstock sprouts. Remove the dried and diseased parts. Intercrops like leguminous crops, short duration vegetables, annual drumstick, Sesamum and Sorghum may be raised in the alley spaces up to four years.
Irrigation	Regular watering should be given once in seven days.
Harvesting	Grafts start yielding from 4 to 5 years. Pods are harvested in January-April every year.
Yield	150- 200 kg/tree/year.

Saffron

Botanical name	*Crocus sativus* L.
Family	Iridaceae
Origin	Mediterranean area, Asia Minor, and Iran
Chromosome number	x=8, 2n=3x=24
Ploidy level	Autotriploid
Type of flowers	Flowers are hermaphrodite (have both male and female organs)
Pollination system	Hand-pollination
Commercial part	Stiga
Main Phytochemical constituent	Picrocrocin

- Saffron (*Crocus sativus*) is a bulbous perennial of the iris family (Iridaceae) treasured for its golden-coloured, pungent stigas, which are dried and used to flavour and colour foods and as a dye.
- The genus *Crocus* (family Iridaceae) comprises some 85–100 species having an old world distribution, primarily in the Mediterranean – Europe and Western Asia.
- Believed native to the Mediterranean area, Asia Minor, and Iran, the saffron crocus has long been cultivated in Iran and India and is supposed to have been introduced into Cathay by the Mongol invasion.
- It is cultivated chiefly in Spain, France, Sicily, Italy (on the lower spurs of the Apennines Range), and in Iran, and India.
- Saffron, the world's most expensive spice, is derived from the stigas of the saffron flowers (*Crocus sativus* L.).
- Commercial saffron comprises the dried red stiga with a small portion of the yellowish style attached.
- It is a prized culinary condiment, widely used in the cuisine of many European and Asian countries.
- It has a strong, exotic aroma and a bitter taste. A golden-coloured, water-soluble fabric dye was distilled from saffron stigas in India in ancient times.
- The stigas in the delicately scented saffron flowers are dried and used in cooking to colour, flavour and add a unique aroma to the meal.
- The flowers are hermaphrodite (have both male and female organs) and are pollinated by bees and butterflies.
- The saffron crocus (*Crocus sativus* L.) is sterile and does not set viable seed. Therefore, the crop must be propagated by corm multiplication.
- *Crocus sativus* is a triploid species with a karyotype similar to diploid *C. thomasii* and *C. cartwrighitanus.*
- Reproductive biology makes it possible to ascertain that saffron is self- and allo sterile although its ovules can be fertilized by pollen of *C. cartwrighitanus* as well of *C. thomasii*. The crossing leads to seed set and fruits.
- The successful pollination in *Crocus* produces a capsule maturing at soil level. *Crocus sativus* hand-pollinated with pollen of *C. cartwrightianus* originated a large capsule maturing in May and visible above ground.

- The average weight of fresh stigas is 0.03 g per flower and dry weight is 0.007 g per flower. About 150 flowers are needed to obtain 1 g of dry stigas.
- The saffron crocus (*Crocus sativus* L.) is sterile and does not set viable seed. Therefore, the crop must be propagated by corm multiplication.
- *Crocus sativus* is a triploid species with a karyotype similar to diploid *C. thomasii* and *C. cartwrighitanus.*
- Reproductive biology makes it possible to ascertain that saffron is self- and allo sterile although its ovules can be fertilized by pollen of *C. cartwrighitanus* as well of *C. thomasii*. The crossing leads to seed set and fruits.
- The successful pollination in *Crocus* produces a capsule maturing at soil level. *Crocus sativus* hand-pollinated with pollen of *C. cartwrightianus* originated a large capsule maturing in May and visible above ground.
- The average weight of fresh stigas is 0.03 g per flower and dry weight is 0.007 g per flower. About 150 flowers are needed to obtain 1 g of dry stigas.
- The flowers have three bright, orange-red stigas which are the true saffron.
- The three stigas are handpicked from each flower, spread on trays, and dried over charcoal fires for use as a food flavouring and colouring.
- Saffron contains 0.5 to 1 percent essential oil, the principal component of which is picrocrocin. The colouring matter is crocin.
- The bitter taste and an iodoform or hay-like fragrance is caused by chemicals picrocrocin and safranal.
- It is used to colour and flavour many Mediterranean and Oriental dishes, particularly rice and fish, and English, Scandinavian, and Balkan breads. It is an important ingredient in bouillabaisse.

Qualities of saffron according to I.S.O.

The I.S.O. (International Standard Organization) defines in its norm 3632-2 1994 the different qualities of saffron in filaments and powder based in their chemical properties. This is gathered in the following table:

Characteristics	Requirements		Test Method
	Saffron threads	Saffron powder	
Moisture and volatile matter, % (m/m), max.	12	10	ISO 3632-2 CLAUSE 9
Total ash, % (m/m) on dry basis, max.	8	8	ISO 928 & ISO 3632-2 CLAUSE 10
Acid-insoluble ash, % (m/m), on dry basis, max.			ISO 930 & ISO 3632-2 CLAUSE 11
Categories I & II	1.0	1.0	
Categories III	1.5	1.5	
Solubility in cold water, %(m/m), on dry basis, max.	65	65	ISO 941
Bitterness, expressed as direct reading of the absorbance of picrocrocine at about 257 nm, on dry basis, min.			ISO 3632-2 CLAUSE 13
Categorie I	70		
Categorie II	55		
Categorie III	40		
Safranal, expressed as direct reading of the absorbance at about 330 nm, on dry basis. All categories			ISO 3632-2 CLAUSE 13
Min.	20		
Max.	50		
Colouring strength, expressed as direct reading of the absorbance of crocine at about 440 nm, on dry basis, min.			ISO 3632-2 CLAUSE 13
Categorie I	190	190	
Categorie II	150	150	
Categorie III	100	100	
Total nitrogen, % (m/m), on dry basis, max.	3.0	3.0	ISO 1871
Crude fibre, % (m/m), on dry basis, max.	6	6	ISO 5498

Production techniques

Climate	Saffron grows well in areas that have cold winters and warm dry summers.
Soil	The plant prefers light (sandy) and medium (loamy) soils, requires well-drained soil, and can grow in nutritionally poor soil.
Propagation	Through corms.
Planting	The corms are placed in ridges of about 20 cm. depth. The distance between the bulbs should be of 10 cm. In India, corms are planted 7.5-10 cm apart, in rows 15-20 cm apart. Recommended planting depths for corms vary from 7.5-10 cm to 15-22 cm.
Seed rate	Optimal corm quantity per hectare is 13-15 tons, which is about 600- 700 thousand corms with an average weight of 20-22 g each (45-48 corms/kg).
Season	The sowing takes place in the months of June and July.
	20-30 tons FYM before planting and 80 kg P/ha and 30 kg K/ha followed by a split application of 20 kg N/ha in autumn and again immediately after flowering is recommended.
Harvesting	The harvesting takes place between the ends of October-beginning of November. The rose of saffron blooms at dawn and should stay the least possible time in the plant because it withers quickly and the stigas loose colour and aroma. This is why they are gathered between dawn and 10 a.m. Once the flowers are gathered, stigas are separated from the rest of the flower.
Yield	2-7 kg/ha.
PHT	The stigas of saffron have a high level of moisture, so it is necessary to dry them for its good preservation. This is the process of roasting, in which the stigas get it definitive aspect: bright red, rigid and without wrinkles.After the process of roasting, the stigas of saffron would have 1/5 of their original size. This means that for one kg of raw stigas we will obtain 200 g of saffron ready for consumption.For its perfect preservation, saffron is stored in big wooden trunks lined with metal plate inside protecting it from heat, cold and specially moisture.

Cinnamon

Botanical name	*Cinnamomum verum* Bercht. and Presl [Syn. *Cinnamomum zeylanicum* Blume]
Family	Lauraceae
Origin	Srilanka
Chromosome number	2n=24
Type of inflorescence	Paniculate cymose
Pollination system	Cross- pollination
Type of fruit	One-seeded berry
Commercial part	Bark, leaf
Main Phytochemical constituent	Cinnamic aldehyde

- Cinnamon (*Cinnamomum verum*) a bushy evergreen tree of family Lauraceae is one of the earliest known spice mainly cultivated for the dried inner bark of the tree.
- It is native to Sri Lanka (Ceylon), the neighbouring Malabar Coast of India, and Myanmar (Burma) and also cultivated in South America and the West Indies for the spice consisting of its dried inner bark.

- The chromosome number of *Cinnamomum verum* is 2n=24.
- It is a bushy evergreen tree (6-8 m tall), cultivated as low bushes to ease the harvesting process.
- The leaves are long (10-18 cm), leathery and shining green on upper surface when mature.
- Inflorescence is paniculate cymose, the initial branching of the inflorescence paniculate, with alternate or opposite branches, while the flowers are arranged in cymes.
- The flowers have a fetid, disagreeable smell.
- The flowering time varies from October to February and the fruit ripens during May-June.
- Flowers exhibit protogynous dichogamy, the male and female phases are separated by almost a day.
- The genus *Cinnamomum* is open-pollinated, and pollination is most probably by insects, especially flies.
- Fruit is a one-seeded berry, ellipsoidal to ovoid, 1-2 cm long, black when ripe, surrounded by the enlarged perianth at the base.
- The spice is light brown in colour and has a delicately fragrant aroma and warm, sweet flavour. It is lighter in colour and milder in flavour than the other related species.
- Cinnamon bark oil is one of the expensive essential oils in the world market.
- Cinnamon contains from 0.5 to 1 present essential oil, the principal component of which is cinnamic aldehyde (about 60%).
- Other components are eugenol, eugenol acetate, and small amounts of aldehydes, ketones, alcohols, esters and terpenes.
- The oil is distilled from fragents for use in food, liqueur, perfume and drugs. The aldehyde can also be synthesized.
- In India, it is grown in one or two locations in Kerala.
- The commercial products of cinnamon are quills, quillings, featherings, chips, cinnamon bark oil and cinnamon leaf oil.
- Quills are the most valuable cinnamon product, which are obtained from the bark of the cinnamon tree.
- 'Quills' are scraped peel of the inner bark of the mature cinnamon shoots, joined together with overlapping tubes, the hollow of which has been filled with smaller pieces of cinnamon peels which is dried first in the sun and thereafter in the shade.

- Featherings are feather-like pieces of inner bark consisting of shavings and small pieces of bark left over from the process of making quills.
- Cinnamon bark is a popular spice with a delicate fragrance and a warm agreeable taste.
- It is used in the form of small pieces or powder. It is widely used in flavouring confectionary, liquors, pharmaceuticals and cosmetics.
- It is found to help diabetics in digestion of sugar. It has astringent; stimulant and carminative properties and can check nausea and vomiting.
- The cinnamon bark oil has anti-fungal properties and cinnamon leaf oil is widely used in perfumery and cosmetics

Production techniques

Climate	Cinnamon is a hardy plant and is cultivated in Sri Lanka under varying conditions ranging from semi dried to wet zone conditions. The ideal temperature for growing cinnamon is between 20-30 degree C and rainfall between 200 to 250 cm. It comes up well from sea level to an elevation of about 1000m.
Soil	Sandy or lateritic soils with high humus are suitable.
Varieties	YCD 1, PPI – 1, Nithyasree, Navasree, Konkan Tej, Sugandhini.
Propagation	Propagation of cinnamon is by rooted cuttings, air layering and seedlings.
Season	Cinnamon planted during June-July.
Planting	Take pits of 50 cm3 at 3 m x 3 m spacing. Fill the pits with top soil and FYM 10 kg. One year old seedlings or rooted cuttings are transplanted under partial shade.
Manuring	20gN, 18g P_2O_5 and 25K_2O/seedling is recommended for the first year. The dose of fertilizers is increased gradually to 200gN, 180g P_2O_5 and 200K_2O for plants of 10 years and above.
Intercultural operations	Provide temporary shade after transplanting. Weeds are removed as and when necessary. Young trees are cut close to the ground to produce side shoots from second or third year onwards. This process is called "Coppicing". By stooling around the stumps, more side shoots are encouraged from the base of the trees.
Irrigation	Protective watering during summer is in beneficial.
Harvesting	The harvest starts from 4th or 5th year after planting. Cut the shoots in May and November for bark harvest. Young shoots spring up from the stump ready for removal in subsequent season within 18 months when they attain one metre long and 1 to 2 cm thick. Terminal ends of shoots are removed. Peeling is done by knives after scraping off the outer bark. Harvested produce is called as 'Quills'. From leaves, Cinnamon oil can be extracted by steam distillation.
Yield	100 g of dried bark/ bush/ year. 35 kg of leaf oil/ ha/ year.

Nutmeg

Botanical name	*Myristica fragrans* Houtt.
Family	Myristicaceae
Origin	Indonesia
Chromosome number	2n=42
Type of inflorescence	Umbellate cymes
Sex expression	Dioecious
Pollination system	Cross- pollination
Type of fruit	One seeded drupe
Commercial part	Seed
Main Phytochemical constituent	Terbein

- The nutmeg plant, ***Myristica fragrans* Houtt.,** is a member of the small primitive family **Myristicaceae,** taxonomically placed between the Annonaceae and Lauraceae (Joseph,1980).
- Genetically the somatic number of chromosomes in *Myristica Fragrans* is **2n=42,** and the basic chromosome number for the genus is suggested to be **7** (Purseglove, et al., 1981).
- Nutmeg is the dried **kernel** of the seed and mace is the **dried aril** surrounding it.
- The nut has 7.1% essential oil, 9.8% oleoresin and 30.9% butter, while the mace has 7.1% essential oil and 13.8% oleoresin
- The nutmeg is a medium-sized, spreading or conical, thickly-leaved evergreen tree that attains the average height of 4-10 m but sometimes may reach heights of 20 m and over.
- These are simple, persistent, alternate, glabrous and exstipulate, elliptic or oblong lanceolate with an apex acute or slightly acuminate and an acute base tapering into a short petiole slightly flattened adaxially and about 1-1.5 cm long.
- The lamina, 5-15 cm long and 2-7 cm broad, is coriaceous, medium to dark green above and shining, light silvery-green or subglaucous beneath.
- The tree is usually dioecious but sometimes there is the occurrence of male and female flowers on the same tree.
- The trees are normally unisexual, bearing either male or female flowers.

- The inflorescences with male and female flowers are structurally similar and are axillary and glabrous, bearing flowers in **umbellate cymes** with the male 1-10 usually outnumbering the female 1-3. It is quite common to find in the male inflorescence flowers in various stages of development.
- The pedicels, 1-1.5 cm long, are pale green with a minute caducous bracteole at the base of the flower.
- The flowers are creamish-yellow in appearance, waxy and fleshy, fragrant and may measure up to 1 cm in length. Petals are absent so the dominant calyx is bell shaped, nectiferous at the base, with 3-reflexed triangular lobes. The female flowers, up to I cm long, exhibit a puberulous, superior, sessile, one celled ovary about 7 mm long and topped by a very short, white two lipped stiga.
- The male flowers consist of an androecium I cm long, glabrous, with a 2 mm stalk, with 8-12 stamens, with anthers adnate to a central column and attached to each other by their sides
- The male flowers are born in clusters, whereas female flowers are often solitary.
- The fruit is a fleshy, pendulous, **one seeded drupe** suspended by a greenish brown fruit stalk about 1.5 cm long. It is broadly pyriform, yellow, smooth, 6-9 cm long and almost as broad.
- Some fruits are obviously bell shaped, longer than there are broad, and in rare cases observed locally some races may bear joined double fruits.
- When the fruit mature it burst open along the groove exposing the bright attractive mace, covering the hard black, shiny shell of the seed called **nutmeg.**
- Nutmeg tree is indigenous to Moluccas. The major nutmeg growing areas are Indonesia and Granada. It also grows on a smaller scale in Sri Lanka, India, China, Malaysia, Zanzibar, Mauritius and Solomon Island.
- The spice in the ground form is mainly used in the food processing industry especially as a standard seasoning in many Dutch dishes.
- Nutmeg oleoresin is used in the preparation of meat products, soups, sauces, baked foods, confectionaries, puddings, seasoning of meat and vegetable etc.
- The fleshy outer cover of the fruit is crystallized or pickled or made into jellies.

Production techniques

Climate	Nutmeg thrives well in places with warm humid climate from sea level up to 1300 m MSL. A well-distributed annual rainfall of 150 cm is ideal for the crop.
Soil	It grows on a variety of soils from sandy to clayey loams and red laterite soils with good drainage.
Varieties	Vishwashree, Konkan Sugandha and Konkan Swad.
Propagation	Nutmeg is commercially propagated through grafts.
Time of planting	Planting is done at the onset of rainy season.
Planting	Plant 12-18 months old seedlings/ grafts/ budded plants in pits of 75 cm3 size. Take pits at a spacing of 9 m x 9 m. For planting plagiotropic graft, spacing 5X5m has to be adopted. A male graft has to be planted for every 20 female grafts in the field.
Manuring	Apply FYM 15 kg, 20:18:50 g of NPK per tree during first year and FYM 50 kg, 500:250:1000 g of NPK for adult trees (15 years onwards) in two splits June-July, September-October. Apply 50 g in each of *Azospirillum* and *Phosphobacteria* one month after manuring.
Intercultural operations	Keep the area around the plant weed free. It requires medium shade especially during the initial stages of growth. Fast growing shade trees or banana are planted in between them a few months prior to planting and can be thinned out later. It can be grown as mixed crop with arecanut and coconut. In Arecanut plantations, Nutmeg can be planted after every third row of Arecanut.
Irrigation	Irrigation is given once in 5-7 days during summer months.
Aftercare	Nutmeg can be grown as an intercrop in coconut, clove, Arecanut and coffee gardens. Regular mulching of plant basins after slashing weeds and shading in summers in the early years are very essential.
Harvesting	The bearing starts from six to seven years after planting. The fruits are ready for harvest in about 9 months after flowering. The fruits are ripe and ready for harvesting when the pericarp splits open. After harvest the outer fleshy portion is removed, and the mace is manually separated from the nut. The aril commonly known 'mace' and 'seed' are separated and dried.
Yield	1000-2000 fruits/tree, Dried nuts: 5-7 kg/tree, Dried mace: 0.5-0.7 kg/tree.

Clove

Botanical name	*Syzygium aromaticum* (L.) Merrill & Perry
Family	Myrtaceae
Origin	Indonesia
Chromosome number	2n=22
Type of inflorescence	Terminal cymes
Type of Flowers	Hermaphrodite
Sex expression	Monoecious

Contd.

Botanical name	*Syzygium aromaticum* (L.) Merrill & Perry
Pollination system	Self-pollination
Type of fruit	Fruit olive-shaped, 1-seeded, popularly
referred to as	'mother of clove'
Commercial part	Unopened flower bud
Main Phytochemical constituent	Eugenol

- *Syzygium aromaticum* (L.) Merrill & Perry is an evergreen tree belonging to the *Myrtaceae* family.
- It is cultivated for the production of cloves, dried unopened flower buds which are used as spices.
- The name clove is believed to be derived from the French word clou meaning nail due to the appearance of this spice.
- The tree grows to a height of 10-12m and start flowering in about 7 years.
- It continues to produce flower buds for 80 or more years. It is a valuable spice of the orient.
- Clove clusters are plucked by hand when the buds are fully developed with a pronounced pink flush and then dried over several days in the sun.
- Cloves vary in length from about 1/2 to 3/4 inch (13 to 19 mm).
- Leaves glabrous, with numerous oil glands on lower surface.
- The inflorescences are lateral and terminal cymes, either lax or tightly congested.
- Flowers small, in terminal cymose clusters, each peduncle bears 3-4 stalked flowers at the end. Sepals minute triangular projections.
- Fruit olive-shaped, 1-seeded, popularly referred to as 'mother of clove'.
- The clove tree is monoecious, flowers are hermaphrodite and self-pollinating.
- Fruits mature approximately 9 months after flower initiation and are considered physiologically mature when the exocarp turns reddish-purple in colour.
- Unopened flower buds, leaves and stalks yield essential oil.
- Cloves contain 14 to 20 percent essential oil, the principal component of which is the aromatic oil eugenol (70-90%).
- Cloves are strongly pungent owing to eugenol, which is extracted by distillation to yield oil of cloves.

- Three essential oils are available from this spice: clove bud oil, clove stem oil and clove leaf oil. Each has different chemical composition and flavour.
- Clove bud oil, the most expensive and the best quality product, contains eugenol (70-90%), eugenol acetate (15%) and beta caryophyllene (5-12%).
- The use of clove in whole or ground form is mainly for culinary purposes and as a flavouring agent in food industry.
- Its flavour blends well with both sweet and savoury dishes. It is highly valued in medicine as carminative, aromatic and stimulant.
- In Indonesia, the lion share of production is consumed in production of 'kretek' cigarettes.
- The antiseptic and antibiotic properties of clove oil are used in medicine especially in dentistry, oral and pharyngeal treatments.
- It has wider applications in preparations of toothpaste and mouthwashes, soaps and perfumes. It is also reported to help diabetics in sugar assimilations.

Production techniques

Climate	High atmospheric temperature (25 to 35 degree C) with heavy sun light, good and well-distributed rainfall (above 150 cm) and high humidity (above 70%) are preferred.
Soil	Deep rich loams with high humus content and lateritic soils are best suited.
Propagation	Through seeds.
Time of sowing	June – July and September-October.
Planting	Seeds are extracted from ripe fruits and sown immediately in poly bags. The seeds germinate in five to six weeks. Dig 75 cm3 pits at a spacing of 6-7m. Fill the pits with top soil, 50 g *Azospirillum* and FYM 10 kg/pit. Two years old seedlings are planted in pits.
Manuring	One year old plants can be applied with FYM 15 kg, 20:20:60 g of NPK per plant in two splits during June-July and September-October. Every year the dosage is correspondingly increased and a bearing tree of 7 years old may be applied with FYM 50 kg, 300:300:960 g of NPK per plant 50 g in each of *Azospirillum* and *Phosphobacteria* to be applied one month after manuring.
Intercropping	Clove prefers partial shade. Under Indian conditions, it is best suited for mixed cropping in older coconut or Arecanut gardens or in coffee estates. Intercropping with banana is very good.
Irrigation	Frequent watering is essential in the initial stages in the absence of rains. Apply 8 litres of water either through drip or through basin during the months of January – May.

Harvesting	Bearing starts from sixth year onwards. The flower buds should be harvested when they are fully mature but before opening. The buds are harvested as clusters and separated and dried in the sun for five to seven days.
Yield	2-3 kg dried buds/tree. 1kg cloves have 11,000-15,000 dried cloves.

Allspice

Botanical name	*Pimenta dioica* (Linn.) Merill
Family	Myrtaceae
Origin	West Indies
Chromosome number	2n = 22
Ploidy level	Diploid
Type of inflorescence	Racemose cymes
Pollination system	Cross- pollination
Pollinator	Honey bees
Type of fruit	Berry
Commercial part	Fruit and seed
Main Phytochemical constituents	Eugenol, 4-hydroxy-3-methoxy cinnamaldehyde, 3,4-dimethoxycinnamaldehyde, vanillin and 3-(4-hydroxy-3-methoxypheny) pro-pane-1,2 diol

- *Pimenta dioica* (Linn.) Merill. Family: Myrtaceae, well known for its berries called Pimento, has been used as an important spice since time immemorial, for its culinary as well as medicinal qualities.
- *P. dioica* is a small evergreen tree native to the West Indies. It has been spread throughout tropical regions through trade as an ornamental species and for its spice known as allspice (Rodriquez, 1969).
- *Pimenta dioica* is native to the Caribbean region, especially Jamaica and Cuba.
- The basic chromosome number for the genus is x = 11 and *Pimenta dioica* is a diploid with 2n = 22.
- It is cultivated in certain isolated pockets of Kerala, Karnataka, West Bengal, Bihar and Odisha. Commercial cultivation is not popular in India.
- Allspice plant, of the myrtaceae family, grows about 9-12 m tall with erect trunk, bark grey; much branched, round topped and foliage dense.
- Leaves opposite, oval-oblong to elliptical, entire, leathery, glandular-pinnetate on lower surface; 12-20 cm in length, deep green and lustrous.
- Flowers are borne on racemose cymes, small, white to greenish-white, each flower having four tiny petals, a single style with one ovary, two ovules and a cluster of anthers.

- As the flower opens, the style straightens, and although the stiga is raised above the anthers, the flower appears to be hermaphrodite.
- Chapman (1966) considered the plant to be cross-pollinated by bees.
- The fruits have two kidney-shaped seeds (Weiss, 2002; Parthasarathy, 2007).
- Berries are globular 4-7mm in diameter, hard with rough surface and a reddish brown colour.
- Pericarp woody, brittle and around 1mm in thickness.
- The berry has two locules separated by a thin partition. Each locule contains a single, reniform, hard and dark brown seed.
- The dried berries range in size (6.5 to 9.5 mm in diameter) and there are 13 to 14 berries per gram.
- The quality of pimento is affected by factors like growing area, stage of maturity of berries at harvest and storage conditions.
- Allspice contains essential oils (2.5-4.5%) in both leaf and berry.
- A phenylpropanoid, *threo*-3-chloro-1-(4-hydroxy-3-methoxyphenyl) propane-1,2-diol, has been isolated from the berries of *Pimenta dioica* together with five known compounds, eugenol, 4-hydroxy-3-methoxy cinnamaldehyde, 3,4-dimethoxycinnamaldehyde, vanillin and 3-(4-hydroxy-3-methoxypheny) pro-pane-1,2 diol was found to show inhibition of autooxidation of linoleic acid in a water alcohol system (Kikiuzaki *et al.*, 1999).
- Aqueous extract of Pimenta berries showed 91.7% DMPO scavenging when determined by ESR spectra (Young *et al.*, 2002).
- However, the leaf oil has a different flavour profile even though the principal component is eugenol.
- The level of volatile oils can vary depending on their origin, weather, and harvest and processing conditions.
- The dried, green-mature fruit is the commercial flavourant and curing agent. Its extract is an essential component in the classic liqueurs Chartreuse and Benedictine, and in the local Jamaican drink Pimento Dram.
- The major use of allspice is in food industry (65 to 70%) in domestic use (5% to 10%), production of berry oil (20% to 25%), extraction of oleoresin (1% to 2%) and pharmaceutical and perfume industry.

- Berry, berry oil, oleoresin, leaf oil are products of economic use.
- It is used mostly in Western cooking and less suitable for Eastern cooking.
- It has medicinal, anti-microbial, insecticidal, nematicidal, anti-oxidant and deodorant properties.

Production techniques

Climate	Altitude : 1000 m above mean sea level, Annual rainfall : 100-200 cm, Ideal temperature : 27°C.
Soil and	Deep rich loams with high humus content and lateritic soils are best suited.
Propagation	Through seeds and vegetative propagation: Air layering in the month of January.
Planting	9-10 months old seedlings of 25-30 cm height are used for field planting.
Spacing	6 m x 6 m. Since the plant is dioecious in nature, it is desirable to keep a male: female.
Manuring	FYM (10 Kg) + 20:180:50 g of NPK per tree in the first year. For grown up trees of 15 years or more, FYM (50 kg) + 300:250:750 g of NPK / tree is applied as two split doses.
Growth regulator	Soil application of Paclobutrazol (1.25g/tree) induces flowering, increased number of panicles, flowers and setting of berries which in turn resulted in higher yield.
Harvesting	Clonally propagated plants start flowering in three years while the seed propagated plants in 6 years. Fully developed green berries are harvested.
Yield	20-25 kg dry berries per year/ tree.

Vanilla

Botanical name	*Vanilla planifolia* Andrews
Family	Orchidaceae
Origin	Brazil
Chromosome number	2n=32
Type of inflorescence	Raceme
Pollination system	Cross- pollination
Type of fruit	Capsule
Commercial part	Pods
Main Phytochemical constituent	Vanillin

- Natural vanillin is obtained from the cured pods (fruits) of the vanilla plant *Vanilla plantifolia* (Family: Orchidaceae).
- Vanilla is a perennial climbing orchid with sessile leaves and succulent green stems, producing aerial roots at the nodes.

- The inflorescence is a raceme with 20 or more flowers.
- Flowers are 6 cm long, 2.5 cm wide, either yellowish green or white.
- It is hermaphroditic (containing both male and female parts) and the flower itself is self-fertile, but incapable of self pollination.
- Vanilla's pollen is also largely inaccessible to most pollinating insects. It requires outside action to either transfer pollen from an anther to the stiga, or lift the thin-membrane preventing self-fertilization, or rostellum, and press the flower's anther to the stiga.
- Fruit popularly known as 'beans' or 'pod' is a capsule, nearly cylindrical and about 20 cm long.
- It is a native of Atlantic Coast from Mexico to Brazil.
- The fragrance and flavour of vanilla is due to numerous compounds produced during the curing process, among which vanillin is the most abundant. Other compounds are vanillic acid and an oleoresin.
- Vanilla is world's most popular flavourant for numerous sweetened foods, several commercial food products, liquors, perfumes etc.
- Vanilla extracts or essence are extracted with alcohol and contains the aroma and flavour principles and sweetening/thickening agents.
- They are widely used as a flavouring par excellence for ice creams, soft drinks, chocolates, confectionary, candy, tobacco, baked foods, puddings, cakes, cookies, liquors, and in perfumery.
- The important vanilla producing countries are Madagascar, Mexico, Tahiti, Malagasy Republic, Comoro, Reunion, Indonesia, Seychelles and India.

Production techniques

Climate	It thrives well in humid tropical climate with an annual rainfall 200 – 300 cm from sea level to 1500 m. A warm humid climate with temperature ranging from 21° C to 32° C is ideal.
Soil	It grows best in light, porous and friable soils with pH. 6-7. Clayey soils and water logged areas are not suitable for the plant. Partial shade is essential for successful cultivation.
Propagation	Through stem cuttings.
Season	Planting of cuttings should be taken up preferably during September-November.
Planting	Pit size 30 cm3 for standards and for planting of vanilla cuttings.
Spacing	Plains: 2.0 to 2.5 m X 1.2 – 1.5 m, Hills: 1.5 x 1.5 m.

Manuring	Apply 40 – 60: 20 – 30: 60 – 100 g of NPK per vine per year. It is given in 2 to 3 splits. Spray 1 % solution of 17: 17: 17 NPK mixture once in a month for boosting growth and flower production.
Training	Train the vines to a height of 1.2 to 1.5 m. Then trained horizontally or allowed to grow downward towards the ground. Horizontally trained vines are coiled round the pole connecting the two supporting trees. Vines trained to grow downward is allowed to touch the soil and allowed to root and again brought back upward on the same supporting tree and the same procedure is repeated
Flowering	Flowering starts in the third year of planting during December – January. Pinching of top 7.5 – 10 cm of vine 6 – 8 months before flowering seasons encourages flower bud initiation. Pruning of older fruiting branches also encourages flower production. Each inflorescence consists of 15 – 20 flowers
Pollination	The artificial pollination is useful in vanilla and pollination must be done on the same day as flowers start opening from 4.00 am and extend upto 1.00 pm. About 10 to 20 inflorescence may be pollinated in a vine. Normally 5 to 6 flowers in the lower side of inflorescence are pollinated. Hand pollination is done using a needle or a piece of pointed wood or a tooth pick to lift the hood covering the anther cap so that the anthers are brought into contact with stiga
Harvesting	The pods are ready for harvest in 6 to 9 months after flowering. The matured beans change colour from green to pale yellow. The right picking stage is when the distal end of the pod turns yellow. Daily picking of matured pod is essential. The pods are harvested by cutting with a knife
Yield	Average cured bean yield is 300 to 600 kg / ha / year. 6 kg of green pods produces 1 kg of cured beans

Black pepper (King of Spices)

Botanical name	*Piper nigrum* L.
Family	Piperaceae
Origin	Western Ghats of India
Chromosome number	x=13, 2n=52
Ploidy level	PolyPloidy level
Type of inflorescence	Pendent spike
Sex expression	Wild: dioecious, Cultivated: bisexual
Pollination system	Self-pollination
Type of fruit	Drupe
Commercial part	Dried mature berry
Main Phytochemical constituent	Piperine

- Black pepper (*Piper nigrum*), the king of spices, is one of the oldest and the most popular spice in the world.
- Black pepper, *Piper nigrum* L, is a native of humid, tropical, moist evergreen forests of Western Ghats of India.

- The name pepper comes from the Sanskrit word pippali meaning berry.
- It is now grown in Indonesia, Malaysia, Sri Lanka, Thailand, China, Vietnam, Cambodia, Brazil, Mexico, and Guatemala apart from the country of origin.
- Black pepper is a perennial woody climber, climbs on support trees by means of ivy like roots for flowering and fruiting.
- Black pepper is a vigorous vine with thick and rough old stem bearing numerous flowering plagiotropic branches. Runner shoots arise from the base.
- Leaves are of variable shape and size but generally ovate – lanceolate, thick, coreaceous, glabrous with 2 to 3 pairs of lateral ribs.
- Spike filiform, pendulous with sessile decurrent bracts formed into shallow cup subtending the anthers and gynoecium stamens 2 with dithecous anthers.
- The carperl is single with spherical ovary. The stiga is 3-5 lobed with no style.
- The inflorescence is a pendent spike, 3-15 cm long with 50-150 flowers.
- The fruit is a drupe, spherical, pungent and red when ripe.
- Wild forms usually dioecious but most cultivated ones are bisexual.
- Many cultivated type's exhibit protogyny with female phase preceding the male phase by few days or weeks (Ravindran *et al*, 2000).
- Black Pepper is the dried mature berry of *Piper nigrum*, a climbing, perennial shrub mostly found in hot, moist region of Southern India.
- Under cultivation pepper vines are trailed over support as columns, 5-6m tall, 1-2 meter diameter.
- The climbing woody stems have swollen nods with clinging roots at each node, which helps in anchoring the vine to the support trees (standards).
- It has straight upward growing main stem and have lateral shoots from the axils of the leaves having shorter inter nodes without adventitious roots.
- In such branches the terminal buds get modified into an inflorescence (Spike) and the auxiliary buds continue further growth.
- The root system confined to 75-100 cm radius and depth.

- Flowers are minute, white pale yellow, arranged spirally on fleshy pedantries.
- The specie is naturally self-pollinated and pollen dispersal is aided by the presence of water droplets.
- Fruit is a single seeded drupe often called berry. It is spherical in shape, green in colour, changing to red on ripping.
- The cultivated black pepper is having the somatic chromosome number of $2n=52$.
- Various studies on chromosome number among the cultivars and species of Piper indicate the existence of polyPloidy level series in the genus. All the species studied from South India and Sri Lanka could be traced to a common basic number of $x=13$.
- Most of the species including the polyploids exhibited normal meiotic behaviour suggestive of their alloploid nature.
- The sparse pairing of chromosomes in the hexaploid *P. betle* is suggestive of its hybrid origin.
- The chromosome data available on the genus show that this is chromosomally a homogeneous group.
- It was also suggested that the various *Piper* spp. represent a homogeneous assemblage where gene mutation or imperceptible chromosome changes have affected the evolution both at inter and intra specific levels.
- *P. nigrum* is a tetraploid. It was indicated that *P. nigrum* might have originated through hybridization between species occurring in Western Ghats.
- Pepper is largely used by meat packers and in canning, pickling, baking, considering for its preservative value.
- Other products in use are pepper oil, oleoresin, micro encapsulated pepper, green pepper in brine, dehydrated green pepper, frozen pepper etc.
- Black pepper is an essential ingredient in Indian system of medicine.
- Black and white pepper have two main components, the volatile oil and the pungent components, commonly known as piperine.
- The major functional properties of pepper are analgesic, anti-pyretic, anti-oxidant and anti-microbial.
- India is a leading producer, consumer and exporter of black pepper in the world.

- Kerala and Karnataka account for a major portion (92 per cent) of production of black pepper in the country.

Production techniques

Climate	Pepper requires hot and humid climate and grows between 20 degree North and South latitudes, from sea level to up to 1500 meters above MSL. The crop tolerates temperatures between 10 degree and 40 degree C. A well-distributed annual rainfall of 125 to 200 cm is considered ideal for pepper.
Soil	Black pepper can be grown in a wide range of soils with a pH of 4.5 to 6.5, though in its naturalhabitat it thrives well in red laterite soils.
Varieties	Panniyur 2, Karimunda, Sreekara, Subhakara, Panchami, Pournami, IISR Thevam, IISR Malabar Excel, IISR Girimunda, IISR Sakthi, PLD-2Lower elevation and less shady areas- Panniyur 1Higher elevation and more shady areas-KarimundaInter cropping in Arecanut – Panniyur 5Self-sterile varieties: Balankotta, KalluvaliHigh yielding clones: Subhakara, SreekaraPanniyur-4: Selected from KuthiravalyPanniyur-1,3: Hybrid between Uthrinkotta and Cherikaniankadam.
Propagation	Through runner shoot cuttings.
Planting	Slopes facing West and South should be avoided. Pits of 50 cm x 50 cm x 50 cm size are dug at a spacing of 2 to 3 m in either direction (Panniyur 1-3 x 3 m). 5 to10 kg of FYM/Compost is mixed with top soil and the pits are filled. Rooted cuttings of black pepper are planted in June-July @ two per standard. Silver oak, Dadap and Jack can be used as standards and should be planted at a spacing of 7–8 m.
Manuring	Apply cattle manure or compost @ 10 kg/vine-before the onset of South West monsoon. 100:40:140 g of NPK per vine in two split doses during May-June and September-October. Slaked lime at 500 g per vine is applied in alternate years during May-June. Apply *Azospirillum* @ 100 g/vine one month after the application of chemical fertilizers.
Irrigation	Protective irrigation in basins during December-May at 10 days interval.
Weeding	Two weeding are given during the months of June-July and October – November.
Growth regulator	To increase the berry size spray NAA @ 40 ppm. Fruit drop: The spike shedding can be reduced by foliar spray of Diammonium Phosphate 1.0 % four times *viz.,* before flower initiation (May), during new leaves and flower emergence (June) before spike initiation (July) and pinhead stage of berries (August).
Harvesting	Harvesting commences from third year onwards. The harvesting season is from November to March. Harvest is done by hand picking the whole spikes when few berries in the spike start turning red. The berries are separated and dipped in hot water (8°C) for one minute and sun dried for 7 to 10 days.
Yield	Yield: 2 to 3 kg/vine/year, 273kg/ha.

Major black pepper cultivars available in India

Cultivar	Important Characters
Aimpiryan	High yielder, high fruit set, excellent in higher elevations, good in quality
Arakkulammunda	Moderate and regular but early bearer, medium in quality.
Balankotta	Cultivar with large drooping leaves, moderate but early irregular bearing medium in quality, bold berries.
Bilimallegesara	Moderate yielder, good for Karnataka State.
Chengannurkodi	Moderate yielder from South Kerala, medium in quality.
Cheppukulamundi	Moderate yielder from central Kerala, medium in quality.
Cheriyakaniakadan	Popular in north Kerala, moderate and early bearing variety.
Jeerakamundi	Cultivar with small leaves, short spikes with high spiking intensity, and small berries; alternate bear.
Kalluvally	A promising north Kerala cultivar, good yielder, medium in quality, high dry recovery, drought tolerant.
Karimunda	Most popular cultivar suitable for most of the black pepper growing areas, high yielder, regular bearer, medium in quality.
Kottan	A cultivar found in north Kerala, moderate in yield, medium in quality.
Kottanadan	A high yielding, high quality cultivar from South Kerala, drought tolerant and regular bearer
Kurimalai	A cultivar from Karnataka, moderate yielder medium in quality.
Kuthiravally	A cultivar with long spikes, high yield and good quality.
Kuttianikodi	A moderate yielder from central Kerala with relatively long spikes and good spiking intensity.
Malamundi	A moderate yielder, medium in quality.
Malligesara	A common cultivar from Karnataka, relatively good in yield.
Manjamundi	A moderate yielder from north Kerala, medium in quality.
Narayakodi	Popular in south Kerala, moderate yielder, medium in quality.
Neelamundi	A good yielder from central Kerala, medium in quality, tolerant to *Phytophthora* .
Nedumchola	A cultivar with small leaves and short spikes, poor yielder.
Neyyattinkaramundi	A cultivar from central Kerala, medium in quality and yield.
Perambramunda	A cultivar from north Kerala, moderate yielder, medium quality.
Perumkodi	A cultivar from central Kerala, moderate in yield and quality.
Poonjaranmunda	A cultivar originally from central Kerala, sporadically found in gardens of north Kerala. Moderately good in yield with long spikes and quality.
Thommankodi	A cultivar from central Kerala, good yielder and quality.
Thulamundi	A central Kerala cultivar, medium in yield and quality.

Black pepper germplasm having specific traits

Traits	Cultivars/accessions/species
1. Tolerant/Resistant to foot rot (*Phytophthora capsici*)	Balankotta, Kalluvally, Narayakkodi, Neelamundi, Uthirancotta, P24, P1534, P339, Coll: 1041, C 1095, C 847, C 1090, HP 780, *P. colubrinum*
2. Tolerant/Resistant to slow decline *(Meloidogyne incognita)* (*Radopholus similes)*	Pournami (Variety), Acc. No. 4163, 4175, 1090, 334 (Cultivated), Acc. No. 3219, 3286, 3287, 3311 (wild) Acc. No. 820, 3141, 3200, 3283, 3291, 3299 (wild), HP 305 (Hybrid)
3. Tolerant/Resistant to 'pollu' (*Longitarsus nigripennis*)	Acc. No. 816, 841, 1084, 1114, 2070 (cultivar); *P. barberi, P. Chaba, P. hymenophyllum, P. longum*
4. Adaptable to high elevation	HP 34, HP 105, HP 812, HP 728, Coll.1041*; HP 105 & HP 813
5. Tolerant to drought	KS 51, KS 69, KS 114; Panniyur-5, Acc. No. 4216, 4226, 1343, 1368, 1226
6. Medicinal value	*P. longum, P.mullesua, P. betle, P.chaba*
7. *High quality* Essential oil	Balankotta, Kottanadan, Kumbhakodi, Acc. No. 41, 43, 49, Culture-5128
Oleoresin	Kottanadan, Kumbhakodi, Acc. No. 41, 164
Piperine	Kottanadan, Kumbhakodi, Acc. No.41, 164
8. Ornamental value	*P. crocatum, P. magnificum*

Crop improvement

Selection

- The selection programmes attempted in black pepper are selection from within germplasm (inter-cultivar selection), within a cultivar (intra cultivar selection), and selection in segregating open pollinated or selfed progenies.
- An elite plant once identified by any of these methods can form the basis of a new variety, which can be multiplied vegetatively and subsequently released after evaluation for yield and quality traits.
- Selection was carried out in the most popular cultivar of Kerala-Karimunda. Out of 216 elite plants evaluated for yield and quality and two lines under the name **Sreekara** and **Subhakara** were identified and released for cultivation (Ratnambal *et a*l, 1990).
- Panniyur-4, a clonal selection from cv. Kuthiravally and Panniyur-6, a clonal selection from local cv. Karimunda were released from Pepper Research Station, Panniyur (KAU) under AICRP on Spices, while PLD-

2 is a selection from cv. Kottanadan by the Central Plantation Crops Research Institute, Research Centre, Palode.

Selection from Germplasm

- Panchami and Pournami are the improved varieties developed through selection from germplasm.
- Panchami is a selection of Aimpiriyan with high yield and good quality and high fruit set, while Pournami was tolerant to root knot nematode – Meloidogyne incognita.
- Coll. 1041, a germplasm accession is found very promising in the experimental trials conducted at Valparai, at high elevation and this is also field tolerant to foot rot disease.
- This line was recommended for release recently for cultivation under the name "IISR Thevam". It is suitable for high altitude areas of South India (upto 3000 ft. MSL).

Selection from Open Pollinated Progenies

- Pepper, being heterozygous segregation is always expected in the open pollinated (OP) and selfed progenies.
- Because of the geitonogamous mode of pollination, the open pollinated progenies are comparable to selfed offsprings.
- Thus, there is a fair chance to locate useful genotypes in open pollinated progenies. Ibrahim *et al.* (1986) reported comparative genetic variability within the open pollinated progenies of a few varieties of black pepper.
- At the Pepper research Station, Panniyur, three varieties namely *Panniyur-2,* 5 and 7 were developed through selection from OP progenies of cvs. *Balankotta, Perumkodi* and *Kalluvally,* respectively.
- *Panniyur-7* is characterized by long spikes and high piperine content. IISR- Shakti, a high yielding variety tolerant to *Phytophthora* was selected (P24) from open pollinated progeny of Perambramundi.

Hybridization

- Wide variability exists among the cultivars of black pepper for yield and quality attributes.
- Hybridization in pepper mainly exploit the inter cultivar variability.
- For hybridization, cultivars are taken as parental units and crossings are

done among them.

- Genetic improvement through hybridization generally involves selection of parents, hybridization and selection of superior genotypes.

Inter-cultivar Hybridization

- Information on the breeding value, general or specific combining ability of pepper cultivars is not available for selection of parental combinations.
- In the absence of such information the available gene pool is used at random for inter-cultivar hybridization with parents selected based on assessing one or few good characters.
- Evaluation of progenies of many crosses at the Pepper Research Station, Panniyur led to the development of Panniyur-1 which is a selection of a cross between cv. Uthirancotta and Cheriyakanikkadan.
- A second variety, Panniyur-3, was also developed from the cross involving the same parents.
- At IISR, a large number of crosses involving many cultivars were made and the progenies tested for yield in preliminary trials.
- The promising F_1 plants were multiplied and planted for comparative yield evaluation. Among them three lines (HP-34, HP-105, HP-813) having good yield and adaptability for higher elevation have been identified and are in advanced stages of evaluation.
- HP 813 (Cholamundi X Panniyur-1) and HP 105 (Narayakodi X Neelamundi) were released under the names "IISR Malabar Excel" and "IISR Girimunda", respectively.
- Apart from these, a large number of inter-cultivar hybrids are under different stages of testing for yield and resistance to *Phytophthora*. Of the hybrids a few have shown high yield as well as tolerance to *Phytophthora.*

Inter-specific Hybridization

- In an attempt to bring resistance genes from wild relatives to cultivated types interspecific hybrids between *P. nigrum x P. attenuatum* and *P. nigrum x P. barberi* was successfully developed (Sasikumar *et al.* 1999).
- The hybrids exhibited distinct morphological and anatomical features but are not useful.

PolyPloidy level breeding

- A natural triploid (2n (3x) =78), Vadakkan bearing large leaves; very bold fruits with low fruit set was identified by IISR from the germplasm.
- The progenies of this cultivar exhibited wide morphological variations and varying chromosome numbers (cytotypes) (2n=52-104), however, none of these chromosomal variants have any horticulturally useful traits (Nair *et al*., 1993).
- An induced tetraploid (2n=104) was developed by treating the seeds of *Panniyur-1* with 0.05% colchicine (Nair and Ravindran, 1992).
- This tetraploid has larger and thicker leaves but the growth was slower than the diploid parents, and is difficult to establish in the field.
- A tetraploid variant having an extra bold fruits from *Panniyur 1,* was also reported by Ibrahim *et al*. (1987).

Mutation breeding

- Attempts were made using gamma rays as source of irradiation.
- Apart from seeds, rooted cuttings were also irradiated and raised. Irulappan *et al.* (1982) and Ravindran *et al.* (1986) used 1-4 kr gamma rays for inducing variability in *Karimunda, Panniyur 1, Kuthiravally, Kalluvally* (*Pulpally*), *Kalluvally* (Malabar), *Thommankodi* and *Aimpiriyan.*
- Irradiation adversely affected the germination of seeds. As the dose increased, germination was delayed. The M_1 population expressed morphological abnormalities such as chlorophyll changes, twinning of seedlings and rosette leaves.
- Chandy *et al*. (1980) did not observe any mutants in the M_1 by treating vegetative buds with EMS and further generations were not studied.
- Programmes using ionizing radiations for generation of variability for selection of improved genotypes are in progress in Sri Lanka and Malaysia with some encouraging results.

Released black pepper varieties and their characteristics

S. No.	Variety	Av. Yield-dry (kg/ha)	Driage (%)	Quality attributes (%)			Remarks
				Piperine	Oleoresin	Essentialoil	
1.	Panniyur 1 (Hybrid between Uthirankotta x Cheriyakaniyakadan)	1242	35.3	5.3	11.8	3.5	Spikes are long with large berries high in oleoresin. Early bearing, performs well under open situations. Suitable to all pepper growing regions. Not suited to heavily shaded areas.
2.	Panniyur 2 (Open pollinated progeny of Balankotta)	2570	35.7	6.6	10.9	3.4	Shade tolerant, rich in oleoresin and piperine. Suited to all pepper growing tracts of Kerala.
3.	Panniyur 3 (Hybrid between Uthirankotta x Cheriyakaniyakadan)	1953	27.8	5.2	12.7	3.1	Vigorous and late maturing, suitable for all pepper growing region, performs well under open situation. Long spikes & bold berries.
4.	Panniyur 4 (Clonal selection from Kuthiravally)	1277	34.7	4.4	9.2	2.1	Stable yielder, performs well under adverse condition including partial shade.
5.	Panniyur 5 (Open pollinated progeny of Perumkodi)	1098	35.7	5.3	12.3	3.8	This is a shade tolerant long spiked variety better suited for mixed cropping in coconut/ arecanut gardens. Tolerant to nursery diseases.
6.	PLD –2 (Selection from Kottanadan)	2475	-	3.3	15.5	3.5	High oleoresin line recommended for Trivandrum and Quilon

Contd.

S. No.	Variety	Av. Yield-dry (kg/ha)	Driage (%)	Quality attributes (%)			Remarks
				Piperine	Oleoresin	Essential oil	
							districts of Kerala.
7.	Subhakara (Selection from Karimunda)	2352	35.5	3.4	12.4	6.0	Adaptable to all pepper growing tracts. High quality line with high oleoresin and essential oil.
8.	Sreekara (Selection from Karimunda)	2677	35.0	5.1	13.0	7.0	Adaptable to all pepper growing tracts. High quality line with high oleoresin and essential oil.
9.	Panchami (Selection from Aimpiriyan)	2828	34.0	4.7	12.5	3.4	Late maturing variety with excellent fruit set.
10.	Pournami (Selection from Germplasm)	2333	31.0	4.1	13.8	3.4	High yielding variety, tolerant to root knot nematode. Suited to all pepper growing regions of Kerala.
11.	Panniyur 6 (Clonal selection from of Perumkodi)	2127	33.0	4.9	8.3	1.3	Steady and stable yielder tolerant to drought and adverse climatic conditions. Suitable for open condition as well as partial shade
12.	Panniyur 7 (Open pollinated progeny of Kalluvally)	1410	33.6	5.6	10.6	1.5	Vigorous, hardy and a regular bearer, long spike, high piperine tolerates adverse climatic condition suitable open and shaded conditions.

Curry leaf

Botanical name	*Murraya koenigii* Spreng.
Family	Rutaceae
Origin	India, Sri Lanka and other south Asian countries
Chromosome number	2n=18
Type of inflorescence	Terminal cyme
Pollination system	Self- pollination
Commercial part	Leaf
Main Phytochemical constituents	Pinene, sabinene, caryophyllene, cadinol and cadinene

- *Murraya Koenigii*, belongs to the family Rutaceae, commonly known as curry-leaf tree, is a native of India, Sri Lanka and other south Asian countries.
- The fresh leaves of *Murraya koenigii* are generally used in Indian cookery for flavouring foodstuffs.
- It is a small tree up to six meters in height and 15-40cm in diameter.
- Flowers, bisexual, white, funnel-shaped, sweetly scented, stalked, complete, ebracteate, regular, actinomorphic, pentamerous, hypogynous, the average diameter of a fully opened flower being 1.12 cm.
- Inflorescence, a terminal cyme, each bearing 60 to 90 flowers; calyx, 5-lobed, persistent, inferior, green; corolla, white, polypetalous, inferior, with 5 petals, lanceolate; length, 5 mm; androecium, polyandrous, inferior, with 10 stamens, dorsifixed, arranged into circles of five each; smaller stamens, 4 mm. long whereas the longer ones, 5 to 6 mm; gynoecium, 5 to 6 mm long; stiga, bright, sticky; style, short; ovary, superior.
- The long slender leaflets are dark green on top and paler underside.
- The leaves have a strong, warm curry aroma when bruised or rubbed.
- Its sensoric quality is fresh and pleasant, remotely reminiscent of tangerines.
- Fruits, round to oblong, 1.4 to 1.6 cm long, 1 to 1.2 cm in diameter; weight, 880 mg; volume, 895 microlitres; fully ripe fruits, black with a very shining surface; pulp, Wistaria blue 640/2; the number of fruits per cluster varying from 32 to 80.
- Seed, one in each fruit, 11 mm long, 8 mm in diameter, colour spinach green 0960/3; weight, 445 mg; volume, 460 microlitres.
- Seeds of curry leaf are chlorophyllous and poor storers. Generally the seeds are classified into 'orthodox' and 'recalcitrant' based on their desiccation and temperature sensitivity.

- Fresh leaves may contain up to 2.6% essential oil.
- The major constituent responsible for the aroma and flavour has been reported as pinene, sabinene, caryophyllene, cadinol and cadinene.
- Curry leaves are extensively used in Southern India and Sri Lanka (and are absolutely necessary for the authentic flavour), but are also of some importance in Northern India.

Production techniques

Soil and climate	Red sandy loam soils with good drainage are ideal for better leaf yield. The optimum temperature requirement is 26° to 37°C.
Varieties	Sen Kaampa, Dharwad-1 and Dharwad-2.
Propagation	Through seeds.
Sowing	With in 3-4 days of collection of fruits, the seeds should be pulped and sown in nursery beds or poly bags.
Planting	The field is ploughed 3-4 times to get a fine tilth. Before last ploughing well decomposed FYM is applied @ 20 t/ha. Pit size of 30 x30x30 cm is dug one to two months before planting at a spacing of 1.2 to 1.5 m. One year old seedlings are suitable for planting. One seedling is planted at the centre of the pit.
Manuring	After each harvest 20 kg of FYM/plant is applied and mixed with soil.
Intercultural operations	Periodical hoeing and in the first year intercrop like pulses can be grown. After attaining 1 m height, the terminal bud is cut off to encourage basal branching. In total 5-6 branches are maintained per bush. Ten to twelve months after planting the first harvest starts.
Irrigation	Immediately after planting the pits are irrigated. On the third day the second irrigation is given and then the irrigation is given once in a week.
Yield	At the end of first year 250-400 kg of leaves/ha can be harvested.

Chilli

Botanical name	*Capsicum annum* L., *Capsicum frutescens* L.
Family	Solanaceae
Origin	South America
Chromosome number	2n=2x=24
Type of inflorescence	Solitary axillary
Pollination system	Often cross pollinated crops
Type of fruit	Berry
Commercial part	Green as ♀♂ well as ripe and dried pod (fruit)
Main Phytochemical constituents	Capsaicin, capsanthin

- *Capsicum annuum* L. is a domesticated species of the plant genus Capsicum in the family Solanaceae, native to South America.
- Genus *Capsicum* includes 22 wild species and three varieties as well as five domesticated species and their wild relatives.
- The domesticated *Capsicum* spp. Can be grouped into two groups depending upon the cross compatibility.
- The white flowered which includes *C, annum, C chinense C. fruitescens* (*C. annum* group) and *C. baccatum var. baccatum* and *C. baccatum var. pendulum* (*baccatum* group).
- The purple flowered comprising of *C. pubescens, C. eximinum* and *C. cardenasii.*
- All the species in the genus have n=12 except *C. ciliatum* and *C. scolnikianum* which have n=13.
- The species of the genus are classified into two groups, according to the number of chromosomes: one with 2n=2x=24 chromosomes and another with 2n=2x=26 chromosomes.
- Domesticated species belong to the first group (Moscone et al. 2007, Pozzobon et al. 2006 and Pozzobon and Schifino-Wittmann 2006).
- Chilli is considered as one of the most important commercial spice crops and is widely used universal spice, named as *wonder spice*.
- Capsicum is derived from the Greek word "Kapsimo" meaning "to bite."
- Genus *Capsicum* is divided into three sections by Hunziker-Monotypic Tubocapsicum , Pseudoacnistus and Capsicum.
- Chilli was introduced into India from Brazil in the 16th Century by the Portuguese
- Chillies are referred to as chillies, chile, hot peppers, bell peppers, red peppers, pod peppers, cayenne peppers, paprika, pimento, and capsicum in different parts of the world.
- The original Mexican term, *chilli* (now *chile* in Mexico) came from the Nahuatl word *chilli* or *xilli*, referring to a larger *Capsicum* variety cultivated since 3000 BC, as evidenced by remains found in pottery from Puebla and Oaxaca.
- The world's hottest chilli "Naga Jolokia" is cultivated in hilly terrain of Assam in a small town Tezpur, India.

- India is the largest producer and exporter of chillies.
- India has immense potential to export different types of chillies required by various markets around the world. It is the leader in export, with 25% share in world trade, followed by China with 24% share in total global export.
- *Capsicum annuum* is an annual sub –shrub, the flowers of which are borne singly and fruits usually pendent, which provide red peppers, cayenne, paprika and chillies and sweet pepper (bell pepper) a mild form with large inflated fruits.
- *Capsicum frutescence* is a perennial chilli with small sized pods which are highly pungent. It is commonly known as 'bird chilly' and 'Tabasco'.
- Capsicum plants are herbaceous or semi-woody annuals or perennials.
- The leaves are ovate, tapering to a sharp point, measuring up to 15 cm, dark green on the upper surface and pale green on the lower surface.
- The glabrous stem is angular and the inflorescence terminal, flowers are axile in placentation borne at nodes. The corolla is whitish and calyx greenish.
- The flowers are small, white and borne singly or in clusters of 2 or 3 in the axils of the leaves.
- Capsicum (*Capsicum annuum*) are generally self-pollinating, although cross-pollination is also common.
- The stiga (♀) is receptive before pollen is released from the anthers (♂). Pollen is released a few hours after the flower opens, but in some cases no pollen is released at all. The stiga can be either shorter or longer than the stamens (♂), and in some cases the stiga may touch the anther, causing self pollination.
- Capsicum flowers produce nectar at the base of the flower, with the quantity of nectar produced varying between varieties.
- The fruit is a berry and may be green, yellow, or red when ripe. The fruits are of diverse shapes and sizes depending upon the variety.
- Dry chilli is extensively used as spice in curried dishes. It is also used as an ingredient in curry powder and in seasonings.
- Bird chilli is used in making hot sauces as pepper sauce and Tabsco sauce.
- Paprika, Bydagi chilly, Warangal chapatta and similar high colour less pungent varieties are widely used for colour extraction.

- The pungency for the red peppers and the colour value for the paprikas are the most important parameters.
- Pungency in chilli is due to chemical compounds known as capsaicinoids, which are alkaloid compounds found only in the plant genus, *Capsicum.* Capsaicin, also known as n-vanillyl-8-methyl-6-(e)-noneamide, is the most pungent of this group.
- It is sparingly soluble in water but highly soluble in fats, oils and alcohol.
- It is a tasteless, odourless white crystalline substance.
- The Scoville organoleptic test is a refined, systematic approach to measure the pungency level.
- It was the first laboratory test reported and used to measure the pungency in chillies .
- One 'Scoville Heat Unit' corresponds to about fifteen parts per million (ppm) of the capsaicin.
- The pure capsaicin has the maximum possible Scoville rating of 15 million units (Scoville, 1912).
- Capsaicin and dihydrocapsaicin together make up 80- 90 per cent of the capsaicinoids found in hot peppers.
- In *Capsicum annuum* species, the total capsaicinoid content ranges from 0. 1 to 1.0 percent and the capsaicin to dihydrocapsaicin ratio is about 1:1.
- In *Capsicum frutescens,* the total content of these compounds ranges from 0.4 to 1.0 percent with the ratio around 2: 1.
- The whole chilli consists of 40 per cent pericarp containing an inner sheath known as dissepiment, 56 percent seeds and 4 per cent stalks.
- The pericarp contains al most all of the pungency, whereas the chilli seeds contain only traces of pungency with a capsaicin content of 0.005 percent.
- The capsaicin content of chilli oleoresin ranged from 2.5 to 3 per cent.
- The pungency of the pericarp is mostly concentrated in the dissepiment.
- The colour value is the principal criterion for assessing the quality of paprika.
- The pigent content ranged from 0.1 to 0.8 per cent. The extractable colour value is measured by a spectrophotometric process, usually expressed in

terms of ASTA colour value (American Spice Trade Association).

- Common extractable colour values present in the industry are 85, 100, 120 and 150 (Anu and Peter, 2000).
- It is reported that approximately 20 carotenoids contribute to the colour of *Capsicum* powder.
- The major red colour develops from capsanthin and capsorubin, while the yellow orange colour is from beta-carotene and violaxanthin.
- Capsanthin, the major carotenoid of ripe fruits, contribute up to 60 per cent of the total carotenoids.
- Other carotenoids present are capsorubin, zeaxanthin, lutein, kryptoxanthin and alpha and beta-carotene.
- This colour is highly popular among food and beverage processors for its use as a colourant, since this being a 'natural plant colour'.
- Capsicum has a tonic and carminative action. Taken inordinately it may cause gastro-enteritis.
- Oleoresin capsicum is used in pain balms and vaporubs.
- Dehydrated green chilly is a good source of vitamin 'c'.

Production techniques

Climate	Long, warm growing season, frost free for 4-5 months is ideal. In tropical southern India the crop is raised during winter months (August – December) is desirable and hence temperature-21°C at night & 27 °C in day. Optimum temperature for seed germination 18-24 °C.
Soil	Well drained, light fertile loam soil with moisture holding capacity is ideal. pH 6-7.
Varieties	Arka Abir, Kt-Pl-19, Byadagi.
Seed rate	450-500g seeds are required to raise seedlings to plant one hectare of land.
Transplanting	Age: 35 – 40 days old seedlings are ready for transplanting. Hills: 45 x 45 cm. Plains: 60 x 45 cm.
Manuring	200-250 q FYM, 120kgN, 60kgp and 50kgK/ha.
Harvesting	The matured fruits are ready for harvest after 70-80 days of planting. Fruits are harvested 3 to 6 times in a crop. The harvesting is spread over a period of 3-4 months.
Yield	25 – 35 q dry chilli/ha.
Storage	Chillies on harvesting have a moisture content of 60-88% depending on whether partially dried on the plant or harvested while still succulent, this must be reduced to 8-10% to prepare dried spice. Out of 100 kg of fresh fruits, 25-35kg of dried fruits may be obtained.

Varieties/Hybrids

Variety/Hybrid	Salient features
IARI, New Delhi	
Pusa Jwala	Plants dwarf, bushy, light green; fruits 9-10 cm long, light green, ripe fruits light red, highly pungent; fairly tolerant to thrips and mites. Average yield 85 q/ha (green) and 18 q/ha (dry).
Pusa Sadabahar	Plants erect, perennial (2-3 years), 60-80 cm tall; fruits 6- 8 cm long, borne in clusters with 6-14 fruits per cluster, ripe fruits dark red, highly pungent, resistant to CMV, TMV and leaf curl complex; first picking in 75-80 days after transplanting. Average yield 95 q/ha (green) and 20 q/ha (dry).
IIHR, Bangalore	
Arka Lohit	Fruits dark green turning deep red on maturity (Capsanthin 0.21%); Fruits highly pungent (Capsaicin 0.71%); Suitable for irrigated and rainfed cultivation. Average yield 25 t/ha (Green Chilli) and 3 t/ha (Dry chilli).
Arka Suphal	Fruits green, smooth, medium long (6-7 cm x 1 cm), pendent, turning deep red on maturity. Suitable for irrigated and rainfed cultivation. Tolerant to powdery mildew. Average yield 25 t/ha (Green Chilli) and 3t/ha (Dry Chilli).
Arka Harita	F1 hybrid, Fruits light 10 cm and width 1cm. Fruits are dark green and turn to red on maturity. Resistant to powdery mildew and viruses. Average yield 38.2 t/ha green chilli or 5.86 t/ha dry chilli.
Arka Meghana	CMS based high Yielding F1, hybrid with resistance to powdery mildew and viruses Fruit length 10.6 cm and width 1.2 cm. Very early types, Fruits are dark green and turn to red on maturity. average yields 33.5 t/ha green chilli or 5 t/ha dry chilli.
Arka Sweta	CMS based high yielding F1, hybrid for fresh market. Suitable for *kharif* & *rabi* seasons under irrigated conditions. Fruit length 11-12 cm, width 1.2-1.5 cm; smooth and medium pungent. Fruits light green and turn red on maturity. Tolerant to viruses Duration 160 days. Yields 33 t/ha (fresh) & 5 t/ha (dry).
IIVR, Varanasi	
Kashi Early	Plants of are tall (100-110 cm height) without nodal pigentation on dull green stems and bear pendant fruits. Fruits are long (8-9 x 1.0-1.2 cm), attractive, dark green and turn bright red at physiological maturity, pungent with smooth surface. First picking of the green fruits starts at about 45 days after transplanting. The average yield of this hybrid is 250 q/ha (red ripe).
Kashi Anmol	Plants are determinate, dwarf (60-70 cm) with nodal pigentation on stem and bear green attractive pendant fruits. First picking starts 55 days after transplanting; yield 200 q/ha.
Kashi Surkh	Plants are semi determinate (1-1.2 m), erect and nodal pigentation on stem. Fruits are light green, straight, length 11-12 cm, suitable for green as well as red fruit production. First harvest starts after 55 days of transplanting. Green fruit yield is 240 q/ha.

Contd.

Variety/Hybrid	Salient features
HAU, Hisar	
Hisar Shakti	Fruits upright in clusters, pointed, thin, medium long, highly pungent, tolerant to leaf curl mosaic virus, fruit rot and powdery mildew. Average yield is 170-190 q/ha.
Hisar Vijay	Early, fruit similar to Hisar Shakti but with blackish tinge on fruit surface under low temperature, tolerance like, Hisar Shakti. Average yield is 150-170 q/ha.
AAU, Anand, Gujarat	
Chilli 101	The fruits are medium pungent, long, compact and strait with green shining colour, more shelf life and high ascorbic acid content. Average green fruit yield of 145 q/ha.
Chilli 111	The fruits are attractive, medium long, straight, pungent and slightly wrinkled surface with good shining. Average yield is 121 q/ha.
Chilli 131	The fruits are elongated, straight, compact, short pedicel length, smooth shining surface and pointed blossom end. The fruits having light green colour and non-pungent with good aroma. Average yield is 89 q/ha.
Chilli 112	The fruit of this variety are pungent, elongated straight, light green colour and semi wrinkled surface. Average yield is 127 q/ha.
Chilli hybrid-1	The fruits of this hybrid are elongated straight with pointed blossom end having semi wrinkle surface and light green colour. Average yield is 273 q/ha.
G.B.U.A.&T, Pant Nagar	
Pant C-1	The green pod yield of this variety is 75 q/ha. The dry pod yield is about 15 q/ha. This variety is easily distinguishable from the other varieties because of the fact that it has upright fruiting pods. The pods are highly pungent, relatively small in size, broader at the base and narrow towards the tip. It is moderately resistant to mosaic and leaf curl virus. The green pod yield of this variety is 75 q/ha. The dry pod yield is about 15 q/ha.
MPKV, Rahuri	
Phule Jyoti	Plants are tall and spreading. Fruits are smooth and medium long. Tolerant to thrips, mites and wilt. Suitable for kharif season, average yield 250 q/ha. (Green).
Phule Mukta	Resistant to powdery mildew and Fusarium wilt. Fruits are small, dark green and smooth. Suitable for kharif season. Average yield 15 q/ha. (red dry).
TANU, Coimbatore	
CO 1	Fruits long and red at ripening; seed content 55%; samba type. Average yield 2 t/ha
CO 2	Fruits oblong and bright red on ripening; seed content 60%, gundu type. Average yield 2 t/ha.
K1	Plants bushy; pods green; long samba type; dark red on ripening. Average yield 2 t/ha.
K2	Plants tall, compact, pods long and bright red on ripening; samba type; high seed content. Average yield 2 t/ha.

Variety/Hybrid	Salient features
PMK.1	Suitable for semi-dry condition. Fruits contain high capsaicin 0.36%, gundu type. Average yield 3 t/ha.
PKM 1	Hybrid derivative of the cross between AC. No. 1797 x CO.1. It is suitable for cultivation under irrigated conditions. Bold pods, which are dark red in colour. The number of pods per plant is 204 with 85 seeds per fruit. Average yield 3 t/ha (Dry pod).
MDU.1	Determinte growth. Fruits borne in clusters of 4-9. Fruits are long, dark red and shiny. Average yield 2 t/ha.
CO 3	Fruits long, slender and dark green. Suitable for export. Average yield 4 t/ha.
PLR1	Suitable for green chilli. Average yield 19 t/ha. (Green).
CO 4	Vegetable chilli, fruits are dark green, less pungent and suitable for high density planting. Average yield 3 t/ha (Green).
TNAU Chilli Hybrid Co1	F1 hybrid developed by crossing Selection 1 x CA97. Fruits are long, smooth and dark green. It possesses 0.58% capsaicin Suitable for green and dry chilli. Moderatelt resistant to fruit rot. Average yield 28 t/ha (Green) and 7 t/ha. (Dry).

Cardamom small (Queen of Spices)

Botanical name	*Elettaria cardamomum* (Maton.)
Family	Zingiberaceae
Origin	India
Chromosome number	x=12, 2n=48
Type of inflorescence	Panicle
Sex expression	Bisexual
Pollination system	Cross-pollination
Pollinators	*Apis cerana* and *Apis dorsata*
Type of fruit	Capsules
Commercial part	Fruit (Capsule)
Main Phytochemical constituent	The volatile oil contains about 25-40% cineole, 30-40% a-terpinyl acetate and about 1-2% limonene

- Cardamom *(Elettaria cardamomum* Maton) acclaimed as the 'Queen of Spices' is the true cardamom belonging to the family Zingiberaceae under the natural order Scitaminae. It is one of the most important and highly priced spices.
- It is one of the highly prized spices of the world. India is considered the native home of *Elettaria cardamomum.*
- Cardamom is a herbaceous perennial (2-5 m in height) with underground (subterranean) rhizomes and aerial leafy stems (tillers) made of leaf sheaths.

- Inflorescence is a long panicle arising from the underground stem, but comes up above the soil.
- Flowers are arranged in clusters (known as *cincinni*) subtended by scale leaves.
- Flowers are bisexual, bracts linear, oblong and persistent, sepals 3, petals 3, unequal, lip longer with violet tinge carpels 3, style 1, ovary-trilocular, axile placentation, ovules-numerous in each carpel.
- The peak flowering is spread over a period of six months from May to October.
- The time required to reach full bloom stage from flower/bud initiation ranges from 26 to 34 days and capsule development takes about 110 to 120 days from the full bloom stage.
- In general maximum number of flowers open during early hours of the day 3.30 – 8.00 AM immediately followed by the anthesis.
- The fruit are tri-locular, ovoid or oblong, greenish-brown capsules containing about 15-20 seeds attached to axile placenta. Light reddish or dark reddish brown seeds are irregularly 3-sided, transversely wrinkled or furrowed and are covered by a membranous aril.
- Cardamom has bisexual flowers, self compatible but cross-pollination is more common.
- *Apis cerana* and *Apis dorsata* are the predominant pollinators.
- Cardamom flowers remain in bloom for 15-18 hours and stiga receptivity and pollen viability were reported to be maximum during morning hours between 8 AM and 10 AM.
- Pollination during this time gives about 72% fruit set. Thereafter, the stiga receptivity decreased gradually resulting in the minimum fruit set of 24%. The active foraging of bees is seen in the morning hours of the day resulting higher fruit set in cardamom.
- The somatic chromosome number of cardamom is reported to be 2n = 48 (Gregory, 1936; Sharma and Bhattacharya, 1959).
- Earlier workers have reported that cardamom is of amphidiploid origin from wild species, which are probably extinct. Allied genera such as *Globa, Balbifera, Phoemaria, Amomum* sp. and *Alpinia* spp also possess 2n=48 and are considered to be evolved from a common basic number, X=12.

- Indian cardamom is offered to the international markets in different grades: 'Alleppey Green Extra Bold' (AGEB), 'Alleppey Green Bold' (AGB) and 'Alleppey Green Superior' (AGS) are names that register instant appeal worldwide.
- Cardamom oil is a precious ingredient in food preparations, perfumery, health foods medicines and beverages.
- Indian cardamom enjoys a premium preference in the Middle East, Japanese and Russians who relish it for its distinct enriching properties.
- Cultivation of cardamom is mostly concentrated in the ever green forests of Western Ghats in South India.
- Besides India, cardamom is grown as a commercial crop in Guatemala and on small scale in Tanzania, Sri Lanka, El Salvador, Vietnam, Laos, Thailand, Cambodia, Honduras, and Papua & New Guinea.
- The optimum altitudinal range on growing cardamom is 600 to 1500 m above MSL. The cardamom growing regions of South India lies within 8-30 degree N latitudes and 75-78 degree longitudes.
- The major use is for the preparation of 'gahwa' – a strong cardamom coffee concoction which is a symbol for hospitality among Arabs.
- Apart from this cardamom is widely used as a flavouring material in whole and ground form.
- Cardamom oil and oleoresin has applications in flavouring processed foods, cordials, and liquors and in perfumery and in Ayurvedic medicines.
- Cardamom contains between 2% and 10% essential oil, out of which the major portion is in seeds.
- The volatile oil contains about 25-40% cineole, 30-40% a-terpinyl acetate and about 1-2% limonene.

Production techniques

Climate	It is a shade loving plant thriving well in elevations up to 600-1200m above MSL under an average annual rainfall of 1500-4000mm and temperature range of 10-35^{0}C.
Soil	Thick shady areas with loamy soil are ideal. Adequate drainage must be provided.
Varieties	Malabar- Mudigree-1 and 2, PV 1, ICRI 1, ICRI 3, TKD 4, IISR Suvarna, IISR Vijetha, IISR Avinash, Mysore-ICRI 2, Vazhukka – PV2, Njallani (Green gold).
Propagation	Through Seeds and suckers.

Seed rate	600g/ha (fresh seeds), Treat with commercial grade Sulphuric acid or Hydrochloric acid for 20 minutes. Wash with water.
Time of planting	June – December.
Planting	Larger types: 2.5 x 2.0 m; Smaller types: 2.0 x 1.5 m.
Manuring	Apply compost 25 t/ha; 75:75:150 kg of NPK per ha in two split doses during June-July and October – November.
Irrigation	Generally Cardamom is grown as a rainfed crop, but sprinkler irrigation may be provided during summer for increased yields.
Shade regulation	Moderate shade – 50-60 %, Rainfed – 40 – 50 %, Irrigated – 55-60 %, Mixed population of medium sized shade trees- Karona, Red cedar, konikonna, Jack, Vellakil, Thempavu, Thambahom, Bolongi, Elangi.
Weeding	Spraying of Paraquat (625 ml in 500 litres of water/ha) may be done only in interspaces between rows leaving 60cm around the plant base. Spraying may be done in pre-and post-monsoon periods.
Harvesting	Harvesting commences from third year onwards. Economic yields from 5th year. Harvesting is done once in a month. Pick only those fruits which are just ripe but not fully ripe. Fully ripe fruits tend to split on drying and do not develop the desirable dark green colour.
Yield	Yield: 200-250 kg/ha.

Crop improvement

- The main focus of Cardamom breeding in addition to high yield are resistance to biotic stress viz., viral diseases such as 'katte' and 'kokke kandu' and fungal diseases such as rhizome rot, clump rot and capsule rot; drought tolerance; plants with bold capsules with more number of seeds/fruit; Higher percentage of capsule dry recovery (>22%), higher percentage of essential oils, α -terpenyl acetate which is responsible for the aroma and flavour and varieties with wide adaptability.
- Cardamom breeding depends on selections from germplasm and from open pollinated progenies of popular cultivars.
- IISR Vijetha is katte virus tolerant line while IISR Avinash and ICRI 4 are relatively tolerant to rhizome rot. PV 1 has long and bold capsules.
- The variety CCS 1 has compact growth habit and is suitable for high density planting.
- Hybridization between NKE, RR, extra bold and Multibranch types are in progress with an aim to evolve desirable types.

Clonal selection

- All the existing improved varieties have been evolved by selection for desirable characters such as higher yield and superior capsule characters.
- Selection in cardamom is based on both qualitative and quantitative characters from preliminary, comparative yield trial and multiplication trials to confirm the superiority of the selected clone.

Hybridization

- Intervarietal hybridization was made between identified superior cultivars for deriving lines with high yield, *'katte'* resistance and drought tolerance.

Intergeneric hybridization

- In an effort to bring **Katte resistance** from wild relatives to cultivated cardamom , inter-generic crosses were made using *Ammomum subulatum*, *Alpinia neutans*, *Hedychium flavascence* and *Hedychium coronarium* as male parents (Parameshwar, 1977).
- Cross with *A. neutans* set a few fruits and in other cases no fruit formation was noticed. Compatibility barriers prevented the formation of fruits in these cross combination (Madhusoodanan *et al.*, 1994).

Mutation breeding

- Effort has been made to develop genotypes tolerant to cardamom mosaic (*katte*) virus, drought, and better quality through treatment of cardamom seeds and rhizomes with different mutagens such as γ-rays and Nitrosomethyl Urea (NMU), Diethyl Sulphate (DES) and Ethyl Methyl Sulphate (EMS) but no desirable mutant could be identified so far.

Poly Ploidy level breeding

- Polyploids were induced in cardamom by treating the sprouting seeds with 0.5 per cent aqueous solution of Colchicine (Sudharshan, 1989). The polyPloidy level lines exhibited increased layer of epidermal cells, thick cuticle and thicker wax coating on the leaves which are the general characters associated with drought tolerance in nature.

Released varieties of cardamom with yield and quality characteristics

S. No.	Variety	Source	Yield (kg/ha)	Essential oil %	1, 8 cineole %	α -Terpenyl acetate %	Capsule shape	Areas recommended for cultivation
1.	IISR Coorg Suvasini	IISR, CRC Appangala	409	8.7	42	37	Oblong	Kodagu and Hassan districts of Karnataka
2.	PV-1	KAU, Pampadumpara	260	6.8	33	46	Long	All cardamom tracts of Kerala & Karnataka
3.	Mudigere 1	UAS, Bangalore	275	8.0	36	42	Oval	Malnad region of Karnataka
4.	Mudigere 2	UAS, Bangalore	476	8.0	45	38	Round	Traditional cardamom growing Tracts of hill zones of Karnataka
5.	ICRI-1	ICRI, Myladumpara	325	8.3	29	38	Round	South Idukki zone of Kerala
6.	ICRI-2	ICRI, Myladumpara	375	9.0	29	36	Oblong	Vandanmettu and Nelliampathi zones of Kerela
7.	ICRI-3	ICRI, Myladumpara	439	6.6	54	24	Oblong	Hill zones of Karnataka
8.	ICRI-4	ICRI, Thadiyankudisai	455	6.4	—	—	Globose	Lower Pulneys in Tamil Nadu
9.	IISR Avinash	IISR, CRC, Appangala	847	6.7	30.4	34.6	Oblong	Rhizome rot infested areas
10.	IISR Vijetha 1	IISR, CRC, Appangala	643	7.9	44.9	23.4	Oblong	Moderate to high shaded mosaic infested areas
11.	PV-2	KAU, Pampadumpara	982	10.45	—	—	Long	Cardamom hill reserves of Kerala
12.	Njallani Green Gold	Farmers Selection	1600	—	—	—	bold	All cardamom growing regions

Ginger

Botanical name	*Zingiber officinale* Rosc.
Family	Zingiberaceae
Origin	South East Asia
Chromosome number	x=11, 2n=22
Type of inflorescence	Spike
Sex expression	Trimerous, bisexual
Type of fruit	Capsule
Commercial part	Underground rhizome
Main Phytochemical constituents	Gingerol, shogoal and zingerone

- Ginger (***Zingiber officinale*** Rosc.) (Family: ***Zingiberaceae***) is an herbaceous perennial, the **rhizomes** of which are used as a spice.
- India is a leading producer of ginger in the world.
- The word ginger is derived from a **Sanskrit** word **singabera** meaning 'shaped like a deer's antlers (horn)'.
- It is believed to be a native of **South East Asia.**
- Ginger is an herbaceous perennial having underground branched rhizome with small scales.
- Leaves are sheathing arranged alternatively, linear lanceolate, gradually acuminate and glabrous.
- Flowers are borne on a **spike** produced in a peduncle different from the aerial leafy stem, arising directly from the rhizome.
- Flowers are many, **trimerous bisexual**, irregular, epigenous, yellow in colour with dark purplish spots.
- Outer periapt is cylindric, shortly three lobed, inner periapt tube is cylindric, lobes are lanceolate, one stamen perfect, two combined into a petaliferous leaf, the labellum.
- Androecium consists of stamens of which the outer 3 are reduced to stamenoids. The inner lateral stamens are united and showy to form a deep purple coloured labellum.
- The posterior stamen of the inner whorl is the only fertile stamen, which is enclosed by the labellum. The filament is flat and short with two prominent anther lobes.
- The style passes through the grove formed by the anther lobes and ends in a capitate stiga. Anther cells are contiguous, produced into a long beak.

- Ovary is inferior, three carpelled, three celled. Ovules are many on axial placentation. Style is long, delicate, lying in a groove in the stamen.
- Stiga is small and sub-globose. Fruit is very rarely produced, which is an oblong capsule. (Ravindran and Babu 2005; Purseglove 1981). But *Zingiber officinale* is not known to set seeds.
- The somatic chromosome number of ginger is 2n = 22 in all species except in *Z. mioga* (2n=55).
- Ginger of commerce is the dried underground stem of the herbaceous tropical plant grown as an annual.
- The whole plant is refreshingly aromatic and the underground rhizome, raw or processed, is valued as spice.
- Ginger is a slender perennial herb, 30-50 cm tall with palmately branched rhizome bearing leafy shoots.
- The leafy shoot is a pseudostem formed by leaf sheath and bears 8 to 12 distichous leaves.
- Ginger contains 1.5-3.0% volatile oil.
- The volatile oil is composed mainly of sesquiterpene hydrocarbons.
- Oxygenated sesquiterpenes are present up to 17% and the remainder is composed of monoterpene hydrocarbons and oxygenated monoterpenes.
- Of the sesquiterpene hydrocarbons, about 20-30% is (-) a -zingeberene, up to 12% (-)-b -bisabolene, up to 19% (+)-ar-curcumene and up to 10% farnasene.
- The pungent components of ginger are three compounds: gingerol, shogoal and zingerone.
- Gingerols are a series of compounds with the general structure, 1-(4'-hydroxy -3' methoxyphenyl) -5-hydroxyalkan-3 ones.
- They are mainly condensation products of zingerone. The presence of zingerone and shogoals in fresh extract is non existent. They are formed during the drying process.
- Fresh ginger, dry ginger powder, oleoresin and oil are used in food processing.
- It is indispensable in the manufacture of ginger bread, confectionary, ginger ale, curry powders, certain curried meats, table sauces, in pickling and in the manufacture of certain cordials, ginger cocktail, carbonate drinks, liquors etc.

- In medicine, it is used as carminative and stimulant. It has wider applications in indigenous medicines. The ginger oil is used as food flavourant in soft drinks.

Production techniques

Climate	Ginger grows well in warm and humid climate and is cultivated from sea level to an altitude of 1500 m above sea level. Ginger can be grown both under rain fed and irrigated conditions.
Soil	A friable loam rich in humus is ideal. However, being an exhausting crop it is not desirable to grow ginger in the same soil year after year.
Varieties	Rio de Janeiro, Maran, Nadan, Suruchi, Surabi, Suprabha, IISR Varadha, IISR Rejatha, IISR Mahima, Athira & KarthikaSurbhi: Mutant cultivarRio de Janeiro: exotic cultivar Himgiri: Tolerant to rhizome rot.
Propagation	Through rhizomes.
Seed rate	15 – 18q of rhizome/ha.
Time of sowing	The best time for planting ginger in the West Coast of India is during the first fortnight of May with the receipt of pre-monsoon showers.
Spacing	Irrigated crop – 40 x 20 cm in ridges and furrows. Rainfed crop – Raised beds of 30 x 30 cm or 25x 25 cm.
Manuring	Basal: FYM 25-30 t + 30 tonnes green leaves as mulch in three splits : 15 tonnes immediately after planting, 7.5 tonnes – 60 days and 120 days after planting, 50: 25 kg of P and K per haTop dressing: 37.5: 12.5 kg of N and K per ha applied on 45th and 90th day after planting.
Intercultural operations	Mulching is done at the time of planting with green leaves. After each top dressing, earthing up of plants is done.
Irrigation	Ginger requires 1300-1500 mm of water during its crop cycle. The critical stages for irrigation are during germination, rhizome initiation (90 DAP) and rhizome development stages (135 DAP). The first irrigation should be done immediately after planting and subsequent irrigations are given at intervals of 7 to 10 days in conventional irrigation.
Weed control	Weeding is done just before fertilizer application and mulching; 2-3 hand weedings are required depending on the intensity of weed growth. Proper drainage channels are to be provided when there is stagnation of water.
Earthing	Earthing up is essential to prevent exposure of rhizomes and provide sufficient soil volume for free development of rhizomes. It is done at 45 and 90 days after planting immediately after weeding and application of fertilizers.
Harvesting	Ginger attains full maturity in 210-240 days after planting when leaves start yellowing and drying.
Yield	12-15 t/ha. The yield of dry ginger is about 19-25 per cent of fresh ginger depending on the variety and climatic zone.
Processing	Processing of ginger to produce dry ginger basically involves two stages-peeling of the ginger rhizomes to remove the outer skin and sun drying to a safe moisture level.

Peeling	Peeling serves to remove the scaly epidermis and facilitate drying.
Drying	The moisture content of fresh ginger at harvest is about 80-82 per cent which is brought down up to 10 per cent for its safe storage.
Storage	Fully dried rhizomes can be stored in airtight containers such as high density polyethylene or similar packaging materials. Long term storage for more than two years would result in deterioration of its aroma, flavour and pungency.
Bleached ginger	Bleached ginger is produced by dipping scrapped fresh ginger in a slurry of slaked lime, Ca(OH)2, (1 kg of slaked lime/120 kg of water) followed by sun drying.

Crop improvement

- The crop improvement in ginger is aimed at to develop high yielding varieties with wide adaptation, high quality parameters (oil, oleoresins) and low fibre, besides resistant to major pest and diseases such as rhizome rot and shoot borer.
- In India attempts were made to develop varieties through introduction, selection mutation and polyPloidy level breeding. Hybridization in ginger is not feasible due to sterility.

Selection

- For crop improvement, major emphasis was given for augumentation of germplasm from different localities, their comparative yield evaluation and selection of superior types based on yield and quality traits.
- The cultivars differ considerably in their rhizome characters and production potential which is influenced by various factors.
- The high yielding capacity of exotic variety Rio-de-Janeiro was proved in many locations. This variety possesses a number of quality attributes.
- The yields of Himachal Pradesh, Kuruppampadi, China and Maran are comparable to Rio-de-Janeiro (Jogi *et al.* 1972, Nybe *et al* 1980, Kumar *et al.* 1980, Mohanty *et al.* 1981, Thangaraj and Muthuswamy (1983).
- Varieties like Nadia, Burdwan are high yielders in certain locations (AICSCIP, 1978, Aiyadurai, 1966, Khan 1959, Nybe *et al* 1980, Arya and Rana, 1990, Saikia and Shadeque, 1992, Chandra and Govind, 1999, Gowda and Mehata, 2000).
- The varieties popular with farmers include Rio-de-Janeiro, Himachal Pradesh, Kuruppampadi, Maran, Nadia and Burdwan.

- Thingpuri has given highest yield in Odisha (Panigrahi and Patro, 1985). Under Nagaland conditions, Thinladium, Nadia and Khasi Local were the best (Singh *et al.,* 1999).
- For Quality Attributes- Rio-de-Janeiro and Maran had the highest oleoresins, Karakkal had the highest essential oil, crude fibre was least in China and Nadia and high in Kuruppampadi, Maran, Jugijan, Ernad Manjeri, Nadia, Poona, Himachal Pradesh, Tura and Arippa (Jogi *et al.* 1972, Nybe *et al* 1980, Kumar *et al.* 1980).
- Most of the improved varieties of ginger are the result of direct selections from germplasm. Seven varieties were so far released in ginger.

Mutation breeding

- Physical and chemical mutagens are employed for creating variability in the sterile vegetatively propagated plants. The induced variability, once fixed, can be maintained through vegetative propagation.
- Use of chemical mutagen, Ethyl methane sulphonate (EMS) resulted in reduced growth and increased cytological irregularities (Rattan 1987). Use of Gamma rays also had similar effects (Rattan,1987).
- Almost all the induced changes appearing in the R_1 generation were in chimeric form and expressed a stunted or semi-dwarfing effect, and were inhibitory on production of rhizomes (Giridharan and Balakrishnan, 1991& 1992, Jayachandran and Mohanachandran, 1992).

Polyploidy level breeding

- Ramachandran (1982) and Ramachandran Nair (1992) reported successful induction of stable **tetraploids having 2n = 44** in ginger (cv. Maran and Mananthody) by treating the sprouts with 0.25 per cent aqueous colchicines. The polyploids were more vigorous than the diploids and flowered during the second year of induction.
- These **autotetraploids** had larger rhizomes and high yield (198.71 g/ plant). However, oil content of these rhizomes was lower (2.3%) than the original diploid cultivar (2.8%).

Promising selections of ginger for quality attributes

Character/ trait	Variety /Cultivar/ Accession	Reference
High dry recovery (%)	Tura local , Thodupuzha, Nadia, Kuruppampadi, Maran, Thinladium, Jorhat, Vengara, Ernad (Chernad), Himachal Pradesh, Sierra-Leone, Suruchi, Assam, China, Mowshom, Thingpui, Varada, Acc. Nos. 27, 117, 204 & 294, SG-685, Narasapattam, Rio-de-Janeiro	Mohanty (1984), Sreekumar *et al.* (1980), Nybe *et al.* (1980), Nair *et al.* (1969), Muralidharan (1972), Thomas (1966), Sasikumar *et al.* (1994 & 1999), Paulose (1973)
High Oleoresin (%)	Assam , Mananthody, Wynad Local Kuruppampadi, Rio-de-Janeiro, Maran, Wynad Kunnamangalam, Ambalavayalan, Santhing Pui, Himachal, Varada , China, Santhing Pin (Manipur –I), Ernad Chernad, Erattupetta, Tamarassery local, PGS-33 & PGS-1, Nadan Pulpally, Nadan Acc. Nos. 3, 6, 14,57, 110, 118, 582, 236, 388, 414,	Krishnamoorthy *et al.* (1970); Natarajan *et al.* (1972) Muralidharan (1972), Nybe *et al.* (1980), Sreekumar *et al.* (1980) Sasikumar *et al.* (1994, 1999), Sarma and Sasikumar *et al.* (2002),Zachariah *et al.* (1993, 1999)
High Essential oil (%)	V_1S_1-8, BDJR-1226, Chanog-II , Mananthody, Karakkal , Rio-de-Janeiro , Vengara, Valluvanad , Elakallan, Sabarimala, Acc. Nos. 118 , 14 , 64,418, 399, 389, 205, 110, 236, 104, 296, Nadan Pulpally, Thodupuzha, BLP-6, SG-723, BDJR-1054, SG-55, Maran, Shilli, Bangi, Himgiri, Acc. No., V_1E_4-4, PGS-23 & SG-706	Krishnamurthy *et al.* (1970), Lewis *et al.* (1972), Nybe *et al.* (1980), Sarma *et al.* (2001), Sarma and Sasikumar (2002), Zachariah *et al.* (1999)
Low crude fibre (%)	China, UP , Himachal Pradesh, Nadia, Tura, Ernad-Chernad, Zahirabad, Kuruppampadi, Mizo, PGS-16, Jamaica, Acc. Nos.15, 27, 287 , 288, 22, 18, 419, 386, 415, 200, 110, 336, Varada	Thomas (1966),Sreekumar *et al.* (1980),Nybe *et al.* (1980), Sasikumar *et al.*(1994, 1999) , Sarma and Sasikumar *et al.* (2002), Zachariah *et al.*, (1999)
High gingerol and Shogaol	Wynad Kunnamangalam, Ambalavayalan, Ernad Chernad, Sawthing Pui, Rio-de-Janeiro, Mizo, Nadia, Maran, Ernad, Chernad, Kada , Narianpara, Santhing Pin (Manipur-I), PGS-37, S-641, Maran, Erattupetta, Nadan Pulpally, Jorhat Local, PGS-16, Baharica & Kunduli	Sasikumar *et al.*(1994, 1999); Sarma *et al.* (2001), Zachariah *et al.* (1993)
High zingiberene and (6) gingerol	Baharica & Amaravathy	Sasikumar *et al.*(1999)
Salted ginger	Acc. Nos. 35 , 27	Sasikumar *et al.* (1999)

Released varieties of ginger and their salient features

S. No.	Cultivar	Institution/Released from	Av. Yield (fresh t/ha)	Maturity (days)	Dry recovery (%)	Crude recovery (%)	Oleoresin (%)	Remarks
1.	Suprabha	Odisha University of Agriculture & Tech. Pottangi	16.6	329	20.5	4.4	8.9	Plumpy rhizome, wider adaptability, suitable for both early and late sowing
2.	Suruchi	Odisha University of Agriculture & Tech. Pottangi.	11.6	218	23.5	3.8	10.0	Bold rhizome, suitable for rainfed and irrigated conditions.
3.	Suravi	Odisha University of Agriculture & Tech. Pottangi.	17.5	225	23.0	4.0	10.2	Plumpy rhizome, suitable for rainfed and irrigated conditions.
4.	Himgiri	Y.S. Parmar University of Horticulture and Forestry, Solan, H.P	13.14	230	20.2	6.05	4.29	Best for green ginger, less susceptible to rhizome rot, suitable for rainfed condition
5.	IISR- Varada	Indian Institute of Spices Research, Calicut, (Kerala)	22.6	200	20.7	3.29-4.5	6.73	Bold rhizome, tolerant to rhizome rot. Wide adaptability
6.	IISR-Mahima	Indian Institute of Spices Research, Calicut, (Kerala)	23.2	300	23.0	3.26	4.48	Plumpy, bold rhizome with low fibre content
7.	IISR-Rejatha	Indian Institute of Spices Research, Calicut, (Kerala)	22.4	200	20.8	4.0	6.34	Plumpy, round and bold rhizome with low fibre and high oil content

Turmeric

Botanical name	*Curcuma longa* L (Syn: *C. domestica* Val.)
Family	Zingiberaceae
Origin	South East Asia
Chromosome number	x=21, 3x = 63
Ploidy level	Triploid
Type of inflorescence	Spike
Pollination system	Cross- pollination
Type of fruit	Capsule
Commercial part	Rhizome or underground stem
Main Phytochemical constituents	Curcuminoids, curcumin

- Turmeric is the dried underground rhizome of perennial herb *Curcuma longa* L (Syn: *C. domestica* Val.) of the family Zingiberaceae.
- India is the single largest producer and exporter of turmeric in the world.
- The country of origin of turmeric is presumed to be South East Asia.
- It is a cross-pollinated, triploid species (2n=3x=63), which can be vegetatively propagated using its underground rhizomes (Sasikumar, 2005)
- Turmeric is an erect perennial herb, but it is grown as an annual.
- The leaves are alternate, obliquely erect or subsessile, lanceolate, green acuminate with a long leaf sheaths forming a pseudo stem.
- The underground stem or rhizome is fleshy at the base of each aerial shoots consisting of an erect, ovoid or ellipsoid structure (mother rhizome), ringed with the bases of old scale leaves, bearing when mature several to many horizontal or curved rhizomes (fingers), which are again branched (secondaries).
- The rhizomes show yellow to bright orange yellow colour inside of the rhizome.
- Rhizomes are rich in **curcumin** for which turmeric is valued.
- Leaf blades usually more or less erect, often with a purple-flushed strip on either side of the midrib. The inflorescence is terminal and borne in between the leaf sheaths.
- **Flowers** are in cincinni of 2-7, each cicinnus in the axil of a bract. Flowers are pale yellow in colour, length equalling those of the bracts.
- **Calyx** short unequally toothed and split nearly half way down one side.

- **Corolla** tube + staminal tube, tubular at the base, the upper half cup shaped, the corolla lobes inserted on the edges of the cup, and the lip, staminodes and stamen just above them.
- Corolla lobes thin, translucent white or pink to purplish, the dorsal are hooded and ending in a hollow hairy point.
- **Staminodes** elliptic-oblong, their inner edges folded under the hood of the dorsal petal.
- Labellum obovate, consisting of a thickened yellow middle band which points straight towards or in some what reflexed, its tip slightly cleft, and thinner pale (white or pale yellow) side-lobes up-curved and overlapping the staminodes.
- Filament of stamen short and broad, constricted at the top, anther versatile, the filament joined to its back, the pollen sacs parallel, with usually a curved spur at the base of each; connective some times produced at the apex into a small crest.
- Stylodes are cylindrical, 4-8 mm long. Ovary trilocular; fruit ellipsoid, thin-walled, dehiscing and liberating the seeds in the mucilage of the bract pouch; seeds ellipsoid; with a lacerate aril of few segents which are free to the base. (Holttum ,1950).
- The tuberous rhizomes or underground stems of turmeric are used from antiquity as condiments, a dye and as an aromatic stimulant in several medicines.
- The **inflorescence** is a **cylindrical spike**, 10-55 cm long, 5-7 cm wide and terminal on the leafy shoot.
- The flowers are yellow or pale yellow, borne in a spike. They arise from two buds situated in the axils of bracts and mature successively. Bracts are greenish-white; the uppermost tinged with pink. The bracteoles are thin and elliptic.
- Fruit is a thick walled trilocular capsule with numerous arillate seeds.
- Rhizomes have a distinctive taste and smell, brownish and scaly outside and the inside is bright orange in colour.
- The roots are fleshy, often ending in a swollen starchy tuber.
- It is a triploid and might have evolved as a hybrid between tetraploid *C. aromatica* and diploid *C. longa* (having 2n=42) types or one of these is evolved from the other by mutational steps, represented by these intermediate types that are known to occur.

- Basic chromosome number of 21 might have been derived either by dibasic amphidiPloidy level (by combination of lower basic numbers of 9 and 12 found in some genera in the family) or by secondary polyPloidy level.
- Detailed cytological investigations by Nambiar (1979) revealed that the earlier reports of chromosome numbers of 2n=32, 62 and 64 for *C. longa* and 42, 63 and 86 for *C. aromatica* were probably exceptional cases and the correct chromosome numbers for these species are 2n=63 and 84 respectively.
- *Turmeric* contains two primary constituents, the colouring matter and the volatile oil.
- The volatile oil of turmeric is about 1.5 -6.0% and is composed of a variety of sesquiterpenes, many of which are specific for the species.
- Most important for the aroma are turmerone (max. 30%), ar-turmerone (25%) and zingiberene (25%).
- Conjugated di-arylheptanoids (1,7-diaryl-hepta-1,6-diene-3,5-diones, e.g. curcumin) are responsible for the orange colour and probably also for the pungent taste (3 to 4%).
- The crystalline colouring matter, curcumin, is a diferuloyl methane. It dissolves in concentrated sulphuric acid giving a yellow-red coloration.
- The rhizomes contain curcuminoids, curcumin , demethoxy curcumin, bis-demethoxycurcumin, 5'- methoxycurcumin and dihydrocurcumin which are found to be natural anti-oxidants.
- The colour curcumin extracted from turmeric is used as a colourant.
- Turmeric is also used as a dye in textile industry. It is used in the preparation of medicinal oils, ointments and poultice.
- It is stomachic, carminative, tonic, blood purifier and an antiseptic. It is used in cosmetics.
- The aqueous extracts has biopesticidal properties.
- It is a very important spice in India, which produces nearly the whole world's crop and uses **80%** of it.
- Presently, it is cultivated in China, Taiwan, Indonesia, Sri Lanka, Australia, Africa, Peru and the West Indies.

Production techniques

Climate	Turmeric is a tropical crop. It grows at all places ranging from sea level to an altitude of 1220m above MSL. It is very sensitive to low atmospheric temperature.
Soil	It grows in light black, black clayey loams and red soils in irrigated and rainfed conditions. The crop cannot stand water logging or alkalinity.
Varieties	CO 1, BSR 1, BSR 2, Roma, Suvarna, Sudarshana, Suguna, Sugandham, Ranga, Rasmi, Rajendra Sonia, Krishna, Suroma and Allepy Supreme, Kedaram, Prabha, PrathibaCo-1: Vegetative mutant selection derived by X-ray irradiation of" Erode localAllepy turmeric: Its quality is the best in the world.
Propagation	Through mother rhizome & finger rhizomes.
Seed rate	Seed rate of finger rhizome-25q/ha.
Time of planting	May-June.
Planting	30X20, 45 x 15 cm. 25-30 g weight rhizomes are to be dibbled at a depth of 4 cm.
Manuring	Basal: FYM-25 t /ha, Neem or groundnut cake-200 kg/ha, 25:60:108 kg of NPK per ha; 30 kg of FeSO4 and 15 kg of ZnSO4, 10 kg in each of *Azospirillum* and *Phosphobacteria* per ha to be applied at the time of plantingTop dressing: 25: 108 kg of N and K/ha applied on 30, 60, 90, 120 and 150 days after planting.
Aftercare	First earthing up should be given 50-60 days after planting and the next after 40 days.
Intercropping	Turmeric comes up well under sparse shade also. It can be grown as an intercrop in coconut gardens like ginger or as mixed crop with red gram, chilli, colocasia, vegetables, maize and ragi.
Harvesting	Turmeric takes 7-9 months for harvesting. The plants will start lodging, yellowing and drying on crop maturity. The rhizomes are dug with spade or digging forks.
Yield	Fresh rhizomes: 25-30 t/ha, Cured rhizomes: 5-6 t/ha.
Storage	Seed rhizomes can be stored in open sand media with partial shade.
Processing	Raw turmeric rhizomes have to be cured for both colour and aroma. For this, the fingers and bulbs are boiled separately in water for 30 to 45 minutes until the rhizomes are soft. For imparting orange yellow colour, the rhizomes are boiled in lime water or sodium bicarbonate solution. The dried produce is cleaned and polished mechanically in a drum rotated by hand or by power.

Crop improvement

Selection

- The varietal improvement work so far attempted is only through clonal selection.
- Most of the improved varieties released so far are selections from germplasm or clonal selections based on location specific comparative

yield evaluation trials mainly to identify turmeric types with high yield potential, high curing percentage and high curcumin content.

- Most of the selections have been made from the landraces collected from various parts of the country.
- Improved selections were developed mainly in *C. longa* and to a lesser extent in aromatica types and *C. amada*.

Seedling selection

- Though clonal selection is the main breeding procedure to obtain improved lines it has the limitation of low variability.
- Hybridisation is very difficult due to sterility and hence seed set is rare.
- This problem is probably compounded by self incompatibility. How ever rare seed set can be utilized for improvement of genotypes and fixing them clonally.

Mutation breeding

- Induced mutations were also tested in turmeric. In Tamil Nadu, two mutants, CO- 1 and BSR-1, identified from Erode local, were released for large scale cultivation (Balashanmugam *et al.* 1986).
- BSR-2 is another induced mutant from same Erode local using x-rays (Chezhiyan and Shanmuga sundaram 2000).
- Suroma is another mutant selection from Tsundur after irradiation with X-rays.
- Preliminary trials using Colchicine, EMS and MNG at 250, 500, 1000 PPM on cv. Mydukar resulted normal spouting in all treatments.
- Plant height and yields were more in colchicine treatments, number of leaves was more in EMS at 1000 ppm, but yield was poor.

Hybridization

- Studies carried out on blossom biology and viable seed set obtained through successful hybridization have opened the way for recombination breeding programmes (Nambiar et al., 1982; Nazeem et al., 1993; Sasikumar et al., 1994, 1996).

Improved varieties of turmeric and their important characters

S. No.	Variety	Pedigree	Crop Duration (days)	Mean Yield (Fresh t/ha)	Dry Recovery (%)	Curcumin (%)	Oleo resin (%)	Essential Oil (%)	Important Characters
1.	CO-1	Mutant (X-ray) selection from Erode local	270	30.5	19.5	3.2	6.7	3.7	Bold, bright orange rhizomes suitable for drought prone areas
2.	BSR-1	Mutant (X-ray) selection from Erode local	285	30.7	20.5	4.2	4.0	3.7	Suitable for drought prone areas
3.	BSR-2	Mutant (X-ray) selection from Erode local	245	32.7	—	—	—	—	Bold rhizomes resistant to scale insects
4.	Krishna	Clonal selection from Tekurpeta	240	9.2	16.4	2.8	3.8	2.0	Moderately tolerant to pests and diseases
5.	Sugandham	Germplasm selection	210	15.0	23.3	3.1	11.0	2.7	Moderately tolerant to pests and diseases
6.	Roma	Clonal selection from Tsundur	250	20.7 areas	31.0	6.1 areas	13.2	4.2	Suitable for hilly areas
7.	Suroma	Mutant (X-ray) selection from Tsundur	253	20.0	26.0	6.1	13.	4.4	Field tolerant to leaf blotch, leaf spot and rhizome scale
8.	Ranga	Clonal selection from Rajpuri local	250	29.0	24.8	6.3	13.5	4.4	Bold rhizomes, moderately resistant to leaf blotch and rhizome scales
9.	Rasmi	Clonal selection from Rajpuri local	240	31.3	23.0	6.4	13.4	4.4	Bold rhizomes
10.	Rajendra Sonia	Local germplasm Selection	225	23.0	18.0	8.4	—	5.0	Bold and plumpy rhizomes
11.	Megha Turmeric	Selection from Lakadong types	300–315	20.0	16.37	6.8	—	—	Bold rhizomes
12.	Pant Peetabh	Selection from germplasm	—	29.0 (potential)	18.5	7.5	—	1.0	Resistant to rhizome rot
13.	Suranjana	Local germplasm Selection	235	—	21.2	5.7	10.9	4.1	Tolerant to rhizome rot and

S. No.	Variety	Pedigree	Crop Duration (days)	Mean Yield (Fresh t/ha)	Dry Recovery (%)	Curcumin (%)	Oleo resin (%)	Essential Oil (%)	Important Characters
									leaf blotch; resistant to rhizome scales and moderately resistant to shoot borer
14.	Suvarana	200	200	17.4	20.0	4.3	13.5	7.0	Bright orange coloured rhizome with slender fingers
15.	Suguna	Germplasm Selection	190	29.3	20.4	7.3	13.5	6.0	Short duration type, field tolerant to rhizome rot
16.	Sudarshana	Germplasm Selection	190	28.8	20.6	5.3	15.0	7.0	Short duration type, field tolerant to rhizome rot
17.	IISR Prabha	Selection from Open pollinated seedlings	205	37.47	19.5	6.5	15.0	6.5	Rhizomes with close internodes
18.	IISR Pratibha	Selection from Open pollinated seedlings	225	39.12	18.5	6.21	16.2	6.2	—
19.	IISR Alleppey Supreme	Selection from Alleppey Finger Turmeric	210	35.4	19.0	5.55	16.0	—	Tolerant to leaf blotch
20.	IISR Kedaram	Germplasm Selection	210	35.5	18.9	5.9	13.6	—	Tolerant to leaf blotch
21.	Kanthi	Clonal selection from Mydukur	240–270	37.65	20.15	7.18	8.25	5.15	Big mother rhizomes and bold fingers with short internodes
22.	Sobha	Germplasm selection	240–270	35.88	19.38	7.39	9.65	4.24	Big mother rhizomes and bold fingers with short internodes
23.	Sona	Germplasm selection	240–270	4.02 t dry	18.88	7.12	10.25	4.4	Field tolerant to leaf blotch
24.	Varna	Germplasm selection	240–270	4.16 t dry	19.05	7.87	10.8	4.56	Bold rhizome with short internodes, field tolerant to leaf blotch

□□□□

4

Recent Initiatives in Crop Improvement

High-yielding crop varieties and hybrids with tolerance to pests /diseases and to various farming constraints such as soil and water salinity, soil acidity, drought, flood and such other factors have been developed.

Eight cardamom hybrids, NKE 12 × MB 5 (1,499 kg/ha), MB 5 × NKE 19 (1,461 kg/ha), GG × NKE 12 (1,350 kg/ha), RR 1 × CCS 1 (1,245 kg/ha), CCS 1 × RR 1 (1,022 kg/ha), ASH (1,930 kg/ha), NKE 12 × GG (1,746 kg/ha), GG × NKE 19 (1,635 kg/ha), were identified for high yield and resistance to mosaic and rhizome rot diseases. Two unique mutants, AFg-3 and AFg-4, for powdery mildew resistance were isolated in fenugreek. In ajwain, AA 93 has been developed which flowers 40 days earlier compared to other varieties. Thirty-two elite saffron clones were selected and evaluated for fresh and dry weight of pistil, stiga and style length, and weight. CITH 125 (4.5 kg/ha), CITH 123 (4.3 kg/ha), CITH 124 (4.3 kg/ha), CITH 122 (4.0 kg/ha), CITH 12 (4.5 kg/ha), CITH 121 (3.9 kg/ha), CITH 107 (3.8 kg/ha), CITH 120 (3.8 kg/ha) and CITH 104 (3.7 kg/ha), having higher saffron yield were identified.

Source: DARE/ICAR ANNUAL REPORT 2012–13

The registration of **nutmeg** accession A 9-71 (IC 537220, INGR 10142) as a source of high sabinene (45.0%) in nutmeg oil and 41.9% in mace oil has been done at the NBPGR, New Delhi. Evaluation of F1hybrids of **cardamom** indicated that three genotypes (IC 584097, IC 584098, and IC 54722) were high-yielding (20-40%) than Appangala 1 and Njalani Gold.

Source: DARE/ICAR ANNUAL REPORT 2011–12

Micro satellites developed for *Piper* species were successfully used to detect polymorphism in its cultivars. The *Piper ornatum,* a wild species related to black pepper, find resistance to *Phytopthora capsici*, *Radopholus similis* and *Meloidogyne incognita,*pathogen involved in quick and slow wilt of black pepper. Cardamom hybrid, GG × NKE 19, was promising with more than 70% of 8 mm capsules and showed field tolerance to rhizome rot and leaf blight. Of the 68 second generation seedling progenies of 5 seedling progenies, most of them showed very high multiplication rate.

Among 13 genotypes analysed for biochemical parameters, oil content varied from 1.2 to 2.8%, maximum being in Acc.162, followed by Acc.95, while oleoresin content was 2–4.6% and fibre 1.87–5.13%. Chromosome number of ginger collection Acc.195, which has high pollen fertility, was found to be 2n=44 in majority of cells, indicating that it is a tetraploid. In **Garcinia**, using the model of GIS, the existence in north-east Himalayan region was assessed. Four species of Garcinia, namely *G. kydia* (Kuji Thekera)*, G. lancifolia* (Rupohi Thekera)*, G. pedunculata* (Bor Thekera) and *G. xanthochymus*(Tepor Tenga) were located in Meghalaya and Assam.

Source: DARE/ICAR ANNUAL REPORT 2010–11

Black pepper: Two advanced lines, INGR 8099-*Piper thomsonii* (IC-398863) - for sex change from male to bisexual plant and INGR 8100- *Piper nigrum*(IC-563950) – a novel spike variant with proliferating spikes were registered with NBPGR for their unique characters.

Cardamom: Evaluation of hybrids under PET 1 and PET 2 led to the identification of five hybrids, namelyCCS-1 × RR-1, RR-1 × CCS-1, MB-5 × GGASH, NKE 19 × GG, GG × NKE 19.

Cassia: The elite line A1 (IC No. 370400) with high cinnamaldehyde content in bark oil (81.5%) and leaf oil (80.5%) has been registered with NBPGR, New Delhi.

Ginger: A high-yielding ginger variety, Subhada, has been recommended for release.

Cumin: A cumin variety, RZ-345, from Jobner was recommended for release.

Fennel: UF-205 from Jobner, LFC-84 from Guntur and HM-219 from Hisar were identified for release.

Fenugreek: The entries RMt-361 from Jobner,LFC-84 from Guntur and HM-219 from Hisar were identified for release.

Coriander: Three varieties, RCr-728, LCC-170 and DH 206, were identified for release.

Cassava: Three cassava hybrid lines, CMR-3,CMR-63 and CMR-70, with extractable starch content of24.6 25.8% were identified as promising under irrigated plains of Tamil Nadu.

Source: DARE/ICAR Annual Report 2009–10

Recent Initiatives by National Research Centre on Seed Spices, Ajmer, (Rajasthan)

Collection, evaluation and documentation of plant genetic resources of seed spices

All the collected and already available germplasm is being maintained at NRCSS. During 2011-12, early flowering coriander, cumin, dill, ajwain, nigella, anise, celery and caraway germplasm was maintained. A total of 133 coriander, 68 cumin, 54 fennel, 208 fenugreek, 94 ajwain, 12 nigella, 14 dill, 18 anise and 36 celery lines were evaluated during the reporting year. NRCSS gene bank have total germplasm collection of 2007 seed spices as on March, 2013.

Apart from this, 54 exotic lines of fenugreek received from ICARDA were also evaluated during this year. The variability present in these exotic lines, however, phenotypically it differed from the available germplasm from India.

Management of plant genetic resources of cumin

In cumin 78 lines were evaluated for Fusarium wilt resistance in sick plot at NRCSS, Ajmer. GKKB- 184, GKKB-190, AC-E-6 and AC-E-7 were found promising which showed tolerance against wilt. Remaining lines showed wilt incidence from 80- 100%. Few survivals were noticed, which were selected for advancement.

Genetic diversity analysis in anise

Eighteen genotypes were evaluated for all the major descriptor traits. Ample amount of variability was found in the germplasm set. The range of variation was high for days to flowering. Genotype Ani-17 flowered in 82 days, whereas the late genotype Ani-1 flowered in 112 days. Likewise high range of variation was observed for plant height (42 to 70 cm), secondary branches (10.6 to 20.4), number of umbel per plant (34.4 to 54.4), and days to maturity (114 to 135 days). Diversity index was estimated among the eighteen lines, Genotype Ani-14, Ani-15 and Ani-16 were most diverse among the set. Nearly 100 single plants were selected based on various traits to enhance the yield potential and to improve other component traits.

Breeding for improvement in major seed spices

Evaluation of fenugreek genotypes

Two entries from NRCSS, namely AFg-3 and AFg-4 were found superior in coordinated trial on fenugreek conducted during 2009-10, 2010-11 and 2011-12 at 12 locations around the country including Guntur, Dholi, Udaipur, Jabalpur, Faizabad, Hisar, Ajmer, Jobner, Coimbatore, Jagudan, Pantnagar and Raigarh. AFg-3 and AFg-4 stood at second and third with 7.26 and 7.58 per cent higher yield than the national check Hisar Sonali.

Ten advanced lines were evaluated with three check varieties namely AM-1 (AFg-1), AM-2 (AFg-2) and RMt-1 using three replications. Performance of these lines for different traits. AM-317 was the highest yielder with 22.44 q/ha as compared to best check i.e. AM-2 with 18.64 q/h seed yield.

On the basis of two year performance, genotype AM-317 gave 12.89 per cent higher yield than best check variety A (15.78 q/ha). AM-317 was identified for incorporation in coordinated trial.

Creation of variability in fenugreek through mutation breeding

Mutation breeding programme was started with three genotypes of fenugreek. This year total 135 lines of M and M were sown in single row of 2 4 5 metre length in Augented design with three parents as check. These lines were selected as variants for plant type, plant height, leaf shape & size and growth behaviour were selected. Extent of variability created for powdery mildew incidence was also recorded.

One station trial was also framed from stabilized mutants with higher productivity and it was taken as fenugreek Station trial II during 2011-12. The mutant A3-28-9 having higher yield along with less powdery mildew incidence (30 PDI) has been submitted for testing in coordinated trial started in 2012-13.

Breeding for high yield and quality in coriander

This project was started from 2007-08 year with the objective of developing genotypes having high yield and good quality i.e. higher essential oil content. The station trial was repeated during the year. Twenty two advance populations were evaluated in RBD with four check varieties namely RCr-435, Ajmer Coriander-1, RCr-41 and Hisar Sugandh. There was severe damage due to frost to all the entries. On the bais of performance during 2010-11 best population namely VDV/GL- 173 and MKSM-1111 have been submitted for evaluation in coordinated trial as ACr-2 and ACr-3,

Breeding for high yield, quality and resistant to biotic and abiotic stress in fennel

This project was started in 2009. In this year 21 selected genotypes were screened in a station trails at NRCSS. Two check varieties namely AF-1 (11.04 q/h & 132 cm) and RF-125 (11.02 q/h & 125 cm) were used in trial. Among the 21 genotypes two genotypes were found high yielding namely AF-05-1- 3 (17.76 q/h &139 cm) and AJ-FNL-2 (13.23 q/h & 122 cm). Both genotypes are included in AICRP for coordinated trails respectively.

Breeding for high yield and quality in dill

One station trial was conducted with 12 test entries and two check varieties namely AD-1 and AD- 2. The best check variety AD-1 gave 1025.0 kg/ha seed yield and it was at par with another check. The test entries AD-S-01-3 and AD-S-01-24 gave higher yield than check varieties.

Creation of variability in dill through gene pool approach

To create variability one gene pool on the basis of yield and umbel characteristics was prepared and planted in rabi, 2010-11. This pool was maintained through closed sib mating for maximum introgression of genes within the gene pool.

Breeding for high yield, quality and resistant to biotic and abiotic stress in celery and caraway

This project was started in 2009. In this year, fourteen selected genotypes were screened in a station trial at NRCSS in RBD of three replication comprising one check variety namely A-Cel-1(2.6 q/h). Among the fourteen genotypes four genotypes were found superior over check variety, namely ACel- 8(Sel) (4.38 q/h), Early celery (4.09 q/h), Cel-08 (3.69 q/h) and A-Cel-5 (3.42q/h).

Three selected genotypes of caraway were screened in a station trails at NRCSS. Among these three genotypes one genotype was found superior over local check namely OPA-CAR-1 (0.53 q/h).

Breeding for high yield and quality in ajwain

On the basis of the previous year's observation one station trial was conducted with 10 test entries and three check varieties namely, Ajmer Ajwain-2, GA-1 and Lam Sel-2 during 2011-12. The flowering in genotype AA-93 was found 35-40 days early as compared to checks. This genotype was identified for earliness.

Breeding for high yield and quality in nigella

Station trial was conducted on 10 advance lines of nigella with 3 check varieties namely Ajmer Nigella-1, Pant Krishna and Azad Kalongi. Plant height, number of branches, days to 50 % flowering, yields attributing traits and seed yield (kg/h) was recorded. The maximum seed yield was recorded in AN-21.

Genetic characterization and documentation of released varieties of cumin, coriander, fenugreek and fennel

Morphological observation of yield and yield contributing characters were taken for all the released varieties available. The extent of genetic variation among the indigenous varieties of fennel (*Foeniculum vulgare*) and cumin (*Cuminum cyminum* L.) was found limited for phenotypic and genetic markers. Multilocus genotyping by RAPD primers detected an average of intraspecific variations amounting to 56.52% and 23.17% polymorphism in banding patterns in fennel and cumin, respectively (Fig. 4 & 5). Six cumin varieties exhibited a uniform ITS length of 600bp with single nucleotide polymorphisms (SNPs). Besides total ITS length variations and SNPs, INDELS were also detected at several sites in ITS-1 region of fennel. Multiple sequence alignment of ITS region is of phylogenetic significance in distinguishing and cataloguing of fennel and cumin germplasm. Eventually, the knowledge of their genetic relationships and intra-specific diversity will be of significance in developing intra-specific crosses of fennel and cumin in ongoing seed spices breeding programme. Ten gene sequences of fennel and 6 gene sequences of cumin have been submitted to NCBI, USA public domain data base. Gen Accession numbers HQ 377206 to HQ 377215 of fennel and HM 176650 to HM 176655 of cumin have been assigned.

Genetic diversity analysis through biotechnological tools in fenugreek

Genetic diversity has been calculated among 75 germplasm lines of fenugreek including released varieties using RAPD markers. Out of twenty primers, 15 amplified a total of 2469 RAPD bands. Out of this 385 bands were polymorphic with 15.59% polymorphism, thus revealing a low degree of polymorphism where as it was high (57.66%) among the released varieties which indicates that there is duplicity in collected germplasm. Cluster analysis of 1 - 0 bivariate data using UPGA method delineated the genotypes into three groups and the data was further used for constructing a dendrogram. Genetic similarity matrix was calculated on the basis of Jaccards algorithm for RAPD data.

Genetic diversity analysis in cumin

Genetic diversity was calculated among 55 germplasm lines of cumin including 6 released varieties using RAPD markers. Out of twenty primers, 15 amplified a total of 1191 RAPD bands. Out of this 218 bands were polymorphic with 18.30% polymorphism, thus revealing a low degree of polymorphism where as it was high (66.66%) among the released varieties which indicates that there is duplicity in collected germplasm. Cluster analysis of 1 - 0 bivariate data using UPGA method delineated the genotypes into three groups and the data was further used for constructing a dendrogram. Genetic similarity matrix was calculated on the basis of Jaccards algorithm for RAPD data.

Standardization of Technologies for crossing in coriander (*Coriandrum sativum* L.)

Three chemicals viz GA 2, 4-D, Maleic 3, hydrazide and absolute alcohol were tested for pollen sterility induction in coriander variety ACr-1. The different concentration of these chemicals were (GA : 3 50,100, 150, 200, 250 & 300 ppm; 2,4-D: 10, 50, 100 150, 200 & 500 ppm; Maleic hydrazide: 75, 100, 125, 150, 200 & 250 ppm and absolue alcohol) sprayed at flower primodia initiation stage @ 25 ml per plant. After full anthesis the pollen sterility were tested by staining with 1% acetocarmine solution. Two chemicals 2,4-D and MH induced sterility up to more than 50%.

Source: Annual Report 2012-13, National Research Centre on Seed Spices, Ajmer.

Recent Initiatives by Indian Institute of Spices Research, Kozhikode, Kerala

Black pepper

Interstock grafting

Since *Piper* accession (Acc. 5815) resistant to *Phytophthora* and nematodes is incompatible with black pepper, *P. colubrinum* and *P. hamiltoni* were tried as interstock for variety Sreekara because these 2 species are not only disease resistant but also compatible with pepper. The survival and growth of Sreekara on these combinations were very poor and hence found not promising. Though *P. longum, P. hymenophyllum, P. attenuatum, P. sermantosum, P. hapnium, P. argyrophyllum* and *P. chaba* were initially successful on Acc. 5815 these subsequently failed. However, the grafts of *P. ornatum* with *P. chaba* had shown excellent growth and fruiting even after 12 months indicating good success. So *P. ornatum*, a resistant species may be used as rootstock for *P. chaba* where foot rot is a problem. Grafts of Sreekara on *P. hamiltoni* died during summer under rainfed conditions. Hence, similar to *P. colubrinum P. hamiltoni* as rootstock is feasible only under irrigated conditions limiting its scope. The resistant species *P. ornatum* and Acc. 5815 cannot be utilized directly for black pepper through grafting but may be useful as material for resistance breeding.

The hybrids (pollu beetle resistant) developed from various cross combinations were multiplied and maintained in the nursery. The hybrids multiplied are (1) Acc. 816 x Subhakara-80 (2) Acc. 1084 x Subhakara-50 (3) Acc. 841 x Subhakara-30 and (4) Acc. 1114 x Subhakara-30.

DNA barcoding to discriminate adulterants

Standardized 8 barcoding loci (mat K, rbcL, trnH-psbA, ITS, atpH-atpF, psbK-psbI, rpob, rpoC1) for *Curcuma* species (*C. longa, C. xanthorrizha, C. zedoria*), *Piper* sp. (*P. nigrum, P. atteunatum, P. galeatum*), chilli (*Capsicum annum*) and *Cinnamon* species (*C. verum, C. cassia, C. malabathrum*) and 4 barcoding loci (mat K, rbcL, trnHpsbA, ITS) were selected based on their discriminatory power.

In the black pepper group comprising 5 accessions, one each of *P. nigrum, P. atteunatum, P. galeatum*, 4 different varieties of papaya and 5 branded market samples, the selected 4 barcoding loci were amplified and the products were gel purified and sequenced bidirectionally. The sequences were aligned using Clustal W, trimmed by Bioedit software and analysed using MEGA 5. Single Nucleotide Polymorphisms (SNPs) were found to be higher in trnH-psbA locus and was found to be the best for adulteration detection.

Detection of adulterants in market sample was done using trnH-psbA locus. Out of 5 branded pepper market samples, an adulterant specific band (~600bp) could be observed in one of the samples. Sequence analysis (NCBI Blast) revealed that this band has high similarity with capsicum. Further validation of the locus was done by checking with simulated samples of chilli: pepper in the ratio 1:10, 1:25, 1: 50, 1:100, 1:200. In all the simulated samples the adulterant specific band (~600bp) was amplified.

Expression profiling of genes induced under water stress

Expression analysis of candidate genes for drought tolerance by qPCR was undertaken in black pepper accession, Acc 4226, tolerant to water deficit conditions by comparing the expression under water deficit and controlled conditions. Among the genes studied, DREB like proteins potentially associated with stress-responsive transcriptional regulation was found to be highly expressed under water deficit. Slight increase in expression of a specific BADH, thought to be involved in osomoprotectant metabolism and the molecular chaperone, HSP 70 was also found. However, the expression of a specific MAPK was found to be reduced under stress and bzip did not show any significant change.

Host pathogen interaction

The transcriptome data analysis derived from *P. nigrum* and *P. colubrinum* leaves challenged with *P. capsici* revealed identification of many defense related genes differentially expressed in these 2 different species of *Piper*.

Sequences with homology to Lectin-like receptor kinase a potential host target for RxLR type of effectors from *Phytophthora* was discovered and over expression of this gene was found to be useful in other crops for enhanced resistance to *Phytophthora*. Other components for induced defense system, the small peptides with antimicrobial activity such as defensins were also discovered. Sequences with homology to Polygalacturonase inhibitor protein (PGIP), a defense gene, potential to be used in the resistance breeding were identified. Defense-associated transcription factors belonging to the family of NAC, BHLH, MYB, MYC, BZIP etc. that can act downstream reactions of defense-related genes were also recovered.

Analysis of allelic variants in selected genes revealed maximum SNPs in MAP kinase gene compared to other genes like PAL, Polyphenol oxidase, etc.

Molecular analyses of the expression of five defense-associated genes were carried out using real-time quantitative reverse transcriptionpolymerase chain reaction (QRT-PCR) assays. The expression profiles of different defense related genes differed significantly in the studies undertaken in *P. colubrinum* leaves challenged with *P. capsici.* The fold expression was very high in case of Polygalacturonase inhibitor protein, followed by Polyphenol oxidase, Thaumatin-like protein, defensin and Cysteine proteinase inhibitor gene.

Host resistance

ISSR profiling of progenies of P1 × Subhakara crosses

Twenty ISSR primers have been screened with parents-Panniyur1 and Subhakara. 10 Primers that showed polymorphism between parents was used to profile 94 selected progenies of Panniyur1 × Subhakara.

Amplification of resistance gene candidates

The primer pair, P6F- 5'-GGACCTGGTGGGGTTGGGAAGACAAC- 3' and P6 R-5'- CAACGCTAGTGGCAATCC-3' was used to amplify R gene candidates. The selected set of primers were designed according to the conserved

P-loop (GGVGKTT) and hydrophobic domain (GLPLAL) from the N, L6 and RPS2 genes of the NBS-LRR class specific against pathogens.

The wild species *P. colubrinum* (Acc. 392) and *P. ornatum* (Acc. 3362) and the moderately resistant varieties IISR Sakthi and P24-O-4 and susceptible varieties, Subhakara and Sreekara were used in this study.

Sequence analysis of *Piper* RGAs

Moderately tolerant varieties (IISR Sakthi, P24-O-4) and susceptible varieties (Subhakara and Sreekara) were amplified using degenerate primers designed from known R genes (N, RPS2). The presence of conserved domains, *viz.*, kinase-1a, kinase-2 and hydrophobic motif provided evidence that the sequences belong to the NBS-LRR class gene family. The presence of tryptophan as the last residue of kinase-2 motif further qualified them to be in the non-TIR NBS-LRR subfamily of resistance genes. *P. nigrum* RGAs were compared with *P. colubrinum* RGA and *P. nigrum* NBS analogs formed 2 groups with *P. colubrinum* resistance gene analogs. *Piper* RGA sequences derived using the Rgene specific degenerate primers have been deposited in the GenBank database under the accession numbers: JX416288-JX416291 and JX898031- JX898033.

Cardamom

Breeding for high yield and disease resistance

Cardamom variety Appangala-1 was registered with Protection of Plant Variety and Farmers Rights Authority, New Delhi. The following five hybrids were shortlisted for Coordinated Varietal Trials (CVT) and are under multiplication.

- NKE12 x MB5 (1499kg ha-1)
- MB5 x NKE19 (1461kg ha-1)
- GG x NKE12 (1350kg ha-1)
- RR1 x CCS1 (1245kg ha-1)
- CCS1 x RR1 (1022kg ha-1)

Under PET 2, 3 entries *viz.*, ASH (1930kg ha-1), NKE-12 x GG (1746kg ha-1) and GG x NKE -19 (1635kg ha-1) with high yield were shortlisted for CVT.

Among the entries in the CVT, IC 349651 recorded the highest yield followed by IC 547185.

Screening hybrids for *Katte* resistance

Hybrids were established under greenhouse to screen against *Katte* disease. 300 F2 mapping population from the cross between NKE 12 x GG were artificially inoculated with *Katte* virus under green house conditions and 240 were found to be susceptible after 3rd round of inoculation. Among 230 F2 mapping population (GG x NKE 12), 45 were found susceptible after first round of aphid inoculation.

Screening hybrids for resistance to thrips

Twenty hybrids from the germplasm block at CRC Appangala were field screened against thrips. Total number of capsules, the total number of capsules damaged, numbers of cardamom capsules with low (<25%), medium (>25% - <50%) and heavy (>50%) thrips attack were recorded. The lowest total capsule damage was recorded on GG x NKE19 cross (27.38%), whereas, the hybrid RR1 x CCS1 recorded the highest total capsule damage of 91.3 %. All the hybrids were found susceptible to thrips.

Screening germplasm for abiotic stress

Physiological parameters such as relative water content (RWC), specific leaf weight and stomatal count were recorded in 30 accessions. Significant variation was recorded for RWC and specific leaf weight.

Screening germplasm for biotic stress

Natural incidence of leaf blight and rhizome rot diseases were recorded in 60 accessions maintained in the field gene bank (FGB). The accessions were further grouped into different categories based on the reaction towards leaf blight and rhizome rot disease.

Ginger

Evaluation of promising nematode resistant accessions

Promising nematode tolerant accessions were evaluated from 2010 to 2013 for yield. Among 4 accessions, maximum yield was recorded in Acc. 219 followed by IISR Varada.

Visual selection for boldness was imposed on the germplasm and 13 lines were selected for its extra bold rhizome character. These lines were further multiplied to generate planting material for yield and quality assessment. Breeder's seeds of IISR released varieties were planted for multiplication.

Induction of variability through induced mutation

The rhizome buds of 3 varieties (IISR Mahima, IISR Varada, IISR Rejatha) were subjected to ^{3} irradiation at different doses of 0.5, 0.75, 1.0, 1.25 and 1.50 kR (1kR = 1000 rad) with a control. Based on probit analysis for mortality percent in ^{3} irradiated ginger, LD50 for 3 varieties was derived. Total of 5.13% chlorophyll mutation frequency was recorded among M1V1 mutants.

Ten shoot bud cultures each of 3 cultivars *viz.*, IISR Varada, IISR Rejatha and IISR Mahima were established *in vitro*. These will be further utilized for establishing callus cultures for *in vitro* mutagenesis. Four *R. solanacearum* resistant mutants were planted for multiplication to take up further yield evaluation. 177 (M1V6) and 87 (M1V7) mutants were maintained in pots. 128 mutants (M1V7 and M1V6) were subjected to preliminary screening for soft rot disease caused by *P. myriotylum* and 35 mutants which escaped infection were short listed. These mutants will be multiplied and subjected to secondary screening during the next season.

Transcriptomics

In order to achieve a broad survey of genes associated with bacterial wilt resistance, mRNA-Seq profiling of *C. amada* and *Z. officinale* leaves following infection with *R. solanacearum* was performed. Leaf samples were collected pooled within a given time point, mRNA was isolated, enriched, sheared into smaller fragents, and reverse-transcribed into cDNA. The cDNA were subjected to Illumina HiSeq™ 2000 sequencing, and the resulting sequencing data were subjected to bioinformatic analysis. A total of 31,845,321x2 (101 base), 24,107,482x2 (101 base) raw reads were generated, accounting for approximately 6.43 Gb and 4.87Gb of sequence data, for *C. amada* and *Z. officinale*, respectively. The contigs were assembled into transcripts using the transcriptome assembly software Bowtie. Transcripts which were shorter than 200 bases in length were filtered out, resulting in 45045 and 65536 transcripts from the *C. amada* and *Z. officinale* libraries, respectively. The transcriptomes of ginger and mango ginger following infection by *R. solanacearum* were compared, and several candidate genes for resistance to bacterial wilt were identified.

Suppression subtractive hybridization

RNA samples derived from plants after 4 and 8h after inoculation (hai) were used to understand the early genes involved in bacterial wit resistance in *C.amada* by subtractive hybridization. Clones containing differentially expressed cDNAs of *C.amada* were sequenced using M13 forward and reverse primers. Of the 150 clones, sequenced, 138 produced suitable sequences and were evaluated by computational analysis. All the sequences were annotated using the PSI-BLAST program. Of the 138 SSH sequences, 15 had no identity with sequences in the databases. Among these differentially expressed genes, sequences coding for putative proteins related to defense response during *C.amada—R.solanacearum* interaction (high homology between 85 and 100% identity) were, glutathione-S-transferase (GST), putative cytosolic ascorbate peroxidase (APX), cell wall associated hydrolase, Xyloglucan transglycosylase (XTH), cytochrome p 450, metallocarboxypeptidase inhibitor, peroxiredoxin and thioredoxin dependent peroxidase.

Turmeric

Evaluation of shortlisted seedling progenies

Thirty five seedling progenies shortlisted based on the performance in the earlier trials were planted with 5 released varieties as control. Highest fresh yield of 12.09kg bed-1 was recorded in released variety Rajendra Sonia, followed by Prabha (10.42) and Sudarsana (10.13). Among the seedling progenies, SLP 414/3 recorded the highest yield of 9.67kg bed-1. High curcumin line SLP 389/1 produced 8.17kg bed-1. Rhizomes from 35 shortlisted seedling progenies and 2 controls harvested from the field trial showed dry recovery of >20% for most of the entries. Four of them showed dry recovery of >24%. Highest curcumin content of 5.99% and oleoresin content of 14.19% was recorded in SLP 389/1 while oil content of 4.9% was recorded in SLP 415/3. Curcumin content in controls *viz.*, Prabha, Kedaram was 4.20% and 5.07% respectively. Nine of the shortlisted progenies were also planted at CRC, Appangala to observe performance at high altitudes with released variety IISR Prathibha as control. IISR Prathibha yielded 20kg fresh rhizomes bed-1. All the seedling progenies tested registered fresh yield above 8kg bed-1. Seedling progeny SLP 138/78 showed 10.33kg bed-1. High curcumin line 389/1 produced 9.33kg bed-1. Highest dry recovery of 23.17% was recorded in SLP 138/43 followed by 138/78 (19.9). High curcumin line 389/1 showed a dry recovery of 18.82% and control IISR Prathibha showed 18.28%.

Highest curcumin content of 5.02 % was recorded in SLP 389/1 which was on par with the control IISR Prathibha (4.79%). All other progenies showed curcumin content <4% only. The last 3 years data revealed that curcumin content in seedling progeny No. 389/1 is constantly above 5%.

Chromosome number analysis of seedling progenies

Chromosome analysis was completed in 96 seedling progenies. All of them showed deviation from the normally reported chromosome number of 2n=63.

Most frequently observed chromosome number was 2n=84. Variations observed were 2n=65, 2n=78, 2n=85, 2n=86, 2n=87, 2n=88 etc.

Mining of DNA markers and genes rom Expressed Sequence Tags

A short and easy protocol suitable for isolation of RNA was standardized. The protocol utilizes 2% SDS, 2% PVP, 100 mM Tris Cl and 25 mM EDTA in the extraction buffer followed by acid phenol extraction in presence of 5M sodium acetate. This protocol can be commonly used for RNA extraction from leaf and rhizomes and gives at least 5 times more yield than available protocols.

Development of SSR enriched libraries and microsatellites

The DNA fragents were enriched for SSR regions by hybridizing with biotinylated microsatellite oligos probe mix [(ACT)12, (AAAC)6,(ACCT)6],(ATC) 8,[(AAT)8, (GCC)8,(ATT)8, (ACC)8, (GATA)6] and captured using streptavidin coated dynabeads. Eluted fragents were again enriched by PCR and ligated to pTZ57R/T (Thermo scientific) and TOPO 2.1 (Invitrogen) cloning vectors and transformed into both TOP 10 competent cells and *E. coli* DH5 α cells (Invitrogen) as described by the manufacturer. A total of 500 colonies were obtained. Out of this plasmid DNA was isolated from 240 colonies and confirmed by colony PCR.After restriction digestion of the plasmid DNA the positive samples were sent for sequencing (Scigenome, Cochin). The sequences were analyzed and primers were designed for the microsatellite regions. Newly designed primers were validated in selected accessions and seven polymorphic markers were validated and added to the microsatellite marker repository.

Source-sink relationship

Samples of varieties IISR Prathibha, IISR Alleppey Supreme and Suguna were taken from 50 DAP. Plant height, number of leaves, leaf area and dry weight (DW) were recorded in these varieties at monthly intervals. All the varieties were on par with regard to leaf area and dry weight at all sampling dates. Starch content increased steadily from 50 to 110 DAP and then there was sudden increase in all the varieties. All the varieties accumulated more biomass in shoot compared to rhizomes up to 110 DAP. After 110 DAP, there was sudden rise in partitioning of biomass to rhizomes compared to shoot. At 150 DAP, rhizomes accumulated >75 % biomass. Endogenous IAA and zeatin riboside levels were quantified in both leaves and rhizomes. Both leaves and rhizomes had similar levels of hormones. IISR Prathibha recorded slightly higher concentration in leaves while Suguna showed higher concentration in rhizomes. Total hormone concentration was more in IISR Alleppey Supreme compared to other varieties. All the varieties showed maximum levels at 110 DAP.

Photosynthetic rate, stomatal conductance and transpiration rate were recorded at 100 and 130 DAP. Photosynthetic values ranged from 5.0-7.8 μ moles. All varieties showed higher photosynthetic rate at 130 DAP compared to 100 DAP. Suguna showed lowest values both at 100 (5.0 μ moles) and 130 DAP (6.5 μ moles).

Vanilla

Morphological and cytological analysis of interspecific hybrids

Morphological characters namely leaf length, leaf breadth and inter node length were recorded from 10 interspecific hybrids involving *V. planifolia* and *V. tahitensis*. Variation in these characters among the interspecific hybrids was evident. Chromosome number analysis of 2 interspecific hybrids showed 2n=30 in one and 2n=32 in other. Interspecific hybrids between *Vanilla* sp. (A&N Islands-White) with wide leaves and *V. aphylla* without leaves showed leaf less nature of hybrid progenies in the juvenile stage.

However, they developed leaves after 1-2 years of growth in the nursery. These leaves also showed variation in size and more similarity in shape to *V. pilifera*. In a few progenies, the flowers were smaller than that of *Vanilla* sp. (A&N) and slightly bigger than that of *V. aphylla*. In general the flower architecture was similar to that of *V. aphylla* but, the frill of labellum was similar to that of *Vanilla* sp. (A&N). These hybrids showed 100% pollen sterility and failed to set fruits on self pollination. However, pollination with *V. aphylla* pollen in these hybrids resulted in fruit set.

Hybrid seedlings between *V. planifolia* and *V. aphylla* showed presence of well developed leaves right from the *in vitro* seedling stage and no hybrids without leaves were observed. These hybrids are at their second year of *ex vitro* establishment and have normal growth.

Nutmeg

Selection of promising clones

Surveys were conducted in farmers' fields in adjoining areas of Pala and Thodupuzha and 6 nutmeg accessions were collected. They are *Cheripurathu* (entire mace), *Madukkakuzhi* (bisexual type, producing large number of clusterted fruits), 2 yellow mace types, *Kinattukara* and *Kochukudi* (both high yielding types). In the trial on evaluation of lines for high yield, significant differences were observed among the lines for all the 4 characters and A9/185 was found to be the best with regard to plant height (7.0m), canopy (5.25m), girth (70.5cm) and number of branches per plant (48).

Recent Achievements of All India Coordinated Research Project on Spices

Genetic resources

Over 7000 accessions of Spices germplasm is maintained at various AICRPS centres. A few promising lines were short listed in the mandate crops for further evaluation. At Panniyur the hybrids PRS160 and PRS 161 were the best with maximum green berry yield of 3.2kg vine-1 and 3.0kg vine-1 respectively. At Kumarganj out of 61 accessions of ginger evaluated, NDG-55 (34t ha-1) was promising with respect to yield.

Crop Improvement

During the year, 9 high yielding high quality varieties of spices were recommended for release. They are ICRI-8 in cardamom, Narendra Haldi-3 in turmeric, PPI (CL)1 in clove, DH 220 and Suguna (LCC-236) in coriander, HF-143 and RF-281 in fennel, Ajmer fenugreek-3 and Rmt-365 in fenugreek and Hisar Ajowain-18 in Ajowan.

In a CVT in cardamom, over the years, genotype CL-722 (289.35kg ha-1) recorded maximum cardamom yield followed by PS-27 (275.58 kg ha^{-1}) than the best check M1 (216.34kg ha-1) at Mudigere. In a CVT on coriander at Jagudan the entries COR-56 and COR-48 gave higher yield (i.e.,1448 and 1405kg ha-1) than best check GCr- 2. At Coimbatore three coriander leafy types *viz.,* CS 1, CS 11 and CS 38 have recorded highest leaf yield of 3250, 4000 and 4275kg ha-1 respectively were identified for off season production. In CVT on fennel FNL-47 and FNL-56 gave higher yield (ie. 1299, 1285 and 1221kg ha-1) than all the checks.

Source: Annual Report 2012-13, Indian Institute of Spices Research, Kozhikode, Kerala.

□□□□

5

Recent Initiatives in Crop Production

Cost-effective farming systems for spices cultivation for increasing production and productivity in different agro ecologies have been developed.

An organic package for production of black pepper, ginger and turmeric by applying farmyard manure, vermicompost, ash, rock phosphate, *Azospirillum* sp. and phosphobacteria and *Trichoderma* sp. and *Pseudomonas* sp. as biocontrol agents for disease control was developed. Further, a foliar spray of 1% solution of complex fertilizer 19 : 19 : 19 (N : P : K) during spike initiation period (April second week, May first week and May fourth week) during lean cropping season under irrigated condition enhanced black pepper (cv. Panniyur I) yield by 20–25%. Soil application of zinc up to 10 kg/ ha or foliar spraying of ZnSO4 (0.25%) and Borax (0.2%) twice (60 and 90 days after planting) was recommended for higher yield and quality for turmeric in zinc and boron deficient soil.

Among all the treatments, sprinkler and drip irrigation methods caused early sprouting; early flowering with increased plant height and more number of leaves and flowers/plant as compared to the control (rainfed). Also, stiga fresh weight, dry weight, length and saffron yield were improved in sprinkler and drip irrigation methods over the control. Raised beds resulted in early sprouting, early flowering with increased plant height and more number of leaves and flowers/plant.

In cumin, irrigation with micro sprinkler and drip irrigation methods not only enhanced the yield but also improved the water productivity by 47.8 and 14.7 kg/ha cm irrigation water than flood. Among land configuration treatments, sowing of 3 rows of cumin on raised beds (75 cm) enhanced grain yield by 38.4 and 12.3 % and water productivity by 16.6 and 8.9 kg grain/ha cm irrigation water than flat bed and wider raised beds (150 cm). Further, coriander grown in plastic walk in tunnel resulted in higher yield over open condition and gave higher net return of '97,700 with a benefit: cost of 1.93. In cumin, low pressure drip irrigation yielded higher by 19.31% (663.28 kg/ha) as compared with flood (555.94 kg/ha), whereas water savings were 30%. In coriander, low pressure drip irrigation produced higher yield (703 kg/ha) as compared with flood (612 kg/ha) with a water saving of 30%.

For multiplication of nutmeg, green chip budding with orthotropic buds on its own rootstock gave 90- 100% success. The ideal time for budding was August-November. In turmeric, organic and integrated systems recorded comparable yields (30.2 and 30.6 tonnes/ ha). The variety Alleppey Supreme recorded higher yield (17%) under organic system compared to inorganic system. Soil organic C,N,P,K,Ca, Mg and Zn contents were higher under the integrated system.

In ginger, highest yield was obtained under organic system (20 tonnes/ha) and varieties Mahima and Varada performed better (up to 19% increased yields). Soil N, K and Ca contents were higher under the integrated system, and Mg content was higher under organic system. The field study on different irrigation methods comprising conventional, furrow and drip irrigation in fennel was conducted in which low pressure drip irrigation using 42% lesser water resulted in 37% higher yield over conventional method, thus resulting in higher water-use efficiency.

In saffron, natural corm multiplication is very slow which takes about four years from small cormlets to produce flower bearing corms of 10-12 g size. The highest corm yield (20.55 tonnes/ha) was obtained with treatment of NAA 1,500 ppm with highest corm weight of 4.8 g, followed by GA3 300 ppm (14.78 tonnes/ha) with corm weight of 4.95 g, but BA 40 ppm resulted in highest number of corms/plant (18.1) with corm weight of 2.3 g. Highest flower bearing corms (65.6%) were obtained in NAA 1,500 ppm, followed by GA3 300 ppm (64.95%).

Intercropping *Vetiveria zizanioides* in juvenile black pepper garden recorded maximum net returns of 46,225, followed by (44,600) *Alpinia calcarata* with a benefit : cost ratio of 2.3 and 2.2, respectively. Alternate year of black pepper bearing was studied and found that carbohydrate to cytokinin ratio was more during the bearing year compared to non-bearing year. Three genotypes of cardamom, IC 349591, IC 349537 and IC 349550, were found relatively tolerant to moisture stress. The mean yield of turmeric was recorded higher (8.1 kg/ 3 m2 bed) under organic system, which was at par to other management systems (7.4– 7.7 kg/3 m2 bed). Organic farming significantly increased biomass-C, P, soil respiration, dehydrogenase, acid phosphatase and beta glucosidase activities.

In saffron, maximum size (2.5 g) of micro-corms were obtained on Linsmaier and Skoog (LS) medium supplemented with 21.6 µM NAA and 22.2 µM BA, followed by 2.2 g on Murashige and Skoog (MS) medium supplemented with 21.6 µM NAA and 22.2 µM B.A. Shoots were generated from ovaries, small corms and eye buds. Shoots when cultured on half strength MS medium containing BAP and NAA developed 2 cormlets/shoot with an average fresh mass of 2.3 g. A total of eight months was taken to harvest a cormlets from eye bud which otherwise takes 22 months under field conditions.

Caradmom

Biofertilizers, *Azospirillum* and P- solubilizers, were found to increase growth and yield of cardamom. Application of 100% inorganic nitrogen + 50 g *Azospirillum* + 5 kg FYM or 100% inorganic phosphorus + 50 g *phosphobacteria* + 5 kg FYM was found very effective in increasing yield.

Source: DARE/ICAR Annual Report 2012-13, 2011-12, 2010-11, 2009–10

Recent Initiatives by National Research Centre on Seed Spices, Ajmer, (Rajasthan)

Standardization of sustainable and profitable cropping system with fruit crops

The study was conducted with six seed spice crops in association with fruit trees and in open space. The highest seed, stover and biological yield of seed spices was obtained in open space followed with Aonla association. The next highest yield of seed spices was achieved with Aonla association. The fruit yield of Ber and Aonla was also recorded. Yields were converted into coriander equivalent yield (kg/ ha). All the seed spices resulted higher coriander equivalent yield with Ber in comparison to Aonla and open space. Among seed spices fenugreek association with fruit trees proved highly beneficial which resulted higher coriander equivalent yield with Ber as well as Aonla.

Evaluation of irrigation methods in fennel under various configuration of crop geometry

Application of irrigation through drip with low pressure or conventional method proved better for getting higher yield (2137 kg/ha), net return (70662 /ha), benefit cost ratio (2.77) and water use efficiency (6.24 kg /ha/mm) followed by all furrow irrigation. Further, it has been observed that beside drip, all furrow irrigation method proved better over conventional and alternate furrow irrigation system. All furrow method of irrigation yielded 15 and 20 per cent higher grain and straw yield, respectively over conventional irrigation.

Standardization of package for organic seed spice production (Coriander and Cumin)

In cumin application of 100 percent N & P through vermi compost and enriched vermi compost is better for realizing higher yield but in coriander the higher yield can be realized with application of 100 percent N & P either through vermi compost and enriched vermi compost, sheep manure and enriched vermi compost or 75 per cent N&P through vermi compost, enriched vermi-compost + seed inoculation with *Azotobactor* and PSB. Seed treatment + Soil application of Neem cake @ 150 kg ha + Foliar spray of Sulphur compound resulted the lowest powdery mildew index in coriander and ST + SA of *Trichoderma viridie* + SA of NC @ 150 kg ha +F.S. of Sulphur compounds resulted lowest wilt incidence cumin. The lowest blight index in cumin was observed with ST+SA of Pseu.Flu.+ SA of NC @ 150 kg ha + F.S. of onion extract. The highest yield of coriander and cumin was obtained with the application of ST + SA of *Trichoderma viridie* + SA of NC @ 150 kg ha +F.S. of Karanj oil . The highest yield attributes and yield of coriander and cumin was recorded with insect pest control by Spray of Sulphur compound of Karanj extract-1% (T) which is at par with application of 7 Allylisothio cynate-1 percent and application of Neemoil-2 percent.

Evaluation and identification of efficient cropping sequences for seed spice growing area

In fennel based cropping sequences the highest fennel yield (1650 kg/ha) was recorded in green gram – fennel – green manuring cropping sequence followed by green gram – fennel – summer fallow and black gram – fennel – green manuring cropping sequence. In fenugreek based cropping sequences the higher fenugreek yield (1525kg /ha) was recorded in groundnut – fenugreek – green manuring cropping sequence being at par with groundnut – fenugreek – summer fallow. In cumin based cropping sequences the highest cumin yield of 690 kg/ha was recorded in pearl millet – cumin summer fellow with summer solarization followed by cluster bean – cumin – summer fallow with summer solarization. Among all the cropping sequences the highest fennel equivalent yield (8316 kg /ha) was recorded in cluster bean – cumin – green manuring sequence followed by cluster bean – cumin – soil solarization. The highest net return of ' 336560/ - per ha was recorded in cluster bean- cumin- green manuring cropping sequence followed by cluster bean – cumin – soil solarization which is statistically at par but the highest benefit cost ratio of 9.17 was obtained in cluster bean – cumin – soil solarization cropping sequence followed by cluster bean- cumin green manuring cropping sequence.

Effect of date of sowing, crop geometry and varieties on performance of seed spices under different agro-climatic conditions

Early sowing of fennel on 10th October was more effective to increase the yield than mid and late sown condition and resistant to frost than other seed spice crops. Other seed spices viz. coriander, fenugreek, ajwain and nigella resulted higher yield on mid sown condition. The vegetative growth and yield of all seed spices under study except fennel was good in mid season sowing but lower in early and late season sowing.

In fennel variety RF-125 sown in early season resulted the highest yield attributes and yield (10.41 q/ha). RCr-436 variety of coriander sown in mid season resulted the highest yield attributes and yield (18.97 q/ha). AFg-2 sown in mid season proved superior over all other varieties and resulted the highest yield attributes and yield (17.93 q/ha). Ajmer Ajwain-1 sown in mid season was better compared to Ajmer Ajwain-2. Ajad Kalongi sown in mid season resulted the highest yield (13.12 q/ha).

Inter-cropping of seed spices in vegetables (AICRP on Vegetables)

In the study three seed spices viz. fennel, ajwain and coriander with three vegetables viz. pea, cabbage and carrot were taken. Results revealed that all the vegetables viz. carrot, cabbage and pea performed better with fennel compared to ajwain and coriander and accordingly exhibited higher vegetable yield. Inter-cropping of seed spices with carrot resulted higher yield of fennel, coriander and ajwain compared with cabbage and pea. Inter cropping of fennel with cabbage resulted the highest fennel equivalent yield (27.37q/ha)and net return (' 134,15/- per ha) followed by with carrot + fennel. The highest benefit cost ratio of 4.76 was obtained in carrot + fennel intercropping followed by with cabbage + fennel.

Identification of critical stage for weed management in seed spices

Weed flora of the experimental field consisted of *Chenopodium murale* L., *Chenopodium album* L., *Cynodon dactylon* L. and *Cyperus rotundus*. Maximum seed yield of cumin, coriander and fenugreek (645.8, 1262.5 and 2387.5 kg/ha) were recorded with the weed free throughout growth period treatment, where as lowest yield was recorded at weedy check plots, respectively. The critical stage for weed competition in cumin, coriander and fenugreek were found 51, 35 and 31 days respectively. The loss in yield due to weeds (weedy throughout growth) in fenugreek, coriander and cumin were 629.5, 919.2 and 1675.0 kg/ha as compared with weed free plots. The loss in yield due to weeds (weedy upto 15 DAS to weedy throughout growth period) were range from 22.57- 97.47% in cumin to 12.27- 72.8 % in coriander and 21.41-70.1 % in fenugreek as compared with weed free plots.

Effect of land configuration and water management techniques on yield of seed spices in maize (baby corn) based cropping system

The baby corn yield was 3625, 3756 and 3014 kg/ha under flood, drip and micro sprinkler irrigations, whereas the yield was 3515, 4055 and 2825 kg/ha with flat bed, raised beds of 75 and 150 cm width. Three irrigation measuring 13.46, 5.41 and 4.08 ha cm water was given under flood, drip and micro sprinkler irrigation. Under land configuration treatments the irrigation water delivered was 7.78, 7.67 and 7.50 ha cm with flat, 75 and 150 cm wider beds. After harvest of the baby corn, cumin variety GC-4 was sown on 11-11-2011. Five irrigations were given to the crop measuring 30.81, 10.84 and 15.55 ha cm water through flood, drip and micro sprinkler systems. Whereas 19.62, 17.68 and 19.49 ha cm water was delivered under flat, raised beds (75 cm) and wider beds (150 cm). Results show that irrigation with drip not only enhanced the yield by 41.07 and 23.05 % than flood and micro sprinkler irrigation methods but also improved the water productivity and gave 41.3 kg seed/ha cm irrigation water. The improvement in water productivity with drip irrigation was 32.85 and 19.55 kg seed/ha cm irrigation water than the flood and micro sprinkler irrigation. Sowing of 3 rows of cumin on raised beds (75 cm) enhanced the grain yield by 42.72 and 41.28 % and water productivity by 15.13 and 13.61 kg grain/ha cm irrigation water than flat bed and wider raised beds (150 cm).

Scaling up Water Productivity in Seed Spices cultivation (Cumin and Nigella)

Application of irrigation through drip irrigation and conservation of moisture by mulching with 20 micron plastic sheet is better for realizing higher growth, yield and water use efficiency in Nigella. In case of Cumin, application of irrigation through low pressure drip irrigation and conservation of moisture by mulching with 20 micron plastic sheet is better for realizing higher growth, yield and water use efficiency.

Standardization of NPK levels for coriander and cumin under protected environment in the scenario of climate change

Among different protected structures, plastic walk in tunnel proves significantly better in terms of yield and its attributing characters. The plant height at harvest (119.20 cm), umbel per plant (55.0), umbellate per umbel (7.42) test weight (8.78 g) and seed yield (1533.71 kg/ha) were recorded in the said treatment.

Whereas, under application of NPK fertilizers, economic seed yield (1454 kg/ ha) with highest test weight was recorded with 80 per cent RDF. In interaction effects economic yield (1582.16 kg/ ha) of coriander was recorded in treatment combination containing plastic walk in tunnel with 80% RDF. Minimum aphid population (20.1) was recorded in the month of January in the plastic walk in tunnel as compared to all other treatments.

Effect of different protected structures and fertilizer doses on growth and yield of cumin

Insect proof net exhibited significantly higher values of plant height at harvest (28 .73 cm), umbels per plant (12.83), seeds per umbellate (6.08), test weight (3.66 g) and seed yield (806.78 kg/ha) as compared to black net and open field in cumin. Similarly under application of NPK fertilizers, 100 percent RDF exhibited maximum plant height at harvest (29.18 cm), seeds per umbellate (6.67), test weight (3.87 g) and seed yield (787.82 kg/ha). In interaction effects maximum yield (898.78 kg/ ha) of cumin was recorded in treatment combination containing insect proof net with 100 % RDF. Minimum aphid population (7.37) in the month of January was also recorded in the insect proof net as compared to all other treatments.

Nutrient Influx Efficiency in *Coriandrum sativum, Cuminum cyminum, Foeniculum vulgare* and *Trigonella foenum-graecum* cultivars

Cumin genotypes differed in nutrient influx efficiency depending upon their growth stages. N, P, K, Mn, Zn and Cu influx rate was highest in GC- 4 at 40 days while at 90 days it was highest in AC- 167. However, Fe influx was highest in GC-4 at all the stages. In fennel N, P, K and Mn influx rate up to 45 days was highest in genotype RF-101 while Cu, Fe and Zn influx rate at 45 days was higher in AF1-87. Nutrient influx rate in fenugreek cultivars Rajendra kranti and UM-35 was highest up to 35- 40 days. However, at the age of 75-80 days nutrient influx efficiency was highest in RMt-305.

Nutrient uptake by coriander, cumin, fennel, fenugreek and anise with various IW:CPE ratio

Nitrogen, phosphorus, potassium and Zn uptake was highest in coriander with IW:CPE ratio 0.6 and least with 0.4 IW:CPE ratio. Copper uptake was highest with IW:CPE ratio 1.2. However, iron and Mn uptake was higher with IW:CPE ratio 1.0. N and P uptake in fennel was highest with IW:CPE ratio 0.8. Fennel removed highest K with IW:CPE ratio 1.2. Cu uptake increased with increase in IW: CPE ratio. While, Mn uptake increased up to 0.8 IW:CPE ratio. NPK and micronutrient uptake in fenugreek increased with increase in IW:CPE ratio. N, K, Zn and Cu uptake in anise increased with increase in IW:CPE ratio up to 1.0, whereas P uptake increased up to 0.8. Fe uptake was highest with IWCPE ratio 0.8 and 1.0.

Screening of plant growth promoting *rhizobacteria* for coriander (*Coriandrum sativum*)

Twenty five coriander plant soil samples have been collected from Ajmer, Baran, Jhalawar and Kota Districts of Rajasthan for isolation of rhizospheric bacteria. These samples have been analysed for their electrical conductance (EC), pH, total viable bacterial count, and total viable fungal count. Microbial count on nitrogen free *Azotobacter* medium, Pikovskaya medium containing Tricalcium Phosphate and Kings B medium have also been done for prospective microbial analysis of soil samples. Total thirty six cultures were isolated and screened for IAA production, acid production and capacity to solubilise the Tricalcium phosphate. Fourteen bacterial cultures were found superior w.r.t. IAA production and these were further screened for their biochemical characteristics to utilize different sources of carbohydrates and production of the specific enzymes.

Isolation and Evaluation of phosphate solubilising microorganisms (PSM) for fennel (*Foeniculum vulgare* L.)

Fennel plant and soil samples have been collected from Ajmer, Pali, Jhalawar, and Kota Districts of Rajasthan for isolation of *rhizospheric* bacteria. Total fourteen soil samples have been analysed for EC, pH, bacterial count and fungal count. Sixth Phosphate solubilizing microorganism were isolated and further screened on National Botanical Research Institutes Phosphate (NBRIP) medium. Maximum Phosphate solubilization in broth assay was observed in bacterial isolates FEN-14(51.66 μg/ml) which was at par with FEN-1(51.30 μg/ml) and FEN-5 (50.68 μg/ml). These isolates were further screened for tolerance to various abiotic stresses such as pH, high temperature and salinity.

Source: Annual Report 2012-13, National Research Centre on Seed Spices, Ajmer

Recent Initiatives by Indian Institute of Spices Research, Kozhikode, Kerala

Black pepper

Tree species-black pepper interaction

In the green house study on the allelopathic effect of tree standards on growth of bush pepper, the leaf and stem extracts of tree species like garuga, erythrina, gliricidia, ailanthus, jack and silver oak at 100% concentration was applied to bush pepper @100mL per plant twice every month.

Observations on growth parameters taken after eight months of application revealed little or no significant differences among the treatments including control. The data on soil parameters revealed that soil pH was near neutral in all treatments. Available nutrient levels including organic C were markedly high in all the treatments. Dehydrogenase level was lowest in the control and silver oak, but markedly higher in soils applied with other tree extracts. Microbial biomass C was lowest with silver oak and ailanthus extracts but greatest with garuga extract.

Microbial consortium for growth promotion

A consortium of three rhizobacterial isolates (BRB3, BRB13 and BRB23) was found to markedly promote the growth of black pepper. However, the PGPR applied in combination with various rates of chemical fertilizers differed in their effects on black pepper growth. The soil microbial activity measured through various parameters was consistently greater in treatments involving PGPR. However, the PGPR treatments also varied in their effects on soil microbial properties. Microbial biomass N and acid phosphatase activity were greatest in the treatment 100%N + 75%P + 100%K + BRB3 + BRB13, microbial biomass C in the treatment 100%N + 100%P + 75%K + BRB3 + BRB23, microbial biomass P in the treatment 75%N + 100%P + 100%K + BRB13 + BRB23 and â glucosidase in the treatment 100%N + 100%P + 100%K + BRB3 + BRB23. Based on a series of experiments, a microbial formulation involving this consortium of PGPR was developed for licensing and commercialization.

Organic farming

An organic package involving FYM, vermicompost, ash and rock phosphate combination and *Azospirillum* and *Phosphobacteria* and *Trichoderma* sp. and *Pseudomonas* sp. was developed and recommended. For black pepper, an organic package consisting of application of 10 kg FYM + 500g Neem cake + 500 g ash+ 2 kg vermicompost with biofertilizers – *Azospirillum* and P solubilizing bacteria (20g) and *Trichoderma* sp. (50g) and *Pseudomonas* sp. (IISR 6) (50 g) per vine and spray of BM and Neem oil has been standardized and demonstrated.

Soil carbon build-up under spices based cropping systems

Soil samples were collected from two depths (0-15 and 15-30 cm) in pure cardamom, cardamom + coffee + black pepper, coffee + black pepper systems in Madikeri and Wayanad districts and also from black pepper maintained under organic, integrated and chemical management systems. The wet oxidizable C ranged from 1.3– 3.1g 100g-1 in 0-15 cm layer and 0.69-2.1g 100g-1 in 15-30 cm layer. The total organic C (TOC) content was 4.31, 3.73 and 3.41g 100g-1 in the 0-15 cm soil layer of black pepper managed under organic, integrated and inorganic systems, respectively. The total organic N (TON) content was higher under organic management (1.61, 1.35g 100g-1) followed by integrated (1.46, 1.19g 100g-1) and inorganic (1.34, 1.35g 100g-1) management systems of black pepper, at surface and sub surface layers. TOC:TON ratio was observed to be in the range of 2.20-2.75.

The particulate organic C content varied from 0.91 – 1.83g 100g-1 in the top soil and 0.53 - 1.25g 100g-1 in the sub surface layer under different management systems. The particulate organic N was higher under organic management (0.83, 0.38g 100g-1) followed by integrated (0.74, 0.21g 100g-1 and inorganic (0.32, 0.32g 100g-1) management systems. The ratio of POC/PON varied from 2.1-3.6. The POC contributed only 23-31% of TOC, where as the wet oxidizable C fraction contributed 59-71% of TOC at different depths and management systems. The POC fraction also accounted for only 35-47% of wet oxidizable C fraction.

Coffee + Pepper system recorded TOC content of 4.29g 100g-1 in the 0-15 cm layer followed by Cardamom alone (3.63g 100g-1) and Coffee + Cardamom + Pepper recorded the least (2.5g 100g-1). The TON content also followed a similar pattern in the surface and sub surface soils. The wet oxidizable C fraction accounted for about 29-71% of the TOC in the systems studied.

Cardamom

Quantifying nutrient removal for targeted yield

To find out the nutrient requirement for targeted yield, graded fertilizer treatments were imposed with four levels – T1- 0,0,0 (control); T2- 75,75,150 T3- 125,125,250 and T4- 150,150,300 kg NPK on two varieties of cardamom (Appangala- 1 - Paka estate and Green Gold - Kauvery estate). In Appangala^{-1}, higher yields were recorded for higher fertilizer levels with the highest in T4 (824kg ha^{-1}) followed by T3 (740kg ha^{-1}), T2 (683kg ha^{-1}) and lowest in control (209kg ha^{-1}). Number of yielding tillers, no. of panicles clump^{-1} and no. of capsules panicle^{-1} were significantly higher in fertilizer treatments compared to control. In Green gold also highest yield was recorded in T4 (883kg ha^{-1}) where highest dose was applied followed by T3 (795kg ha^{-1}) and T2 (739kg ha^{-1}) with the lowest in the control (344kg ha^{-1}). The fertilizer treatments recorded significantly higher no. of panicles clump^{-1} and no. of capsules panicle-1 as compared to control. The average weight of dry matter removed by thrashing (dried tillers, leaves and panicle) during a year and the nutrient removal by this dry matter were quantified. The nutrient removal by the capsule was quantified as 6.1, 1.9 and 4.2kg of N, P2O5 and K2O by 100kg of the produce. From these estimates, the nutrient required for production of 100kg of capsule has been worked out as 11.3, 1.6, 9.1kg of N, P2O5 and K2O by Green gold and 12.2, 1.7, 11.9kg of N, P2O5 and K2O by Appangala^{-1}.

Developing electronic nose for monitoring aroma

This project aims at developing electronic nose for monitoring the quality of cardamom. Essential oil content and composition of 11 genotypes were determined and the Normaroma index was determined using *e*-nose, developed for assessing the quality of tea. In the preliminary studies, quantity of sample (cardamom seeds 5g) and size of sample holder (10g) were standardized.

Ginger

Organic farming

An organic package consisting of 20t FYM + 2t Neem cake + 1t ash + 4t vermicompost ha^{-1}, *Azospirillum* and P Solubilising bacteria (20g bed^{-1}) for nutrient supplement, GRB35 as seed treatment and application of Panchagavya as drench and spray at 45 and 90 DAP has been standardized. Drenching of GRB35 or 57 and GEB17 was helpful in keeping the disease incidence to <10-15%. Spray of BM (1%) to contain foliar diseases and Neem oil (5mL L^{-1} of water) in combination with the cultural control is recommended for shoot borer management. Results showed that organic management system yielded on par with integrated system, whereas the quality was found to be superior under organic systems.

PGPR as biocontrol

The rhizobacterial strain, GRB35 confirmed its plant growth promoting efficiency besides reducing the soft rot and bacterial wilt in ginger (var. Mahima). The field experiment consisted of treatments involving GRB35 delivered through various modes. Among the treatments, activated capsule (2 capsule 5kg^{-1} seed)-T2 followed by non activated capsule (1 capsule 5kg^{-1} seed)-T1 registered markedly lower soft rot incidence (11 and 12%, respectively).
Source: sink relationship

Varieties IISR Varada, IISR Rejatha and IISR Mahima were planted and sampling was done at monthly interval starting from 50 days after planting

(DAP). Plant height, number of leaves, leaf area and dry weight were recorded at monthly intervals starting from 50 DAP. IISR Varada accumulated more leaf area and dry weight at all sampling dates.

Starch content increased steadily from 50 to 110 DAP and then there was sudden rise in all the varieties. IISR Varada showed highest starch content. Dry matter partitioning studies showed that up to 80 DAP, shoot had more biomass than rhizomes and at 110 DAP, and rhizome had more biomass compared to shoot. At 150 DAP, rhizomes accumulated >75% biomass in all the varieties.

Endogenous IAA and zeatin riboside levels were quantified in both leaves and rhizomes. Rhizomes and leaves recorded similar levels and the maximum levels at 80 DAP. IISR Varada had slightly higher levels. All varieties showed higher photosynthetic rate at 120 DAP compared to 90 DAP and all the varieties had similar levels of photosynthesis. Photosynthetic rate ranged from 10.4-13.9ìmoles.

Weed management practices on growth, yield and quality

An experiment was initiated to study effect of weed management practices on weed control efficiency and to study growth, yield, quality parameters and disease incidence with 10 different green and polythene mulch combinations. Height, number of tiller production and yield were higher for the treatments with *Glycosmis pentaphylla* leaves (recommended practices), *Lantana camara* leaves, paddy straw, coir pith compost and application of coconut leaves as mulch in the beds. Maximum weed control efficiency (97%) was recorded by spreading of black colour polythene mulch followed by application mod of ash colour polythene mulch. Weed control efficiency was also higher in the treatment involving coir pith compost (94%) followed by coconut leaves. Shoot borer incidence varied from 10-20% and maximum was observed in beds in which coir pith compost was applied. Percentage of clump infected with *Rhizoctonia* was more in the treatment in which ash colour polythene mulch was used.

Turmeric

Organic farming

An organic package consisting of 20t FYM + 2t Neem cake + 1t Ash + 4t Vermicompost ha^{-1}, *Azospirillum* and P solubilising bacteria (20g bed^{-1}) for nutrient supplement, GRB35 as seed treatment and application of Panchagavya as drench and spray at 45 and 90 DAP has been standardized. Drench of GRB35 or 57 and GEB17 are helpful in keeping the disease incidence to <10-15%. Spray of BM (1%) to contain foliar diseases and Neem oil (5mL L^{-1} of water) in combination with the cultural control is recommended for shoot borer.

Results showed that organic management system yielded on par with integrated systems and curcumin content was found to be higher under organic systems.

Micronutrients on yield and quality

Based on the field studies on effect of Zn and B on the quality of variety. IISR Prathibha for 3 years, soil application of Zn up to 10kg ha^{-1}or foliar spraying of ZnSO4 (0.25%) and Borax (0.2%) twice (60 and 90 DAP) is recommended for high yield and quality in Zn and B deficient soil.

Recent Achievements of All India Coordinated Research Project on Spices

At Sirsi, initial trials in spices based cropping system black pepper, recommended package of practices gave higher dry berry yield of the black pepper (12 kg vine^{-1}) compared to those of organics (0.78kg vine^{-1}).

At Mudigere application of irrigation 9L clump^{-1} day^{-1} along with 100% recommended dose of fertilizer through drip recorded highest cardamom capsule yield.

The results of fertigation experiments in turmeric at Coimbatore revealed application of NPK@150:60:108kg ha^{-1}with urea and potash as straight fertilizers and phosphorus as water soluble fertilizer weekly once recorded a yield of 49.11t ha^{-1}. In a micronutrient trial of turmeric at Pundibari, soil application of boron gave the maximum yield 9.50kg plot^{-1} and 19.15t ha^{-1}, respectively. Soil application of FYM @30t ha^{-1}+ vermi compost @20q ha^{-1}+ Neem oil cake @ 8q ha^{-1} is recommended for higher yield (48.82t ha^{-1}) of organic turmeric at Dholi. However integrated nutrient management involving soil application of inorganic N @150kg ha^{-1}+ *Azospirillum* @1.5kg ha^{-1}+ FYM @5t ha^{-1}is recommended for higher yield (56.6t ha^{-1}) in turmeric.

In a trial conducted for 3 years to standardize nutrient management for off season production of coriander for leafy purpose at Coimbatore showed that application of 30:40:20 kg NPK ha^{-1} combined with spraying of GA 15 ppm at 20 DAS recorded the highest leaf yield per ha (5870 kg ha^{-1}) with high benefit cost ratio of 2.37. Nutrient supplementation through organic manure using FYM 5.0t ha^{-1} + vermi compost 2.5t ha^{-1} or vermin compost @ 5t ha^{-1} or FYM 2.5t ha^{-1} + vermi compost 3.75t ha^{-1} or FYM 10t ha^{-1} is recommended to increase the seed yield of organic coriander at Coimbatore and Guntur. For the saline soils where there is Zn deficiency (less than 2 ppm) spraying of 0.5% of zinc sulphate (2 sprays at 45 and 60 DAS) is recommended for higher seed yield. To promote micro irrigation and saving water Irrigation with 'Raingun Sprinkler' at either 30 & 60 DAS or 30 & 45 DAS is recommended to improve the productivity of coriander. If water is available for only one irrigation, irrigation with Raingun/Sprinkler at 45 DAS is beneficial and water saving is (58.7%) and 45.3% respectively in Guntur region. Integrated nutrient management with soil application of inorganic N @ 33kg ha^{-1} + Azospirillum @1.5kg ha^{-1} + FYM @ 5t ha^{-1} is recommended for higher yield (1.98t ha^{-1}) of coriander at Dholi region.

The experiment on integrated nutrient management of fenugreek at Dholi, Soil application of inorganic N @13kg ha^{-1} + *Azospirillum* @1.5kg ha^{-1} + FYM @ 5t ha^{-1} is the best and resulted in higher yield of 2.29t ha^{-1}.

Source: Annual Report 2012-13, Indian Institute of Spices Research, Kozhikode, Kerala.

□□□□

6

Recent Initiatives in Crop Protection

A protocol for SYBR green based real-time RT-PCRfor detection of Piper yellow mottle virus and Cucumber mosaic virus infecting black pepper and Cardamom mosaic virus (CdMV) and Banana bract mosaic virus (BBrMV) infecting cardamom was developed. A native isolate of *Phytophthora capsici* (Is. No. 98-93) infecting black pepper was completely sequenced using next generation sequencing platform, Illumina-Solexa GA II. The sequence data was assembled by taking Joint Genome Institute's *P. capsici* as reference genome. The treatment of black pepper cuttings with Carbendazim + Mancozeb (0.1%) was effective for the management of anthracnose disease in nursery.

The black pepper accession, 1114 (Kumbhachola), was tolerant to *P. capsici*. The accession was earlier found resistant to pollu beetle and drought. A new bacterial wilt disease on small cardamom was recorded in Wayanad, Kerala. The phenotypic and genetic characterization revealed that causative organism is *Ralstonia solanacearum* (Biovar 3 phylotype 1) and was 100% similar to ginger strain of *R. solanacearum*. *Bacillus amyloliquefaciens* (GRB 35) and *Serratia marcescens* (GRB 68) were promising for enhancing sprouting of rhizomes, growth promotion and reducing the incidence of soft rot and bacterial wilt diseases in ginger in the field.

Bioagents *Trichoderma viride* and *Aspergillus versicolor* applied as seed treatment (10 g/kg seed) and soil application (2.5 kg/ha) and soil amendment mustard residue (2.5 tonnes/ha) + mustard cake (0.5 tonne/ha) + Neem cake (0.5 tonne/ha) are effective eco-friendly approaches for the management of cumin wilt caused by *Fusarium oxysporum* f.sp. *cumini*. Seed wasp, an important pest of coriander, can be successfully controlled by application of botanicals Neem oil (2%) and insecticides like Thiomethoxam (0.025%) and Imidachlorprid (0.005%). Application of Neem oil (2%), *Metarrhizium anisopliae* and *Verticillium lacanii* at 108 spores/ml gives good control of cumin and fennel aphids.

The efficacy of endophytic bacteria (BP-35,25,17 and TC 10) and rhizobacteria (IISR 853 and IISR 6) against *Phytophthora capsici*, *Radopholus similis* and *Meloidogyne incognita* infecting black pepper was evaluated in field conditions. Application of TC 10 +Metalaxyl-Mancozeb and IISR 853 + Metalaxyl- Mancozeb were found effective for growth and canopy. A model for estimating risk of downy mildew to support the decision on spraying in vineyards was prepared. The model considers daily data on temperature and relative humidity. The model was tested using last two years' data on weather parameters and disease incidence and was found reasonably accurate in estimating the risk. Enzyme linked immunosorbent assay (ELISA) and reverse transcription–polymerase chain reaction (RT-PCR)-based methods were standardized for detection of cardamom mosaic virus.

A virtual screening of phytochemicals from Dr Duke's phytochemical and Ethnobotanical database, literature search, PASS prediction and ADME/Tox screening has identified 56 promising lead compounds with potential nematicidal activity. Flexible docking studies revealed that six of these phytochemicals, ie curcumin, brucine-n-oxide, colubrine, brucine, vanillin, genostrychnine and strychnine, had good binding score (MolDock score) and favourable hydrogen bond interaction with glutathione-S-transferase (GST) of *Brugia malayi*. A new database, GST Lead Base (www.spicebioinfo.res.in/gstleadbase) was developed and hosted. The database includes chemical properties and other details of potential nematicidal compounds that inhibit glutathione-S-transferase in nematodes.*SpiceStat*, the database on vital statistics of spice cultivation, production and export was relaunched with additional features. A new accounts module was developed and integrated in ARISoft, the existing office automation software. An interactive CD on ginger and turmeric package of practices was developed.

Cardamom: Fifty-eight cardamom accessions were screened for leaf blight, rhizome rot and leaf blotch resistance under natural field conditions. The IC-349646 was found to be resistant with an average leaf blight incidence of 20%. Twenty- three accessions exhibited highly resistant reaction to rhizome rot disease.

Turmeric: Maximum disease reduction of rhizome rot was observed with Metalaxyl- Mancozeb (44.5%) followed by Copper oxychloride (36.5 %) compared to the control. *Apanteles taragamme*, an un-identified hymenopterous parasitoid and earwigs, were documented as natural enemies of shoot-borer (*Conogethes punctiferalis*) infesting turmeric. Four strains of entomopathogenic nematodes were isolated from 71 soil samples collected from rhizosphere of ginger and turmeric. Of these, one strain each was tentatively identified as *Heterorhabditis* sp. And *Steinernema* sp.

Ginger: Soft rot of ginger could be controlled by bio-fumigation using cabbage and mustard plant refuses. The bio-fumigation using cabbage increased germination and yield (8.5 kg/plot) by decreasing soft rot (2.25%) and bacterial wilt incidence.

Cumin: Spraying of Mancozeb @ 0.25% at 40, 50, 60 and 70 DAS was found to be effective in controlling blight. Soil solarization + soil application of *Trichoderma harzianum* + spraying of Mancozeb @ 0.25% at 60 DAS and application of vermicompost + soil application of *T. harzianum* + spraying of Mancozeb @ 0.25% at 60 DAS were also effective for controlling the disease.

Source: DARE/ICAR Annual Report 2012-13, 2011-12, 2010-11, 2009–10

Recent Initiatives by National Research Centre on Seed Spices, Ajmer

Management of wilt disease in cumin (*Cuminum cyminum*) through soil solarization

The maximum Percent Disease Index (PDI- 32.0) was found in the treatment (T10) with no soil solarization. The minimum Percent Disease Index (PDI-6.0) was found in the treatment 'T8' (20 days soil solarization + Neem cake + Trichoderma viride) followed by treatment 'T9' (30 days soil solarization + Neem cake + *Trichoderma virede*) with PDI-6.5. Similarly, the maximum yield (173 g/plot) was found in the treatment 'T8' followed by in the treatment 'T9' with (168 g/plot). The minimum yield was recorded in the treatment T10 (control) with 60 g/plot.

Development of plant protection schedule for cumin cultivation

Field experiments were conducted during rabi seasons of 2009-10 and 2010-11 with cumin varieties RZ 209 and GC 4 to evaluate four disease and pest management schedules. The overall results revealed that schedule II [Soil application of Neem cake + mustard cake@ 0.5 t / ha, Seed treatment with Tebucanazole @ 2.5 g/ kg, Soil drenching with Metalaxyl (0.1%) at 25-30 DAS, foliar spray with Mancozeb (0.2%) at 60 DAS and Propiconazole (0.1%) at 75 and 90 DAS, foliar spray with Karathane (0.1%) at 70, 85 and 100 DAS, FS acetamiprid 0.005% 60-70 DAS and imidachlorpid 0.005% 70-80 DAS] reduced diseases PDI and insect population over untreated control.

Evaluation of Trichoderma isolates for the management of wilt (*Fusarium oxysporum* f.sp. *cumini*) disease of cumin

The native Trichoderma spp. isolated from cumin rhizosphere showed variable reaction in colony growth, sporulation, production of volatile and non volatile inhibitors. The isolates CuTk 7-01, CuTa 7-02, CuTh 9-02 and CuTh 3-03 showed more than 40% growth inhibition against cumin wilt and nigella root rot pathogens in dual culture under in vitro conditions.

Efficacy of Picoxystrobin 25 SC on cumin blight caused by *Alternaria burnsii*

Picoxystrobin 25 SC, was applied as foliar spray after initiation of disease at 12 days interval in different concentrations along with Mancozeb (0.2%) and Azoxystrobin (0.1%). The results revealed that all treatments reduced blight disease as compared to untreated control. Picoxystrobin was found effective for blight management even at lower doses; however, there was more reduction in disease with the increase in fungicide concentration. Maximum blight was reduced in the plots treated with Picoxystrobin (0.12%) followed by Picoxystrobin (0.1%) and Azoxystrobin (0.1%). No phyto-toxicity symptoms were observed when Picoxystrobin was sprayed up to 4 ml/lit, which permit the use of Picoxystrobin on the cumin plants even at higher concentration.

Biorational Management of aphids in cumin and fennel crops

Different biorational products viz., entomopathogen (*Beauveria bassiana, Metarhizium anisopliae* and *Verticillium lacanii*), and Neem oil alone and in combination with entomopathogen, *Coccinella septempunctata* (predator) and a standard insecticide were tested for control of aphids *Myzus persicae* and *Apis gossypii* in cumin and fennel crop. In cumin field level colonization of *V. lecanii* was found higher than other entomopathogen and gave maximum control for both the species of aphids. Among plant products Neem oil, allyl–iso–thio cyanate and sulphur extract of Karanj gave good control. Overall, insecticidal check gave maximum protection but all other treatments were significantly superior over the control in term of pest control and seed yield. In fennel crop similar trend was observed.

Source: Annual Report 2012-13, National Research Centre on Seed Spices, Ajmer

Recent Initiatives by Indian Institute of Spices Research, Kozhikode, Kerala

Black pepper

Development of Real-time PCR for viral detection

A SYBR Green based quantitative real time polymerase chain reaction (qPCR) for *Piper yellow mottle virus* (PYMoV) and reverse transcription (RT) qPCR for *Cucumber mosaic virus* (CMV) infecting black pepper was developed for sensitive and specific detection and quantification of viruses. The primers were designed to conserved regions in both the viruses (coat protein in CMV and ORF III in PYMoV) identified by Clustal x alignment. The qPCR and RT-qPCR was performed on a Rotor Gene Q real-time system (Qiagen, Hilden, Germany). The qPCR reaction for PYMoV was carried out in a final volume of 25μL each containing: 12.5μL of 2x QuantiFastTM SYBR Green PCR Master mix, 1.0μL of each forward and reverse primer (1μM μL) and 1μl template (about 60ng). Thermocycling conditions used were an initial denaturation step at 95°C for 5 min, followed by 35 cycles of 95°C for 15s and 60°C for 45s. The components of RT-qPCR reaction for CMV consisted of 50U of Revert Aid reverse transcriptase in addition to the above components and thermocycling conditions included an additional step of cDNA synthesis at 42°C for 45 min at the beginning. Healthy black pepper total nucleic acid and water control (without template) were used as negative controls to monitor specificity of the primers and potential contamination within the real time PCR reagents. Following qPCR and RT-qPCR, the amplicons were subjected to melt analysis from 60-95°C. The specificity of the qPCR and RT-qPCRproduct was verified (apart from melt curve analysis) periodically using a 1.5% agarose gel electrophoresis.

The qPCR and RT-qPCR primers amplified expected product from known infected plants that were previously analysed by PCR and RT-PCR while no amplification was seen for healthy sample and water control. The melt curve of qPCR and RT-qPCR products for infected black pepper samples had a single peak at 82.5- 83.0°C indicating that only the target fragent was amplified. The healthy and water control showed no melting peak. The specificity of the product was also confirmed by agarose gel electrophoresis which showed single band at expected size.

In order to develop standard curve for quantification, the target regions from both the viruses were amplified and cloned in plasmid vector. Yield of recombinant plasmid DNA was calculated using a spectrophotometer and copy number of recombinant plasmid was calculated. The recombinant plasmid was then diluted serially and subjected to qPCR as explained above. The output data were analyzed with the software available in Rotor Gene Q system (Qiagen). A standard curve for each virus was constructed by plotting Ct values from the assays with standard dilution versus the logarithm of the concentration for the exponential phase of the reaction.

The sensitivity limit and range of the SYBR Green qPCR and RT-qPCR assays were deduced from the standard curves for each virus using plasmid standards. The amplification of the standard dilution series yielded in linear relations with reliable results (R2 = 0.99984 for CMV and R2 = 0.99968- for PYMoV) in the range of 10 – 108 copies of target DNA locus per PCR. The detection limit was as low as 26 copies for CMV and 24 copies for PYMoV as estimated from the standard curve equation.

In order to know detection limit of the viruses in plants, total nucleic acid from infected plant was diluted 10-fold and was subjected to qPCR and RT-qPCR using 1 μL^{-1} of each dilution. The detection limit (copy number) of CMV and PYMoV was determined from the standard curve using Ct values obtained. The detection limit was as low as 14 and 7 copies (dilution 10-4 and 10-3) for CMV and PYMoV respectively as estimated from the standard curve equation. The developed qPCR and RT-qPCR assays were used to detect virus in 174 symptomatic and non-symptomatic field samples of black pepper, long pepper and betel vine collected from different agro-climatic regions. All the symptomatic samples were either infected by one or both the viruses while non symptomatic plants tested were infected with a single virus.

Development of LAMP for detection of viruses

A quick detection assay based on loop mediated isothermal amplification (LAMP) was developed for detection of PYMoV and reverse transcription (RT) LAMP for detection of CMV infecting black pepper. Each LAMP and RT-LAMP assay used a set of five primers designed against conserved sequences located in the viral genome. Total nucleic acids containing both RNA and DNA isolated from black pepper was used as template for assays. The LAMP and RT-LAMP were carried out at 65°C for 60 min. The results of LAMP and RT-LAMP reaction was assessed visually by checking turbidity, pellet formation and green fluorescence in the reaction tube and also by gel electrophoresis. The assay successfully detected both the viruses in infected plants while no cross-reactions seen with healthy plants. The optimized LAMP and RT-LAMP were validated by testing field samples collected from different regions.

In vitro evaluation of fungicides against *C. gloeosporioides*

In vitro evaluation of fungicides on the growth of *C. gloeosporioides* from necrotic lesion bits obtained from the runner shoots showed that the combination product of Carbendazim and Mancozeb was more inhibitory to the growth of the fungus compared to Bordeaux mixture and Carbendazim.

Management of anthracnose disease in nursery

The technology consisting of pre-planting treatment of three node cuttings prepared from the runner shoots with fungicides to manage anthracnose disease under nursery conditions was validated. The combination product of Carbendazim+ Mancozeb was found to be superior in delaying disease initiation and development

Characterization of endophytic bacteria

Endophytic bacteria such as spontaneous Rifampicin resistant *Bacillus megaterium* BP-17R and *Pseudomonas putida* BP-25R were characterized by Biolog, biochemical and antibiotic sensitivity assays. Reactions of these bacteria to 17 antibiotics indicated that *P. putida* BP-25R is resistant to several antibiotics while *B. megaterium BP-17R* was resistant only to Rifampicin.

Screening of bacterial metabolites

The secondary metabolites of *P. putida* and *B. megaterium* were screened for their activity against fungi, protozoa and oomycetes and for their antihelminthic activity, respectively, using in silico tools, *in vitro* and *in planta* bioassays. Eighty nine compounds from *P. putida* and 131 compounds from *B. megaterium* were predicted to possess anti-oomycete and nematicidal properties, respectively. The predicted compounds from *P. putida* were further docked with Glucanase Inhibitor Protein of *P. capsici* while the compounds of *B. megaterium* were docked with 1, 4 beta endoglucanase enzyme of *R. similis*. The most promising compounds in *P. putida* after the docking study using the above target are 2-octaprenylphenol, 2-methoxy-6-(all-trans-octaprenyl) phenol, canavanino-succinate and 5-amino-1-(5- phospho-D-ribosyl) imidazole. In *B. megaterium*, the highest docking results were obtained with (S)-methylmalonate-semialdehyde, 2-methoxy-6- (all-trans-octaprenyl) phenol, phenyl acetate, 1-palmitoyl glycerol 3-phosphate, tributryin and 5-[(1- phenylcyclohexyl)amino] pentanoic acid. The presence of these compounds was confirmed by LC-MS analysis and genomic approaches in *P. putida*. However, whole metabolome extracted from BP-17R and BP-25R on assaying against *P. capsici, P. tropicalis, P. myriotylum, Rhizoctonia, Fusarium, R. solanacearum* and nematodes by *in vitro* bioassays did not show any activity against any of the tested pathogens, including nematodes. However, at a concentration of 20mg mL^{-1} these extracts inhibited germination of *P.capsic*i zoospores. Preliminary *in planta* assays against *P. capsici* using the crude metabolites were also negative.

Colonization of *P. putida* in black pepper

Colonization studies using different concentrations of gfp tagged *P. putida* (BP-25R::gfp 58) proved the endophytic nature of the bacterium. The bacterial cell density in cut shoots increased as the initial bacterial concentration was increased from 106-1010 cfu mL^{-1}, the population being lower at the distal end. However, on inoculating black pepper rooted cuttings, the bacteria colonized mostly roots, the population directly proportional to the duration of inoculation. Bacterial colonization on stem region was noticed only at the 28th day. Below detectable populations of the bacteria ware observed in leaves of treated rooted cuttings.

Extraction and purification of phenylpropanoids

Qualitative and quantitative assays were carried out for anthocyanidins, proanthocyanidins and their precursors, leucoanthocyanidins, in root extracts by specific extraction procedures and butanol/HCl assay. Levels of leucoanthocyanidins are found to be higher in the susceptible cultivar (Sreekara) while levels of condensed tannins (proanthocyanidins and anthocyanidins) and phenols were higher in the resistant line, Hp 39. Butanol/HCl assay was carried out in a few more selected lines reported as resistant to plant parasitic nematodes. The results have indicated that the level of proanthocyanidins was comparatively high in resistant lines *viz*., Acc. 3219, C.1090 and *P. colubrinum* compared to susceptible lines like Sreekara or Panniyur 1.

***In silico* and *in vitro* assays**

The *in silico* results were further refined by selecting additional targets like calreticulin 1, xylanase, transthyretin-like protein 3 precursor, cathepsin B, cytochrome c oxidase subunit III and cathepsin S-like cysteine proteinase. Syringin, sinapaldehyde, sinapic acid, scopolin and caffeoylquinic acid consistently showed good potential with high mol dock score and higher number of hydrogen bonds. Subsequently, *in vitro* bioassays were conducted with 6 new compounds *viz.*, catechol, gallic acid, phloroglucinol, salicylic acid, syringaldehyde and tannic acid. Out of the 6 compounds, syringaldehyde and salicylic acid caused >80% mortality of *R. similis* within 48h of exposure.

Localization of phenyl propanoids

Phenolic cells and lignified walls were visualized in root sections of both HP39 and Sreekara by applying histochemical staining. Phenyl propanoids were localized in roots through staining with diphenylboric acid 2 aminoethyl ester (DPBA) coupled with fluorescence microscopy. Increased presence of phenyl propanoids was observed in the cortical region of HP39 roots, consequent to nematode inoculation. Lignin was spotted in root sections through two methods *viz.*, Maule staining and Weisner staining. In general, lignification is significantly high in HP39 roots compared to Sreekara indirectly supporting the inhibition of nematodes by phenyl propanoids in the lignin pathway.

Evaluation of new chemicals against *P. capsici*

Two new chemicals namely RIL-070/FI (72WP) and Ergon 44.3% (w/w) [Kresoxim methyl 500 g/L SC] were evaluated under *in vitro* and *in planta* conditions at different concentrations against *P.capsici* and *C. gloeosporioides.* RIL-070/ FI (72WP) when tested at different concentrations from 10-500 ppm showed 100% mycelial inhibition at 50ppm, sporulation at 100ppm and zoospore germination at 200ppm. The average LD50 for *in vitro* inhibition was 29.23 and LD90 was 54.43ppm.

Foliar spraying of the chemical at concentrations from 100-600ppm under challenge inoculation of *P. capsici* showed that immediately after spraying of the chemical, 500 and 600ppm concentrations showed 92% and 96% inhibition of lesion development, respectively. However the effect was reduced (64.44% and 79.25%, respectively) after 10 days of application. Soil drenching of the chemical resulted in significant reduction in mortality of the plants when compared to control and 400 and 600ppm was found effective and showed 100% reduction in mortality. The *P. capsici* population was also reduced to 77.58%, but >600ppm was found to be phytotoxic.

Ergon 44.3% (w/w) [Kresoxim methyl 500 g/L SC] was tested at different concentrations from 10 -6000ppm of the product *in vitro* against *P. capsici,* showed complete inhibition of mycelial growth and sporulation at 6000ppm whereas zoospore germination was fully inhibited at 2000ppm. The average LD50 and LD90 values were 845.51ppm and 1740.71ppm, respectively. Foliar spray of the chemical followed by challenge inoculation showed inhibition of 44.83% over control. Maximum inhibition (57.12%) was observed at 5 days after spraying at a concentration of 10000ppm and was slightly reduced afterwards and almost maintained up to 20 days. Soil application of the chemical showed 50.12% mortality at 6000ppm when compared to control, while no infection or mortality could be observed at 7000 ppm.

Evaluation of new chemicals against *R. similis*

Nematicidal activity of eight chemicals *viz.*, Fipronil, Thiamethexan, Acephate, Cartap hydrochloride, Quinalphos, Flubendamide, Carbosulfan, and Chloropyriphos were evaluated against the burrowing nematode, *R. similis*. Out of eight, five chemicals *viz.*, Fipronil, Thiamethexan, Cartap hydrochloride, Flubendamide, an Carbosulfan showed good nematicidal activity.

Varietal response to biocontrol agents

Two moderately resistant lines to *P. capsici* (C1090 and IISR Shakti) and one line resistant to *R. similis* (Hp 39) were evaluated in presence of biocontrol agents *T. harzianum, P. fluorescens* (IISR 6), *P. aeruginosa* (IISR 853) and *Pochonia chlamydosporia* in comparison with susceptible variety IISR Sreekara. The establishment and growth of C1090 and IISR Shakti were highly promising in the presence of *T. harzianum* with 100% establishment without any disease incidence. Hp 39 was found to be well established in the presence of *P. chlamydosporia*. The susceptible Sreekara grew well in presence of *T. harzianum* when compared to its control where 50% plants died due to *Phytophthora* infection.

Evaluation of consortia of endophytic bacteria against *P. capsici*

Promising endophytic bacteria in different combinations were tested against *P. capsici*. The growth and establishment of the plants showed promising results with consortium containing BP 25 + BP17 and TC10 + *P. chlamydosporia* and Metalaxyl- Mz + *P. chlamydosporia*. However, on challenge inoculation Metalaxyl -Mz + BP17 showed maximum reduction in disease incidence (10%) followed by BP25 + *P. chlamydosporia*.

Evaluation of actinomycetes against *P. capsici*

Nine Actinomycetes belonging to the genera *Kitasatospora* and *Streptomyces* were evaluated in the green house for disease suppression and growth promotion. The actinomycetes were grown as broth culture and inoculated to the base of the plants @100mL plant^{-1} containing a CFU of 108 mL^{-1}. The plants were challenged with the highly virulent isolate of *P. capsici* after one month. The results clearly revealed the efficacy of four isolates (ACt5, Act2, Act9) in checking the infection. The isolates also showed better root development and growth in rooted cuttings. The potential isolates were identified as *Kitasatospora setae* (Act2), *Streptomyces* sp. (Act5) and *S. tauricus* (Act9-VC11).

Development of liquid formulation for biological control agents

Experiments were conducted to formulate a liquid formulation for *Trichoderma harzianum*. The viability of *T. harzianum* conidia in different cryoprotectants, effect of different storage conditions and the effectiveness of stored product for the biological control were tested. Of the different formulations tested *viz.*, Glycerol, Glucose and DMSO and sterile deionized water, mature conidial suspension in sterile deionized water was found as a promising medium for the long term storage and preservation of viable conidia of *Trichoderma* with biocontrol potency for field application.

An encapsulated bead formulation of *Trichoderma* and *P. chlamydosporia* was also developed. These encapsulated formulations when mixed with vermicompost or FYM can release the organisms. The encapsulated beads can be stored in water for long term preservation. Ten encapsulated beads in 100g vermicompost or FYM gives a population of 10^{11} cfu g^{-1} in 10 days which is equivalent to or more than 10kg of talc based formulation.

Multi location testing of *Trichoderma* isolates

Two pot culture experiments were done with geographically different isolates of *Trichoderma* in varieties Panchami and Sreekara with challenge inoculation with *P. capsici* (05-06). The fungal population did not show much difference whereas there was a substantial increase in native *Trichoderma* population. The treatment T10 (PhytoFuRa10) showed minimum disease incidence (8.175%) followed by T3 (PhytoFuRa3) with (10.68%). PhytoFuRa4, 5, 6 & 7 showed more than 80% disease incidence. The treatment with PhytoFuRa13 showed maximum plant height. In case of var. Sreekara, treatment T10 showed less disease incidence followed by PhytoFuRa11, 12 &13 (less than 20% disease incidence). The treatment with PhytoFuRa2 showed maximum plant height.

Cardamom

Evaluation for thrips resistance

Two hundred and ninety six accession were screened against thrips. There was wide variation in the reaction of the accessions to thrips and percentage of infested capsules ranged from 3.0-96.2%. Fifty three accessions recorded <20% capsule damage and 5 accessions recorded <10% capsule damage. Four of the 5 accessions with <10% capsule damage belonged to *Malabar* types and one to *Vazhukka* type. All the 9 highly susceptible accessions belonged to either *Mysore* or *Vazhukka* types.

Plant morphological characters such as nature of leaf sheath (loose or tight), flower bract (persistent or non-persistent) and panicle type (erect, semi erect or prostrate) were recorded in 69 accessions for characterization against thrips attack. The mean capsule damage was higher (57.2%) in *Mysore* types compared to *Malabar* types (32.8%). Plant characters such as erect panicle and persistence of flower bract contributed to higher thrips incidence.

Evaluation of insecticides and organic products for thrips control

Eleven insecticide molecules and organic products *viz.,* Neem soap (10g L^{-1}), Spinosad (0.3mL L^{-1}), Vertemec (0.4mL L^{-1}), Thiamethoxam (0.3mL L^{-1}), Thiacloprid (0.5mL L^{-1}), Imidacloprid (0.5mL L^{-1}), L-cyhalothrin (0.5mL L^{-1}), Zolone (2.0mL L^{-1}), Fipronil (1.0mL L^{-1}) and Quinalphos (2.0mL L^{-1}) were evaluated for their efficacy in management of thrips. The sprays were imposed during March, April, May, August and September. The trial indicated that Fipronil was the most effective in controlling the thrips population throughout the cropping season followed by Imidacloprid, Quinalphos, Thiacloprid and Thiamethoxam which were on par with each other.

Documentation of endosymbionts and entomopathogens on thrips

The bacterial endosymbiont *Wolbachia* was identified from thrips collected from Kodagu (Karnataka), Wayanad, Palakkad, Idukki (Kerala), Yercaud, Nilgiris and Kodaikanal (Tamil Nadu) districts. All the thrips populations were infected with only sub group Con belonging to super group B. Both male and female thrips were infected with the same *Wolbachia* subgroup. An entomopathogenic fungus was isolated from thrips during a survey of cardamom plantations in Wayanad District in Kerala.

Diversity of rhizome- root rot pathogens and their antagonists

Surveys were repeated in Wayanad and Idukki districts of Kerala, Hassan and Kodagu districts of Karnataka to study the seasonal variation of rhizome and root rot diseases. Eighty fungal isolates were isolated from rhizome and root rot disease affected samples. Among the isolates, *Rhizoctonia solani, Pythium vexans* and *Fusarium* species were found to be dominant. Artificial inoculation studies proved that among the different fungi isolated, *R. solani, P. vexans* and *Fusarium oxysporum* were pathogenic.

Different species of *Fusarium* were found associated with root rot disease, of which *F. oxysporum* was found to be the predominant species. Morphological characterization revealed the association of different species *viz., F. oxysporum, F. solani, F. pallidoroseum* and *F. verticillioides. Rhizoctonia* isolates from all the locations surveyed were identified as *R. solani.*

Protocols for DNA isolation of *F. oxysporum* and *Trichoderma* have been standardized. *In vitro* screening of respective *Trichoderma* spp. against the most virulent root rot pathogen, *F. oxysporum* isolates from Kerala, Karnataka and Tamil Nadu led to the identification of WYDT6, RT 7B and RT2A, respectively as the most effective isolates.

Endophytic and rhizospheric microflora

Isolations made from different plant parts including leaves, petioles, pseudostem, roots and rhizomes of *Amomum* sp. and *Alpinia* sp. Yielded 50 fungal isolates and 5 bacterial isolates. Among the plant parts, the isolates had more preference towards pseudostem followed by rhizomes and least for the roots. Endophytic fungi and bacteria were also isolated from 3 ecotypes *viz., Malabar, Mysore* and *Vazhukka.*

Studies on mycelial compatibility groups among isolates of *Colletotrichum gloeosporioides*

Mycelial compatibility studies among the selected isolates of *C. gloeosporioides* (3 each from Kerala and Tamil Nadu and 8 isolates from Karnataka) revealed that majority of the isolates showed incompatible reaction when isolates from different geographical locations were paired, while, isolates from same geographical regions exhibited compatible reaction when paired together.

IDM strategy for nursery

Evaluation of fungicides, a Neem based product and an isolate of *T. harzianum* revealed that spraying the combination product of Carbendazim + Mancozeb in combination with soil application of *T. harzianum* was promising in managing leaf spot disease under nursery conditions.

Real time RT-PCR for detection of viruses

A method for the detection of Cardamom mosaic virus (CdMV) and Banana bract mosaic virus (BBrMV) was developed with the use of SYBR green one step RT-PCR. Primers were designed from conserved coat protein gene sequences available in GenBank database to amplify 127 bp and 117 bp fragents of CdMV and BBrMV, respectively. The method involved isolation of total RNA and subjecting to real time RT-PCR using qPCR master mix, specific primers and reverse transcriptase for each of the virus. The assay amplified all the isolates of CdMV and BBrMV but no amplification was obtained with the RNA ofhealthy host plants. For SYBR Green based virus detection, it is important to run a melt curve analysis followed by real time PCR, because SYBR Green real time RT-PCR detects each double stranded DNA including primer dimmers and PCR products from mis-annealed primer. Examining the melt curve ensures that the desired amplicon was detected. The melting temperature of the amplicon starts at the point of inflection of the melt curve profile. The melt peak occurs at 83ÚC±0.5 for CdMV and BBrMV and no evidence of nonspecific amplification or primer-dimerization. The healthy and water control showed no melting peak. The specificity of the product was further confirmed by agarose gel electrophoresis which showed a single band of the expected size.

Recombinant plasmids carrying target virus regions corresponding to both the viruses were quantified, serially diluted and used as standards for real time PCR to develop standard curve to enable quantification. The amplification of the standard dilution series yielded in linear relations with reliable results (R2 = 0.99906 for CdMV and R2 = 0.99902 for BBrMV) in the range of 10–108 copies of the target DNA locus per PCR. The detection limit for CdMV was as low as 23 copies and 24 copies for BBrMV as estimated from the standard curve. Similarly the detection limits of the viruses in plants were also studied by using 10 fold dilutions of RNA from infected cardamom plant (1 to 10-5). By taking the Ct values for each dilution the detection limit (copy number) of CdMV and BBrMV was determined from the standard curve. It was found that the detection limit is as low as 16 and 10 copies (10-4 dilution) for CdMV and BBrMV respectively, which was approximately 1000 fold higher than conventional RT-PCR. The RT-qPCR method for CdMV and BBrMV were further validated by testing 30 CdMV and 20 BBrMV isolates from Kerala and Karnataka states. The new method offers a fast, reliable, specific and sensitive tool for the detection of CdMV and BBrMV in cardamom.

Development of RT-LAMP for detection of virus

A reverse transcription loop-mediated isothermal amplification (RT-LAMP) method was developed for quick detection of Banana bract mosaic virus (BBrMV). Five sets of primers were designed based on conserved sequences in coat protein gene of the virus. The RT-LAMP reaction components were optimized by using total RNA extracted from a known infected plant (positive control), a known healthy plant (negative control) and water control (negative control containing water instead of RNA). The LAMP reactions were then carried out at 65°C for 75 min. For the visualization of RT-LAMP products, three different methods were used. During LAMP reaction a large amount of DNA is synthesized, which yields a large amount of pyrophosphate ion by-product (magnesium pyrophosphate) which is insoluble thus increasing turbidity of the reaction tube which can be visualized by different means. In the first method, the presence or absence of turbidity in the reaction tube was taken as the criteria for detection. In the second method, addition of 1mM MnCl2 and

50μM calcein to the reaction mix resulted in the formation of a fluorescent metal indicator which can be viewed under UV light. In the third method, amplification products were visualized by 2% agarose gel electrophoresis. Presence of turbidity or green florescence allowed easy detection of RT-LAMP products without gel electrophoresis. Detection limit of RT-LAMP was up to 100 times higher than RT-PCR and on par with real time RT-PCR. In order to validate the method, total RNA isolated from 24 test samples of cardamom and banana collected from Karnataka and Kerala were subjected to RT-LAMP using 1μL of template RNA along with known positive and negative controls. Of the 24 samples tested, 22 samples showed positive reaction, indicating successful detection of BBrMV in different isolates of virus.

Ginger

Isolation and characterization of phages infecting *R. solanacearum*

Rhizosphere soil samples were collected from ginger growing areas of Kerala and Karnataka and phages were isolated, using *R. solanacearum* isolate from Wayanad as the host. The host range of this phage was checked using seventeen different isolates of *R. solanacearum*. This phage was found to be infecting all the isolates collected from Wayanad. The phage was PEG purified and titrated. The stock was found to contain 107 pfu mL^{-1}.

The phage DNA attP site was amplified using the primer pairs attP - L 5'-CAGTATGTGTCCTGGGTGTTTGTCTACCG- 3' and attP - R 5'-CTCTTATCAGAACGCCCCACCTCCC- 3'. But there was no amplification indicating the absence of attP region. So this phage can be effectively utilized for the biocontrol of *R. solanacearum*.

Isolation of bacteria from apoplastic fluid

The pseudostem and leaves were used for the apoplastic fluid extraction. The apoplastic fluid obtained was serially diluted and plated on tryptic soy agar medium. The distinct colonies obtained were purified and stored. Hundred and fifty bacterial isolates were maintained as glycerol stocks. These bacteria will be checked for their biocontrol properties and will be shortlisted for pot culture and field evaluation.

Volatile oil and 3 solvent extracts (hexane, methanol and chloroform) from the rhizome of *C. amada* were used to examine the antimicrobial activity *in vitro*. The extracts showed varying levels of inhibitory effect on the test pathogens. Among the solvent extracts chloroform extract exhibited greatest antibacterial activity against *R. solanacearum*, followed by hexane and methanol extracts.

In case of *P. myriotylum*, hexane extracts showed greatest activity followed by chloroform extract. Methanol extract exhibited least effect. Essential oil at 1.5% (v/v) caused the complete death of bacteria showing greatest antimicrobial effect. Essential oil was subjected to GC-MS analysis and the major components detected were β -Myrcene and β -Pinene. These were tested for their inhibitory activity against *R. solanacearum*. Both the components showed antibacterial activity with myrcene showing greater activity.

Biochemical characterization of resistant and susceptible accessions to shoot borer *(Conogethes punctiferalis)*

Total fibre and lignin contents in mature shoots and total carbohydrates, lignin and proteins in mature leaves were estimated in moderately resistant and susceptible accessions to shoot borer. Lignin, total proteins, phenols, carbohydrates and fibre contents in immature leaf tissues and immature shoots were also determined in susceptible as well as in moderately resistant accessions. Epicuticular wax was estimated only in immature leaf tissues.

Evaluation of EPNs

The infectivity of 4 promising EPNs *viz., Heterorhabditis* sp. (IISR-EPN 01), *Steinernema* sp. (IISR-EPN 02), *Oscheius gingeri* and *Oscheius* sp. (IISR-EPN 08) was tested against shoot borer larva in the greenhouse. The treatments included application of the EPNs @ 25000Ijs pot^{-1} at 21 days interval during August to October, spraying of Malathion 0.1% (present recommendation) and control. Shoot damage was recorded at the end of crop season. Among the EPNs, *Steinernema* sp. (IISR-EPN 02) and *O. gingeri* treated plants showed minimum shoot damage (15.8% and 16.6%, respectively) which was significantly superior over control (36.0%) and was on par with insecticide application.

Turmeric

Biochemical characterization of moderately resistant and susceptible accessions to shoot borer

Total fibre and lignin contents in mature shoots and total carbohydrates, lignin and proteins in mature leaves were estimated in both shoot borer resistant and susceptible accessions. Lignin, total proteins, phenols, carbohydrates and fibre contents in immature leaf tissues and shoots were determined in susceptible and moderately resistant accessions. Epicuticular wax was estimated only in immature leaf tissues.

Evaluation of EPNs

The infectivity of 4 promising EPNs *viz., Heterorhabditis* sp. (IISR-EPN 01), *Steinernema* sp. (IISR-EPN 02), *Oscheius gingeri* and *Oscheius* sp. (IISR-EPN 08) was tested against shoot borer larva in the greenhouse. The treatments included application of the EPNs @ 25000IJs pot^{-1} at 21 days interval during August to October, spraying of Malathion 0.1% (present recommendation) and control. Shoot damage was recorded at the end of crop season. Among the test EPNs, *Steinernema* sp. (IISR-EPN 02) and *O. gingeri* treated plants showed minimum shoot damage (26.1% and 26.6%, respectively) which was significantly superior over control (44.4%) and was on par with insecticide application.

Mass production of promising EPNs

In vitro multiplication of 4 promising EPNs *viz., Heterorhabditids* sp. (IISR-EPN 01), *Steinernema* sp. (IISR-EPN 02), *O. gingeri* and *Oscheius* sp. (IISR-EPN 08) in 3 modified Wouts medium (MWM) was tested. The ingredients of the modified media were mixed and coated on to foam pieces. The flasks were filled with foam media mixture and sterilized. After cooling at room temperature, 1000IJs were introduced into each flask and sealed. The flasks were incubated at 30oC for 25 days. Colonies of nematodes started appearing on the walls of the flasks 2 weeks after inoculation. The IJs were harvested through sieving and decantation method and EPNs counted with the help of Syracuse counting dish. Among the test media, all test EPNs were produced in MWM-I. However, maximum numbers of *O. gingeri* were produced in MWM-II. No multiplication of EPNs was observed in MWM-III.

Recent Achievements of All India Coordinated Research Project on Spices

At Sirsi, among the new fungi toxicant molecules @ 0.1 % Fenamidone (10%) + Mancozeb (50 %) (Sectin) alone and Fenamidone (10%) + Mancozeb (50 %) (Sectin) as spraying (@ 2L vine^{-1}) and drenching (3L vine^{-1}) along with bioagent *T. harzianum* (MTCC 5179) 50g with 1.0kg of Neem cake as soil application separately during first week of June, 2012 and third week August, 2012 recorded statistically significant reduction in *Phytophthora* foot rot disease of black pepper with regard to leaf infection (6.06% and 4.24%), reduced yellowing of vines (7.88 PDI and 6.67 PDI), least defoliation (8.50 PDI and 6.06 PDI), least death of vines (7.27% and 4.86%) and highest green berry yield (2.93kg vine^{-1} and 3.16 kg vine^{-1}) respectively.

Scale insects of black pepper could be controlled by spraying Neem formulation containing 1% Azadirachtin @0.5%, at fortnightly intervals from September onwards (4 applications)

Application of entomopathogenic nematode (EPN) *Heterorhabditis indica* @ 100 Infective Juvenile (IJ) grub was found to be effective in reducing the population of cardamom root grub, when adequate moisture is available in the soil.

Biofumigation using cabbage was found to be the best treatment with highest yield of 6.03 kg plot^{-1} (12.16t ha^{-1})and the lowest bacterial wilt disease incidence (8%). Minimum incidence of soft rot, bacterial wilt, leaf spot and maximum ginger yield was found due to cabbage biofumigation at Solan and Pundibari.

Seed treatment of *Pseudomononas fluorescens* @10g kg^{-1} seed + soil application of *T. harzianum* @2.5kg ha^{-1} and *P. fluorescens* (IISR 6) 108 cfu as a spray at 60 DAS is recommended for management of cumin wilt and blight

At Jagudan, application of Thiamethoxam 25WG @ 0.0084% had recorded the least wasp damage (7.44%) and maximum seed yield (2029 kg ha^{-1}) of fennel followed by acetamiprid 20SP@ 0.004% (1906kg ha^{-1}).

Source: Annual Report 2012-13, Indian Institute of Spices Research, Kozhikode, Kerala.

□□□□

7

Recent Initiatives in Post Harvest Technology

Recent Initiatives by National Research Centre on Seed Spices, Ajmer

Standardization of the drying techniques

In fennel five methods of grain drying were employed viz., shade dryer (30°C), sun dryer (35°C to 37°C), solar dryer (40°Cto 42°C), vegetable dryer (50°C) and multipurpose grain dryer (60°C). Fennel grains which were dried under solar dryer, gave maximum test weight (5.17 g), recovery (39.82 %) with lowest moisture content (61.30%) and highest taste marks (9.3) and colour marks (9.2).

Packaging and storage

Aluminium bag vacuum packaging was proved to be the best as compared to other packaging methods for all the parameters studied in both Fennel and cumin after 8 months of storage.

Recent Initiatives by Indian Institute of Spices Research, Kozhikode, Kerala

Cryogenic grinding of spices

Black pepper, turmeric and cinnamon were powdered using 3 different means such as ambient ground, freshly ground and cryo-ground. The powders obtained from CIPHET, Ludhiana were subjected to quality evaluation and antioxidant studies. The total phenol, antioxidant activity on the basis of DPPH assay,

Phosphomolybdenum method and FRP method using alcohol, water and petroleum ether extracts of turmeric variety Prathiba did not show any variation. This shows that grinding method does not influence the antioxidant property.

The cryo ground black pepper Panniyur-1 yield high oleoresin compared to other methods. Freshly ground sample gave high oil and comparable oleoresin. Piperine content did not show any variation. GC MS analysis of essential oil from black pepper Panniyur-1 indicated that major sesquiterpenes like β–Caryophyllene gave better yield in cryoground sample. Except for β–Phellandrene, no major difference was observed in oil constituents.

Spice extracts for anticancer effect

Telomerase activity is negative in human normal somatic cells but detected in more than 85% of tumor cells. Therefore, targeting telomerase enzyme is an attractive proposition in the treatment of cancer. Thus, in this study, *in silico* methods were used to identify inhibitors of telomerase from major spice phytochemicals with ADMET satisfaction, used as ligands, *viz.*, curcumin, demethoxycurcumin, bisdemethoxycurcumin, tetrahydrocurcumin, piperine, cinnamaldehyde, 1,8-cineol, myristicin, beta-caryophyllene and alpha-pinene. The targets sourced from the NCBI database were:

- 3KYL, the catalytic subunit of telomerase bound to its RNA template and telomeric DNA
- 2R4G, RNA-binding domain of telomerase and
- 2B2A, the essential N-terminal domain of telomerase reverse transcriptase.

The ligands and the targets were optimized using ArgusLab 4.0.1. The phytochemicals were docked to the active-sites of the above targets to assess binding affinity and consequently their inhibitory activity using Molegro Virtual Docker Version 5.5.0. The post-docking analyses revealed that the turmeric bioactive principle - curcuminoids I, II and III (curcumin, demethoxycurcumin and bisdemethoxycurcumin respectively) and tetrahydrocurcumin showed consistently superior docking ability to the telomerase targets, with high scoring functions that bind to the selected targets with strong binding efficiency and hydrogen bond interactions. Piperine and myristicin, though less efficient, had greater interaction than the remaining ligands. These phytochemicals can be potentially developed as lead drug molecules against cancer, by targeting telomerase.

Curing of turmeric in an improved boiler

Studies on curing and drying of turmeric were conducted at Central Institute of Fisheries Technology, (CIFT), Kochi. Experiments on cooking of turmeric at three varying steam pressures (0, 0.5 and 1kg cm-2) corresponding to steam temperatures of 100, 112 and 121°C for four different cooking durations (5, 10, 15 and 20 min) was performed in a pilot-scale rotary retorting system (Model 24, John Fraser and Sons Ltd., Newcastleupon- Tyne, UK) to determine the time required for complete drying and to evaluate the quality of the dry turmeric obtained. The cooked samples were dried in CIFT-CRYER-SDL 250 model solar drier from 9.30am to 4.30pm. The samples were left inside the drier with fans on overnight and the drying process continued till the samples were uniformly dry. Studies indicated that irrespective of the treatment combinations, the drying time required for complete drying was 7 days.

Studies on bulk curing and drying of turmeric were also conducted at CIFT, Cochin. About 75kg was cured in a horizontal retort at 1kg cm-2 pressure for 15 min and dried in CIFT-CRYER-SDL 250 model solar drier from 9.30 am to 4.30 pm. The study indicated that drying was completed in 7 days. An experiment on curing of turmeric (variety IISR Prathibha) was conducted in improved turmeric boiler (TNAU model) and conventional water boiling methods for various curing durations. Cured turmeric was sun dried in cemented yard to reduce the moisture content of fresh rhizomes from 80.50% to less than 10%. Turmeric cured by traditional water boiling method for 45, 60 and 90 min and those cured in improved steam boiler for 30, 45, 60 and 90 min took 10 days for complete drying. But in case of rhizomes dipped in boiling water for 10 min and then dried, the drying time increased to 13 days. When the rhizomes were sliced and dried, the drying time reduced to 8 days. Quality analysis of the dry turmeric samples indicated a marginal reduction in curcumin content from 4.66-4.44% when the curing time increased from 30 min to 90 min in case of improved curing method and from 4.95-4.62% when the curing time increased from 30 min to 90min in case of water boiling method.

Production of food extrudates

The cassava flour was mixed with cardamom power in the ratio of 96:4 and conditioned at 4°C for 15 days. The blend was extruded in a single screw stand alone Brabender extruder at a die temperature ranging from 170-190°C at a screw speed ranging from 70-90rpm and by using Response Surface Methodology, the machine parameters of the extruder were optimized to get optimum quality of the extrudates based on the physical, functional, textural, biochemical and sensory properties of the extrudates from the flour blend of cassava and cardamom. From the regression models obtained it was found that the expansion ratio would be at a maximum of 3.85 at a die temperature of 170.1°C and at a screw speed of 79.1rpm, the bulk density would be at a minimum of 0.19g mm-3 at a die temperature of 172°C and at a screw speed of 76rpm, the porosity would be at the maximum of 62.83% and at a die temperature of 170.2°C at a screw speed of 71.4rpm, the water absorption index would be at the minimum of 5.2 at a die temperature of 173.6°C and at a screw speed of 70.7 rpm, the water solubility index would be at the maximum of 2% at a die temperature of 170.1°C and at a screw speed of 82.3rpm, the oil absorption index would be at a minimum of 2.3 at a die temperature of 172.4°C and at a screw speed of 73.6rpm, the hardness would be at a minimum of 38N at a die temperature of 170.1°C and at a screw speed of 79.3rpm, the toughness would be at a minimum of 34 Nm and at a die temperature of 170.3°C at a screw speed of 79.67rpm, the crispness would be at a maximum of 31.7 numbers of peaks and at a die temperature of 179.9°C at a screw speed of 89.9rpm, the retention of cineol would be a maximum of 23.82mg $100g^{-1}$ of sample at a die temperature of 177.2°C and at a screw speed of 89.0rpm, retention of terpinyl acetate would be at a maximum of 34.74mg 100g-1 of sample at a die temperature of 183.9°C and at a screw speed of 89.1rpm and the overall acceptability scores would be at a maximum of 6.75 at a die temperature of 170.6°C and at a screw speed of 76.7rpm.

Chemoprofiling of turmeric and ginger varieties

Essential oil profile of rhizhomes of seven varieties of turmeric, *viz.*, Sona, Sobha, Varna, Rajendra Sonia, Narendra Haldi, Sugandham and Co-1 was determined by GC-MS and these contained turmerone (5.6-25.8%), ar-turmerone (3.5-20.8%) and curlone (5.4-15.6%) as the chief components. ar-turmerone ranged from 14 -20.8 % in all tested varieties except Sobha with 3.5%. Turmerone level was 20-25% in Rajendra Sonia, Sugandham, Narendra Haldi, and Co-1; 5-7% in Sobha and Sona and 13% in Varna. Varna, Rajendra Sonia, Sobha and Sona contained 5.0-6.8% curlone and Sugandham, Narendra Haldi and Co- I with slightly higher level, (12.5-15.6%). There was not much variation in the contents of the minor and trace components. In ginger, volatile oil profile of rhizomes of seven varieties *viz.*, Varada, Mahima, Rejatha, Suprabha, Surabhi, Himgiri and Rio de Janeiro was determined. The major constituents were zingiberene (16-23%), farnesene (9-12%), β-sesquiphellandrene (11-12%) and ar-curcumene (8.9-10.3%). Zingiberene varied from 20-23% in all varieties except Rio-de Janeiro (16.4%); ar-curcumene was comparatively low in Varada (5.7%) whereas it ranged from 8.9-10.3% in others.

Source: Annual Report 2012-13, Indian Institute of Spices Research, Kozhikode, Kerala

□□□□

8

Diseases of Major Spices and Condiments

Common name, symptoms of damage and management of diseases of major spices and condiments

Name of disease and its causal organism	Symptoms of damage	Management
Cardamom		
Azhukal disease / capsule rot / fruit rot: *Phytophthora parasitica car. nicotianae/ Phytophthora palmivora*	• Symptoms can appear on tender and matured leaves. Large circular, irregular, water soaked spots with black colour appear on leaves. • The exposed portion of the unopened leaves my rot. • Grey patches of irregular spots with brown margin are formed at the base of the leaf sheath. • The basal portion rots and the pseudo stem break away at the collar region. • The infection spreads to the underground plants and the rhizomes become rots. • Small light brown lesions appear in the green tender fruits which fall of in 3-6 days leaving the small fruit stalk. The tip of inflorescence also rots.	• Removal and burning of infected plants. • Avoid moving of rhizomes from diseased areas to healthy area for planting. • Provide proper drainage. • Three sprays with Bordeaux mixture 1% in May, June, and July. • Soil drench with Bordeaux mixture 1 % (or) Copper oxychloride 0.25%.
Chenthal disease / leaf blight: *Colletotrichum gloeosporioides*	• Elongated, water soaked lesions of varying size appear on the upper surface of the leaf. • The spots becomes brown to dark with pale yellow hole. Leaves wither and pseudo stem wilts. • New shoots which develop are reduced in size. Flowers fail to develop. • The inflorescences dry up stating from tip downwards. The affected garden shows burnt appearance.	• Removal and destruction of affected leaves • Three sprays with Carbendazim 0.1% (or) Mancozeb 0.2% (or) Copper oxychloride 0.25% at 30 days interval.

Contd.

Name of disease and its causal organism	Symptoms of damage	Management
Damping off / rhizome rot / clamp rot: *Pythium vexans, Fusarium oxysporum, Rhizoctonia solani, Phytophthora sp.*	• Infected leaves become pale, yellow and ultimately the young leaves die. • Older leaves die prematurely and new shoots that arise are weak, decay and the rhizomes rot at the base of the stem. • The diseased shoot can be pulled out easily.	• Destruction of diseased clumps. • Providing proper drainage. • Changing the nursery site. • Drenching the nursery beds with Copper oxychloride 0.25% or Bordeaux mixture 0.5%, before 15 days of sowing.
Mosaic: Cardamom moxaic virus (CDMV)	• General chlorosis of young leaves – parallel streaks of pale green tissues running along the veins from midrib to the margins. • Leaf sheath also shows stitch stripes. • In advanced stage the whole plant shows mosaic symptom. • Rhizome shrivels and plants dies. If young clumps are attached they die before flowering.	• Collection and removal of infected clumps along with rhizomes and burning. • Raising of nursery in diseases free areas. • Spray with Dimethoate to kill the vector. • Vector: *Pentalonia nigronervosa*.
Clove		
Last Spot, Twig Blight and Flower Bud shedding: *Gloeosporium gleosporioides*	• Necrotic spots of variable size and shapes are noticed on the leaves. • Severely affected leaves wither, droop down and dry up. • In nursery seedlings die back symptoms are seen. Twigs are infected as the symptoms extend from the leaves through petioles. • The affected branches stand without leaves or only with young leaves at tips. Flower buds are attacked by spreading infection from the twigs. • Shedding of flower buds occurs during periods of heavy and continuous rainfall.	• Spraying 0.25% Copper oxychloride at monthly interval reduces disease intensity, defoliation and flower bud shedding. • Initial spray is given just prior to flower bud formation and continued till the harvest of buds.

Contd.

Name of disease and its causal organism	Symptoms of damage	Management
Leaf rot: *Cylindrocladium quinquiseptatum*	• It is noticed in the nurseries as well as in the main field both at young and mature stages. • Infection starts as darkspots at the leaf margin and spreads sometimes with no definite pattern. • Rotting may be in the whole leaf or at the tip resulting in defoliation.	• Seedling and young plants can be sprayed with systemic fungicides Carbendazim 0.1%.
Seedling Wilt: *Cylindro-cladium* sp., *Fusarium* sp., *Colletotrichum* sp., *Rhizoctonia* sp., and *Trichoderma* sp	• Seedling wilt is found mainly in nurseries and causes fiveto 40% death of seedlings. • Leaves of affected seedlings loose natural lustre, tend to droop and ultimately die.	• Since the infected seedlings promote spread of the disease they are to be removed and destroyed and the nursery is drenched with Copper oxychloride 0.25 %.
Coriander		
Powdery Mildew: *Erysiphe polygoni*	• It appears as small, white circular patches on young parts of stems and leaves. These increases in size, often coalesce to cover extensive areas of leaf surface. • Affected leaves are reduced in size and distorted. Premature sterility is also common. In serve cases, the umbels dry up. • Sometimes, the seeds are attached. • Attack of this disease is seen during cloudy weather condition. • White powdery growth appears on the leaves and buds during its primary stage. Seed formation may not take place in affected plants due to this disease.	• To control this disease, dusting of Sulphur dust @ 20-25 Kg per hectare should be done. • Spraying of Wettable sulphur 0.25% or Karathane 0.05 % can also control this disease.
Stem gall: *Protomyces macrosporus*	• Galls appear on the leaves and stems of the plants affected by this disease.	• To control the disease, sowing may be done only after treating the seeds with 2g Thiram and 1g

Contd.

Name of disease and its causal organism	Symptoms of damage	Management
	• Shape of coriander seeds change due to effect of the disease.	Carbendazim / Kg. seeds. • Spray 0.1% solution of Carbendazim when the symptoms start appearing and repeat the spraying at an interval of 20 days till the disease is completely controlled.
Wilt: *Fusarium Oxysporum* f. sp. *corianderii*	• The disease can be easily recognized in the field by drooping of the terminal portions, followed by withering and drying up of leaves, eventually resulting in death. • Discoloration of vascular system of the root is observed. Partial wilting is also found. In partially wilted plants growth is arrested. • The leaves become pinkish yellow to yellow. Sterility is often noticed in such plants. Seeds, if formed are immature and light. • Severe infection in the early stage results in total failure of the crop.	• Deep ploughing should be done during summer season. • Crop rotation may also be followed. In those plots, where effect of this disease has been noticed, coriander crop should not be taken for 2-3 years. • Sowing should be done after seed treatment with Carbendazim 2 g per kg seed or Thiram 3 g per Kg seed.
Cumin		
Fusarium wilt: *Fusarium oxysporum* f.sp. *cumini*	• Infected plants show peculiar symptoms of dropping leaves, leading to mortality of the entire plant. • Attack of wilt is severe in younger plants.	• There is no chemical control for this disease. • Crop rotation and use of Neem cake are helpful in checking spread of the fungus disease. • Seeds collected from disease free plots should only be used for sowing.
Alternaria Blight: *Alternaria burnsii*	• The blight affected plants show very minute brownish necrotic spots, which later turn to blackish. • Mostly diseased plants fail to produce seeds.	• Seed treatment with Thiram 3 g/kg of seed • Spraying of 0.2% solution of Mancozeb 4 times at 10 days interval commencing from 40 days after

Contd.

Common and scientific name of disease and its causal organism	Symptoms of damage	Management
	• If seeds are produced they remain shrivelled, light in weight and dark in colour.	sowing is recommended. • Add 1 ml soap solution / lit water for better efficiency of fungicide.
Powdery mildew: *Erysiphe polygoni*	• Affected plants show whitish powdery growth on leaves, petiole, stem pedicel and seeds. • At later stages of attack seeds become white and shrivelled and light in weight.	• Crop should be dusted with Sulphur dust @ 25 kg/ha to control this disease as soon as the symptoms are noticed. • Spraying of Wettable Sulphur 0.25% or Dinocap (Karathan 0.05%) can also be used to control the disease. • If needed second spray should be given 15-20 day after first spray.
Fennel		
Collar rot: *Sclerotium sclerotiorum*	• This disease appears in those plots where water stagnation near the plant stem is more. • Collar portion of the plant start decaying and the plants turn to yellow colour & die later on.	• Drenching by 1% Bordeaux mixture or spray with 0.25% solution of any Copper fungicides should be done to control the disease. • Avoid water stagnation in the field.
Damping of seedlings: *Phytophthora syringae*	• Infected seedlings are toppled on the ground surface. (Post emergence damping off).	• Drenching of nursery beds with Copper oxychloride @ 0.25 % should be done to control this disease effectively.
Ramularia Blight: *Ramularia foeniculi*	• At the beginning of this disease small ashy spots are observed on the leaves. • These spots get enlarged and turn into brown colour. In case of severe attack whole plant turns to brown colour, resulting in drying up of plant.	• At the initial stage spraying of 0.25 % solution of Mancozeb is effective to control this disease. • Add 1 ml soap solution / liter of water for better efficacy.

Contd.

Name of disease and its causal organism	Symptoms of damage	Management
		• Subsequently, 2-3 sprayings should be done at an interval of 10 to 15 days on need basis.
Fenugreek		
Powdery mildew: *Erysiphe polygoni* and *Leveillula taurica*	• It usually appears late in the season. Powdery white growth can be seen on both the surfaces of the leaves. • It also appears on above ground plant parts.	• The disease is controlled by dusting inorganic sulphur at 25 kg/ha or spraying Wettable sulphur 0.25 per cent at 20 days interval.
Ginger		
Bacterial wilt: *Ralstonia solanacearum*	• Water soaked spots appear at the collar region of the pseudostem and progresses upwards and downwards. • The first conspicuous symptom is mild drooping and curling of leaf margins of the lower leaves which spreads upwards. • Yellowing starts from the lowermost leaves and gradually progresses to the upper leaves. In the advanced stage, the plants exhibit severe yellowing and wilting symptoms. • The vascular tissues of the affected pseudostems show dark streaks. The affected pseudostem and rhizome when pressed gently extrudes milky ooze from the vascular strands.	• The cultural practices adopted for managing soft rot are also to be adopted for bacterial wilt. • The seed rhizomes may be treated with Streptocycline 200ppm for 30 minutes and shade dried before planting. • Once the disease is noticed in the field all beds should be drenched with Bordeaux mixture 1% or Copper oxychloride 0.2%.
Leaf spot: *Phyllosticta zingiberi*	• The disease starts as water soaked spot and later turns as a white spot surrounded by dark brown margins and yellow halo. The lesions enlarge and adjacent lesions coalesce to form necrotic areas. • The disease spreads through rain splashes during intermittent showers.	• The disease can be controlled by spraying Copper oxychloride 0.25 % or Mancozeb 0.2%.

Contd.

Name of disease and its causal organism	Symptoms of damage	Management
	• The incidence of the disease is severe in ginger grown under exposed conditions.	
Soft rot or rhizome rot: *Pythium aphanidermatum / P. vexans / P. myriotylum*	• The infection starts at the collar region of the pseudostems and progresses upwards as well as downwards. The collar region of the affected pseudostem becomes water soaked and the rotting spreads to the rhizome resulting in soft rot. • At a later stage root infection is also noticed. • Foliar symptoms appear as light yellowing of the tips of lower leaves which gradually spreads to the leaf blades. • In early stages of the disease, the middle portion of the leaves remain green while the margins become yellow. • The yellowing spreads to all leaves of the plant from the lower region upwards and is followed by drooping, withering and drying of pseudostem.	• Treatment of seed rhizomes with Mancozeb 0.3% for 30 minutes before storage and once again before planting reduces the incidence of the disease. • Cultural practices such as selection of well drained soils for planting is important for managing the disease, since stagnation of water predisposes the plant to infection. • Seed rhizomes are to be selected from disease free gardens, since the disease is also seed borne. • Application of *Trichoderma harzianum* along with Neem cake @ 1kg/bed helps in preventing the disease. • Once the disease is located in the field, removal of affected clumps and drenching the affected and surrounding beds with Mancozeb 0.2% checks the spread of the disease.
Nutmeg		
Fruit rot: *Colletotrichum gloeosporioides* and *Botryodiplodia theobromae*	• Water soaked lesions are seen on the fruits, the tissues of which become discoloured and disintegrated. • Premature splitting of the pericarp and rotting of mace and seed are the main symptoms of the disease. The internal tissues are found rotten. The fallen fruits become enveloped with the growth of the organism.	• The disease can be controlled by spraying 1% Bordeaux mixture.

Contd.

Name of disease and its causal organism	Symptoms of damage	Management
Leaf spot and shot hole: *Colletotrichum gloeosporioides*	• Sunken spots surrounded by a yellow halo are the initial symptoms. • Subsequently the central portion of the necrotic region drops off resulting in shot hole symptoms. • Dieback symptoms are also observed in some of the mature branches. • On young seedlings drying of the leaves and subsequent defoliation are seen.	• The disease can be controlled by spraying 1% Bordeaux mixture two or three times during rainy season.
Black Pepper		
Anthracnose/ Pollu disease: *Gloeosporiumgloeosporioides*	• Symptoms appear in leaves and stems. Circular or irregular grey spots appear in the leaves. • Concentric rings of acervuli appear on upper surface of the leaves. On the stem the infection appear at the tips spreads downwards and kill the entire vine. • When infections occur in leaf axils the spikes become affected and fall off. • Berry becomes brown in colour and extends downwards from upper portion. And become chaffy (Pollu means hallow thing).	Three rounds of spray with Bordeaux mixture 1%.
Phytophthora foot rot / quick wilt: *Phytopthora capsici*	**A. Die back:** • The aerial branches get infected at any point. At the site of infection of branch, the discoloration occurs and rotting progress upwards and downwards resulting in die-back symptoms.The lateral branches of the affected vines break off at the nodes and fall off.	• Selection of healthy nursery material. • Provide good drainage. • Soil drenching with 1% Bordeaux mixture after removal of the affected plant. • Spraying with 1% Bordeaux mixture (or) COC 0.25% (or) Alitte 0.3%.

Contd.

Name of disease and its causal organism	Symptoms of damage	Management
	B. Foot rot or Collar rot: • The stem near the ground level get infected and the rotting and death of vine occurs within 2-3 weeks. • The affected portion emits bad odour. The necrosis progress down wards to the underground stem and to the root system. **C. Root rot:** • The infection starts at main root or at feeder root. The leaves become yellow and defoliate.	• Soil application of Neem cake and *Trichoderma viride or P. fluorescens.*
Slow decline / slow wilt: F*usarium, Rhizoctonia* and *Diplodia* along with nematode infection	• General yellowing of lower leaves which progress upwards. • The affected leaves become flaccid and fall off. • Tip burn of leaves and die-back of twigs appear. Root knot, can be seen in the vine showing the yellowing of leaves. The affected vines gradually die. • Vascular browning can be seen.	• Soil drenching with Copper oxychloride in May – June and September – October.
Stump rot: *Rosellonia deusta*	• Pepper vines die in patches of 5-15 meter square scattered in the plantation. • The fungus attacks the roots as brown strand mixed with soil. • Leaves turn yellow and fall off and the plant dries up.	• Soil drenching with 0.25% Copper oxychloride.
Vanilla		
Root rot: *Fusarium* sp., *Sclerotium* sp.	• Browning of roots leading to rotting and decay. • Yellowing of leaves, stem and leaves becoming flaccid, shriveled and giving a drooping appearance.	• Cut and remove the affected roots, reduce mulch. • Soil drench and spray the plant with 0.2 % Copper oxychloride @ 2 or 3 litres per plant.

Contd.

Name of disease and its causal organism	Symptoms of damage	Management
Stem blight: *Phytophthora* sp.	• Appearance of brown spots on leaves and stems followed by yellowing of the leaves and rotting of leaves and stem.	• Spray 0.25% Copper oxychloride or Allite 0.1%.
Stem rot: *Fusarium* sp.	• Water soaked lesions giving stem a brown coloured appearance resulting in rotting of tissues, leaves turning yellowish and drying off.	• Cut and remove the affected portion. • Spray 0.1% Carbendazim , repeat spray after one month.
Bean rot: *Phytophthora* sp.	• Rotting and shedding of beans.	• Spray 0.1% Carbendazim and 1 % Bordeaux mixture alternatively at weekly intervals.
Shoot tip rot: *Fusarium, Sclerotium*	• Decaying of shoot tip and leaves at the tip.	• Spray 0.1% Carbendazim at 15 days interval.
Vanilla mosaic: *Vanilla mosaic virus*	• Mosaic pattern or mottling on foliage. • Stunting sterility and leaf distortion.	• Uproot and destroy the affected plant. **Important Prophylactic Measures** • Timely shade management • Avoid excess moisture • Allow free movement of air • Avoid overcrowding of vines by keeping adequate spacing • Avoid excessive mulching during rainy season and mulching with materials that are not easily decomposed • Avoid excessive manuring and use of fresh cow dung. • Apply recommended doses of bio agents like *Trichoderma*, *Pseudomonas*, *Bacillus* etc. • Collect and destroy the parts of plants showing disease symptoms.

9

Insect Pests of Major Spices and Condiments

Common name, symptoms of damage and management of insect pests of major spices and condiments

Common and scientific name of pest	Symptoms of damage	Management
Black Pepper		
White tailed mealy bug: *Ferrisia virgata*	• Mealy bug found in cluster on the terminal shoots, leaves, berries and suck the sap. • Yellowing, withering and drying of plants or shedding of berries.	• Removal of heavily infested plant parts and the destruction will minimize the inoculum. • Spray Dimethoate 2 ml /lit.
Pollu beetle: *Longitarsus nigripennis*	• Grub feeds on the growing tip, tender stem, leaf petiole, spikes and berry. • Tunneling of tender stem causes drying up of the shoot. • The portions of the spikes tunnelled by grubs turn black and cause dropping of spikes due to the subsequent rotting. • Grubs bore into the berries and feed on entire internal contents. • Exit holes could be seen on dry berries. • As a result of grub boring, the berries dry up and turn dark in colour. • They are hollow and crumble when pressed. • Such hollow berries are called 'Pollu' berries.	• Regulation of shade in the plantation reduces the population of the pest in the field.. • Spraying Quinalphos (0.05% each) during June/ July and September/October or Quinalphos (0.05% each) during July and Neemgold (0.6%) (Neem-based insecticide) during August, September. • The undersides of leaves (where adults are generally seen) and spikes are to be sprayed thoroughly.
Pepper mussel scale: *Lepidosaphes piperis*	• Infects the plant by sucking the leaf tip, leaf and stem of the suckers • Infected parts become yellowish and later dry up.	• Remove and destroy the infected plant parts. • Spray Dimethoate 0.05 % followed by one more spray on 21st day of first spray.
Marginal gall thrips: *Liothrips karnyi*	• Downward and inward curling of leaves resulted in the formation of marginal leaf galls. • Later the infested leaves become crinkled and malformed • In severe cases of attack, the whole plant may become	• Spray Dimethoate (0.05%) during emergence of new flushes in young vines in the field and cuttings in the nursery.

Contd.

Common and scientific name of pest	Symptoms of damage	Management
	stunted, affecting adversely the formation of spikes.	
Top shoot borer: *Laspeyresia hemidoxa*	• Larvae bore into tender terminal shoots and feed on internal tissues resulting in blackening and decaying of affected shoots. • The pest infestation is higher during July to October when numerous succulent shoots are available in the vines.	• Spray Quinalphos (0.05% each) on tender terminal shoots; repeat spraying at monthly intervals (during July - October) to protect emerging new shoots.
Cardamom		
Cardamom aphid: *Pentalonia nigronervosa* f.*Caladii*	• Nymphs and adults suck up plant sap. • Colonies of aphids are seen under concealed conditions inside leaf sheaths of the older pseudo stems.	• Removal of *Colocasia* and other hosts in the vicinity of cardamom plantation is recommended. • Remove partly dried and decayed pseudostems which harbour the colonies of aphids to reduce aphid population. • Foliar spraying with Dimethoate 2 ml/lit at an interval of 2 weeks in November and April gives adequate control. • The aphids act as a vector of the mosaic or 'Katte' virus of cardamom.
Beetle borer: *Onthophgus* spp.	• Larva makes the bore hole which is neat, circular tiny and at right angles to the pericarp mostly on the ventral side. • When attack is on young immature capsules, they put forth matured appearance, decay and drop off. • Adults beetle bores the flowers and feed on tissues.	• Judicious shade regulation. • Spraying of contact insecticide viz., Quinalphos @2ml/lit.
Capsule borers: *Jamides* sp. *cardamomi, E. canarica,*	• Caterpillars bore and feed on the inflorescence, flower buds, flowers and capsules.	

Contd.

Common and scientific name of pest	Symptoms of damage	Management
E. fabia Eupterote cardamomi, E. canarica, E. fabia	• Affected capsules become empty with a big circular hole, turn yellowish – brown which decay and drop off in the rainy season.	
Hairy caterpillars: *Eupterote*	• These appear sporadically in enormous populations at intervals of several years and cause heavy damage to the foliage • Caterpillars are gregarious in habit and they congregate on the trunks of shade trees during day time and drop down on the cardamom plants during night time. • They fed voraciously on leaves and defoliating the whole cardamom plants. • Usually the damage is observed during October-December.	• Hairy caterpillar incidence can be checked by lopping off and burning branches of shade trees in which eggs are laid. • Gradual elimination of shade trees favoured for egg laying will reduce the incidence. • Caterpillars that congregate at the base of shade trees can be collected and killed. • Moths can be attracted to light during night and by installing light traps, the moths can be collected and killed. • Spray Quinalphos 2ml/lit.
Root grubs: *Basilepta fulvicorne, Holotrichia serrata*	• The grubs feed on the roots in the form of irregular scraping. • In advanced stages entire root system is found damaged resulting in drying and rotting depending on the season of attack. • In the severely infested plants, leaves turn yellow and dry.	• Collect the beetles using hand nets during peak period of emergence i.e., during April-May and kill them. • Beetles can be seen resting on the leaves during day time • The time of insecticide application is to be synchronized with the emergence of adults and egg laying. • Applying Chlorpyriphos 0.075% or Phorate 2-4 g a.i./plant in May and in September-October is effective in reducing the population of grubs. • Light raking of soil before the insecticide application is essential for effective control of root

Contd.

Common and scientific name of pest	Symptoms of damage	Management
		grubs.
Shoot, panicle and capsule borer: *Dichocrocis (Conogethes) punctiferalis*	• Early stage of the larva bores the unopened leaf buds and feeds on the leaf tissue. • They also bore the panicles leading to drying up of the portion from the affected spot • Immature capsules and feed on the young seeds inside rendering the capsules empty. • Late stage larvae bore the pseudostem and feed the central core of the stem • Resulting in drying of the terminal leaf and thus produce characteristic 'dead heart' symptom. • Oozing out of frass material at the point of tunnelling is the indication for the presence of larva inside the plant parts • The incidence of this pest is noticed throughout the year but they occur in enormous number in four periods, December-January, March-April, May-June and September-October and their abundance synchronizes with the panicle production, fruit formation and new tiller production.	• During day time adult moths rest on the lower surface of the cardamom. They may be collected with insect net and killed. • The practice of removing the tillers showing 'dead heart' symptoms should be carried with due care. • Tillers may be removed if the attack is fresh as indicated by extrusion of frass • The infestation by early stages of larva of this pest in emerging panicle, immature capsule and leaf bud can be controlled effectively with insecticide application. • Once the late larvae bore and go deep inside the pseudostem, the chemical spray even in its higher dose becomes ineffective.
Shoot fly: *Formosina flavipes*	• Larvae feed on the growing shoot of the young cardamom suckers • Dead heart.	• Remove the affected shoots at ground level and destroy them. • Spray Dimethoate or Quinalphos (0.05% each).
Thrips: Sciothrips cardamom	• Panicles become stunted • Shedding of flowers and immature capsules thus reducing the total number of capsules formed. • Infestation causes formation of corky encrustations on	• Regulate shade in thickly shaded areas. • Spray insecticides like Quinalphos 0.025% during March, April, May, August and September.

Contd.

Common and scientific name of pest	Symptoms of damage	Management
	pods resulting in their malformed and shrivelled condition. • Such pods lack their fine aroma and the seeds within are also poorly developed.	
White flies: *Dialeurodes cardamom*	• Nymphs and adults remain in colonies on lower leaf surface and suck the sap. • Chlorotic patches appear on leaves, • Gradual yellowing and get dried. • Sooty mould development occurs due to honeydew secretion.	• Spray Dimethoate (0.05% each) on the foliage.
Cinnamon		
Cinnamon butterfly: *Chilasa clytia*	• Larvae feed on tender and slightly mature leaves. • In severe cases of infestation, the entire plant is defoliated and only midribs of leaves with portions of veins are left.	• Spraying Quinalphos 0.05% on tender and partly mature leaves.
Leaf miners: *Conopomorpha civica*	• Larvae feed on the tissues between the upper and lower epidermis of tender leaves resulting in linear mines that end in 'blister' like patches. • The infested leaves become crinkled and the mined areas dry up leading to the formation of large holes on the leaves.	• Spraying Quinalphos 0.05% during emergence of new flushes is effective in preventing the pest infestation. • Many other leaf feeding caterpillars and beetles also occur sporadically on cinnamon feeding on tender flushes. • Application of Quinalphos 0.05% would keep them under check.
Clove		
Mealy bug: *Pluvinaria psidii*	• Soft green mealy bug with a white mealy ovisac is seen on the under surface of tender leaves. • The feeding causes yellowing and withering of leaves. • Sooty mould developing on the honey dew secreted	

Contd.

Common and scientific name of pest	Symptoms of damage	Management
	by the bug.	
Black Scale: *Saissetia nigra*	• It infests tender leaves, shoots and twigs. • Sooty mould fungus is observed on leaves due to honey dew secretion.	• Removal of affected leaves and branches will prevent the spread.
Masked scale: *Mycetaspis personata*	• Dome shaped and greyish brown scale will be seen on the under surface of leaves especially of young clove plants. • The infested leaves become discoloured and completely yellowish and finally may fall down.	
Stem borer: *Sahyadrassus malabaricus*	• The stem borer infests the main stem of young trees at the basal region. • The larva of the pest girdles the stem and bores downward into it. • The girdled portion and bore-hole is covered with a mat- like frass material. • The infested trees wilt and succumb to the pest attack subsequently.	• Spray Quinalphos 0.1% around the bore-hole and inject the same into the bore-hole after removing the frass. • Swab the basal region of the main stem with Carbaryll and keep the basins free of weeds.
Chilli thrips: *Scirtothirps dorsalis*	• The affected leaves show a pinkish discolouration. • Flower shedding is also noticed.	• Removal of affected leaves and branches will prevent the spread. • Spray Dimethoate 2ml/lit.
Turmeric and Ginger		
Leaf roller: *Udaspes folus*	• Leaves become folded or rolled longitudinally. • Complete defoliation.	• A spray with Carbaryll (0.1%) or Dimethoate (0.05%) may be undertaken when the severe infestation. • Spray the following dissolved in 1 lit of water,

Contd.

Common and scientific name of pest	Symptoms of damage	Management
		Carbendazim 2g +Mancozeb 2g + Teefol 1 ml.
Rhizome scale: *Aspidiella hartii*	**Field infestation** • Plants look devitalized, pale and withered before drying completely. · In such cases at the time of harvest minute yellowish crawlers can be seen moving in large numbers and this is the potential stage of dissemination. **Storage infestation** • In Initial stage of infestation in, the white coloured scales are seen scattered on rhizomes and latter they congregate near the growing buds. • When the infestation is severe the rhizome and buds shrivel and ultimately the entire rhizome dries.	• Discard and do not store severely infested rhizomes • Collect and destroy damaged leaves. • Select healthy rhizomes free from scale infestation for seed materials. • Treat seed material with Quinalphos 0.075% (for 20-30 minutes) before storage and also before sowing in case the infestation persists. • Apply well rotten sheep manure @ 10 t/ ha in two splits or poultry manure in 2 splits followed by drenching Dimethoate 30 EC @ 2ml /lit of water.
Shoot borer: *Conogethes punctiferalis*	• Yellowing and drying of leaves of infested pseudostems. • The presence of a bore-hole on the pseudostem through which frass is extruded and the withered and yellow central shoot.	• Spraying Malathion 0.1% at 30 day intervals during July to October is effective in controlling the pest infestation. • The spraying has to be initiated when the first symptom of pest attack is seen on the inner most leaves on the pseudostem.
Thrips: *Panchaetothrips indicus*	• Leaves become rolled up. • Turn pale and gradually dry-up.	• Spray insecticides like Quinalphos 0.025%.

□□□□

2. Plantation Crops

10

Importance of Plantation Crops

- The term plantation crops refer to those crops which are cultivated on an extensive scale in contiguous area, owned and managed by an individual or a company.
- There are two groups of plantation crops namely coconut, arecanut, oil palm, cocoa, cashewnut dealt by Ministry of Agriculture and tea, coffee, rubber dealt by the Ministry of Commerce.
- These are high value commercial crops of greater economic importance and play a vital role in improving Indian economy, especially in view of their export potential, employment generation and poverty alleviation particularly in rural sector.

Coconut

- Coconut is grown is grown in 2137 thousand ha with an annual production of nearly 15609 thousand MT with average productivity of 7.3 MT/ha.
- **Indonesia**, Philippines and India are the major producers of coconut in the world.
- **India** is the third largest producer, of coconut in the world.

Cashewnut

- India is the third largest producer, of cashew in the world.
- Maharashtra ranks first in area and production of cashew in the country.
- Cashew is grown mainly in Maharashtra, Goa, Karnataka and Kerala along the west coast and Tamil Nadu, Andhra Pradesh, Orissa and West Bengal along the east coast.

Cocoa

- The world's largest cocoa bean producer is Coste d'Ivoire.
- Cocoa is cultivated in India over an area of 66,000 ha with an annual production of 13000 MT with a productivity of 200 kg/ha during 2012-13.
- Kerala is the leading producer of coco a with 6000.10 MT of production.
- It is mainly grown in the states of Kerala, Karnataka, Tamil Nadu and Andhra Pradesh and cultivated mainly as an intercrop in arecanut and coconut gardens.
- It has great potential to be introduced in other states where coconut and arecanut are grown.

Arecanut

- India is leading producer of arecanut in the world.
- Production of arecanut in the country which stood at 609 thousand MT from an estimated area of 446 thousand ha during 2012-13.
- Karnataka ranks first in area and production of arecanut in the country.
- Karnataka, Kerala and Assam are the three major states producing arecanut.
- Arecanut industry forms economic backbone of nearly 10 million people of India and for many of them it is the sole means of livelihood.

Coffee

- The major coffee producers in the world are Brazil, Vietnam, Columbia, Indonesia and India.
- India's rank 5th in coffee production in the world.
- Arabica and Robusta coffee are the two species that are commercially cultivated in the world. Of these, nearly 70% is Arabica and the balance 30% Robusta.

Area, production and productivity of major plantation crops (2012-13)

S. No.	Crop	India					Leading country in the world in production	India's rank in country in the world in production
		Area in 000' ha	Production in 000' MT	PDY MT/ha	Leading state in area (000' ha)	Leading state in production (000' MT)		
1.	Coconut	2137	15609	7.3	Kerala 798.2	Tamil Nadu 4760.7	Indonesia	III
2.	Cashewnut	992	753	0.8	Maharashtra 184.2	Maharashtra 224.6	Viet Nam	III
3.	Arecanut	446	609	1.4	Karnataka 221.4	Karnataka 358.6	India	I
4.	Cocoa	66	13	0.2	Tamil Nadu 22.4	Kerala 6.1	Coste d'Ivoire	-
	Total	3641	16985	4.7				

Source: FAO Website, December 2014 and for India Data Indian Horticulture Database 2013

- The Indian coffee production in 2012-13 was 318.20 thousand MT comprising 98.60 thousand MT of Arabica.

Rubber

- The world's largest rubber producer is **Thailand.** Among all rubber producing countries India position is 5th in production.
- Most of the rubber plantations are located in far-flung villages and they play a major role in rural development of the country. Women are employed in plenty in rubber plantations and this contributes to their empowerment. In employment generation, rubber-manufacturing industry also plays an equally important role.
- Though rubber is traditionally grown in the southern states of Kerala and parts of Tamil Nadu and Karnataka, the crop has now gained acceptance among the people of non traditional regions as well, especially the North-east. Rubber plantations are very helpful in protecting the environment.

Tea

- **India** is the second largest producer, of tea in the world after China.
- During 2012-13, tea production was 1135.07 thousand MT. Assam is the leading producer of tea.
- Tea industry provides direct employment to more than one million workers directly, nearly 50% are women.

Area, production and productivity of coffee, tea, rubber

Crop	Production (000' MT	Leading state in production	Leading country in the world in production	India's rank in the world in production
Coffee- Arabica (2013-14)	102.20	Karnataka	Brazil	
Coffee- Robusta(2013-14)	202.30	Karnataka		
Total coffee (2013-14)	**304.50**	**Karnataka**	**Brazil**	**V**
Tea (2012-13)	1135.07	Assam	China	II
Rubber (2010-11)	861.950	Kerala	Thailand	V

Source: FAO Website, December 2014 and for India Data Coffee Board, Tea Board, Rubber Board of India

- **Agricultural and Processed Food Products Export Development Authority (APEDA)** was established by the Government of India under the Agricultural and Processed Food Products Export Development Authority Act passed by the Parliament in December, 1985.
- **Coffee Board:** The Coffee Board was established by the Government of India through a constitutional act "Coffee Act VII of 1942" under the administrative control of Ministry of Commerce and Industry.
- **Directorate of Arecanut and Spices Development:** The Directorate of Arecanut and Spices Development was established on 1st April, 1966 at Calicut in Kerala as a subordinate office under Ministry of Agriculture, Government of India to look after the development of spices and arecanut at National level. This responsibility was earlier shouldered by Central Spices and Cashewnut Committee, which was abolished in 1965 consequent to the recommendations of Agricultural Research Review Team and the Indian Central Arecanut Committee which was dissolved in the same time.
- **Directorate of Cashewnut Development:** The Directorate of Cashew nut Development was established in 1966 as a primary field functionary under the Union Ministry of Agriculture. The Directorate of Cashew nut Development which was handling only cashew got the mandate for development of cocoa in 1997. 1997-98 the Directorate of Cashew nut and Cocoa Development started driving both the crops on similar tracks.
- **National Research Centre for Oil Palm** was established by the Indian Council of Agricultural Research at Pedavegi in West Godavari district of Andhra Pradesh on February 19, 1995. In February, 2009 the institute was upgraded as Directorate of Oil Palm Research (DOPR) to cater to the location specific programmes across the country.
- **Rubber Board:** The Rubber Board is a statutory body constituted by the Government of India, under the Rubber Act 1947, for the overall development of the rubber industry in the country.
- **Rubber Research Institute of India (RRII):** The Rubber Research Institute of India established in 1955 is located on a hillock in the eastern suburb of Kottayam, Kerala.
- **Tea Board:** The Tea Board of India was set up under section 4 of the Tea Act 1953 was constituted on 1st April 1954.

- The Act (2 of 1986) came into effect from 13th February, 1986 by a notification issued in the Gazette of India: Extraordinary: Part-II [Sec. 3(ii): 13.2.1986). The Authority replaced the Processed Food Export Promotion Council (PFEPC).

11

Technical Information on Plantation Crops

Common name, botanical name, family, origin, chromosome number, Ploidy level, sex expression, pollination system, type of inflorescence, type of fruit, and method of propagation of plantation crops

S. No	Common name	Botanical name	Family	Centre of origin	Chromosome number 2n =	Sex expression	Pollination system	Type of inflorescence	Type of fruit	Method of propagation
1.	Cashewnut	*Anacardium occidentale* L.	Anacardiaceae	Brazil	42	Allogamous	Cross pollination	Panicle	Nut	Soft wood grafting
2.	Palmyrah	*Borassus flabellifer*	Arecaceae	Tropical Africa	18	Dioecious	Cross pollination	Spadix	Fleshy drupe	
3.	Arecanut	*Areca catechu* Linn.	Arecaceae	Malaya Peninsula, Thailand	32	Monoecious	Cross pollination	Spadix	Single seeded berry	Seeds
4.	Coconut	*Cocus nucifera*	Arecaceae	South east Asia	32	Monoecious	Dwarf –Self pollination Tall- Cross pollination	Spadix	Drupe	Seeds
5.	Oil palm	*Ealias guinensis*	Arecaceae	Central and Western Africa	32	Monoecious	Cross pollination		Drupe	Seeds
6.	Rubber	*Hevea brasiliensis*	Euphorbiaceae	Brazil	36	Monoecious	Cross pollination	Panicle	Capsule	Bud grafting
7.	Betelvine	*Piper betle*	Piperaceae	Malaysia	X=13, 2n=26 and 52	Dioecious	Cross pollination			Terminal stem cuttings
8.	Coffee (Arabica)	*Coffea arabica*	Rubiaceae	Ethiopia and central Africa	4X=44		Self pollination	Cymose fascicle	Pea Berry	Seeds
9.	Coffee (Robusta)	*Coffea canephora*	Rubiaceae	Western and central Subsaharan Africa	22		Cross pollination	Cymose fascicle	Pea Berry	Cuttings
10.	Cocoa (Food of god)	*Theobroma cocoa*	Sterculiaceae	South America	20	Dioecious	Cross pollination	Cauliflory	Drupe	Soft wood grafting
11.	Tea	*Camellia sinensis* (L.) Kuntze	Theaceae	China	30		Cross pollination	Terminal or axillary	Capsule	soft wood cutting

□□□□

12

Production Techniques of Plantation Crops

Cashewnut

Botanical name	*Anacardium occidentale* L.
Family	Anacardiaceae
Origin	Brazil
Chromosome number	X=21, 2n=42
Type of inflorescence	Panicle
Sex expression	Polygamomonoecious
Pollination system	Cross pollination
Type of fruit	Nut
Edible portion	Cotyledon and fleshy peduncle
Nutritive value	Fat (47%), protein (21%) and carbohydrates (22%)
Nature of crop	Evergreen
Leading country in production	Viet Nam
Leading state in production	Maharashtra
Leading state in area	Maharashtra

- Cashew (*Anacardium occidentale* L.) is native of Brazil introduced to India just as other commercial crops like Rubber, Coffee, and Tea etc. by the Portuguese nearly five centuries back.
- Among the Agri-Horticultural commodities getting exported from India, cashew ranks the 2nd position.

- The **cashew tree** is a tropical evergreen that produces the cashew nut and the cashew apple.
- The tree is large and evergreen, growing to 10-12m (32 ft) tall, with a short, often irregularly shaped trunk. The leaves are spirally arranged, leathery textured, elliptic to obovate, 4 to 22 cm long and 2 to 15 cm broad, with smooth margins.
- The flowers are produced in a panicle up to 26 cm long; each flower is small, pale green at first, then turning reddish, with five slender, acute petals 7 to 15 mm long.
- The true fruit of the cashew tree is a kidney or boxing-glove shaped drupe that grows at the end of the cashew apple.

Production technology

Climate	The distribution of cashew is restricted to altitude below 700m where the temperature does not fall below 20°C for prolonged periods. However, temperature above 36°C between the flowering and fruiting period could adversely affect the production.
Soil	It is very sensitive to water logging and hence heavy clay soils with poor drainage conditions are unsuitable for its cultivation. Excessive alkaline and saline soils also do not support its growth. More than 8 pH is not suitable for its commercial cultivation.
Propagation	Through soft wood grafting.
Planting	7.5 X 7.5 or 8X8 m.
High Density Planting	Spacing of 5 x 4 m accommodating 500 plants per hectare is recommended prune the interlocking branches during the July-August to maintain the frame.
Planting time	July-August.
Harvesting	February- March.
Yield	10kg/tree at 16-20 years old tree.

Varieties

Variety	Salient features
Bhubaneswar-1	It is a selection of WBDC-5 (V36/3) having medium compact canopy.
BPP-4	It is a selection of Epurupalem having open canopy.
BPP-6	Trees having open/spreading canopy.
Chintamani-1	It is a selection of 8/46 Thaliparmba having open canopy.
Goa-1	It is a selection of Balli-2 having semi spreading compact canopy.

Jhargram-1	It is a selection of Tenor 16 of Bapatla having medium compact canopy.
K-22-1	It is a selection of Kottarakkara 22 having compact canopy.
Madakkathara-1	It is a selection of Bapatla having compact canopy.
Madakkathara-2	It is a selection of Neduvellur having open canopy.
NRCC-2	It is a selection of 2/9 Dicherla having medium compact canopy.
Ullal-1	It is a selection of Thaliparamba having medium spreading canopy.
Ullal-2	It is a selection of 3/67 Guntur having medium spreading canopy.
Ullal-3	It is a selection of 5/37 Manchery having open canopy.
Ullal-4	It is a selection of 2/77 Tuni having open canopy.
UN-50	It is a selection of 2/27 Nileswar having medium canopy.
Vridhachalam-3 (M 26/2)	It is a selection of Edayanchavadi having compact canopy.

Hybrids

Hybrid	Parents	Salient features
Amrutha	BLA-139-1 x H-3-13	Average nut yield 18kg/ tree having nut weight of 7.18 g, kernel weight of 2.24 g with 31.58% shelling rate and its export grade is W210.
BPP-8 (H2/16)	T1 x T3	Average nut yield 14kg/tree having nut weight of 8.2 g, kernel weight of 1.89 g with 29% shelling rate and its export grade is W210.
Dhana	ALGD-1×K-30-1	Average nut yield 11kg/tree having nut weight of 8.2 g, kernel weight of 2.44 g with 29.8% shelling rate and its export grade is W210.
Kanaka	BLA-139-1×h3-13	Average nut yield 13kg/tree having nut weight of 6.8 g, kernel weight of 2.08 g with 30.58% shelling rate and its export grade is W280.
Priyanka	BLA-139-1 x H-30-1	Average nut yield 17kg/ tree having nut weight of 7.40 g kernel weight of 2.87 g with 26.57% shelling rate and its export grade is W180.
Vengurla-4	Midnapur red x Vetore-56	Average nut yield 17kg/tree having nut weight of 7.7 g kernel weight of 1.91 g with 31% shelling rate and its export grade is W210.
Vengurla-6	Vetore-56 x Ansur-1	Average nut yield 14kg/tree having nut weight of 8 g kernel weight of 1.91 g with 28% shelling rate and its export grade is W210.
Vengurla-7	Vengurla-3 x M-10/4 (Vri-1)	Average nut yield 18kg/ tree having nut weight of 10 g Kernel weight of 2.9 g with 30.5% shelling rate and its export grade is W180.

Arecanut

Botanical name	*Areca catechu* Linn.
Family	Arecaceae
Origin	Malaya Peninsula
Chromosome number	2n=32
Type of inflorescence	Spadix
Pollination system	Cross pollination
Type of Sex expression	Monoecious
Type of fruit	Single seeded berry
Edible portion	Endosperm
Astrigent taste of nuts due to	Polyphenols and tannins
Value added products	**Kalipak:** Processed green nut
Leading country in production	India
Leading state in production	Karnataka
Leading state in area	Karnataka

- *Areca catechu* Linn. is commonly known as Arecanut.
- The Malaya, Archipelago, the Philippines and other East Indies islands are likely to be the centre of origin since majority of the species of Areca have been reported from the said region.
- It is a very widely cultivated plant in eastern countries like India, Bangladesh, Ceylon, Malaya, the Philippines and Japan.
- Arecanut palm is monoecious with male and female flowers occurring on the same spadix.
- The chromosome number of arecanut is 2n=32 and is a secondary allotetraploid.
- Seed is the only propagule of arecanut as in the case of many other palm species.
- The genus *Areca* belongs to the family Arecaceae (Palme) under the tribe Areceaea and is reported to contain about 74 species.
- Among several species *A. catechu* is the only cultivated species, the nuts of which are chewed as a mild stimulant.
- Nuts of *A. triandra* are also used as a mastigatory nut.
- Arecanut is a graceful erect growing palm with a single unbranched stem and grows to a height of 60 to 70 feet depending on the variety and environmental conditions.

- The stem is cylindrical throughout and it results from a single terminal growing point situated at the top and any severe injury to buds result in the death of the palm.
- The "nut" (actually the seed endosperm) is chewed as a stimulant masticatory by 5% of the world's population, making it more popular than chewing gum but not as popular as tobacco.
- The areca nut contains the tannins arecatannin and gallic acid; a fixed oil gum; a little terpineol; lignin; various saline substances; and three main alkaloids — arecoline, arecaidine and guvacine — all of which have vasoconstricting properties.
- More than half the arecanut production in the world is from India (53%), followed by Indonesia, China, Bangladesh, Thailand and Malaysia.

Production technology

Climate	Arecanut grows at an altitude up to 1,000 m above sea-level. The crop flourishes well at a temperature range of 14°-36°C.
Soil	The largest area of Arecanut is found in gravelly laterite soils of red clay. In parts of Karnataka, arecanut is planted in fertile clay soils with an admixture of tank silt. Sticky clay, sandy, brakish and calcarious soils are not favourable for its cultivation.
Propagation	By **seeds,** weight of nut 35g minimum, germination 40 DAS.
Planting	2.75 X 2.75 m.
Planting time	May -June
Manuring	Apply to each bearing palm (5 years and above) 10 - 15 kg of FYM or green leaf. 100:40:140 g of NPK/ tree/ year.
Harvest	The bearing starts after 5 years of planting. Nuts are harvested when they are three quarters ripe.
Yield	200-250 q. ha ripe nuts, 30-60q/ha chali.

Improved varieties

Variety	Salient features
Kahikuchi	It is a tall cultivar with medium thick stem possessing comparatively shorter internodes.
Mangala	It is high yielding semi tall palm with partially drooping crown, earliness in bearing, more number of female flowers/inflorescence, higher nut set, quicker stabilisation, round and medium sized yellow coloured nuts.
Mohitnagar	It is high yielding tall palm with medium thick stem, partially drooping crown, and orange yellow coloured oval to round shaped nuts.

Variety	Salient features
Shreewardhanee	It is a selection from Shriwardha local. The yield ranges from 7.5 to 8.0 kg/tree/ year.
Sreemangala	It is high yielding tall palm with sturdy stem, partially drooping crown, round and bold with deep yellow coloured nuts. The average yield 44 q/ha (dry kernel).
Sumangala	It is high yielding tall palm with partially drooping crown, oval to round shaped deep yellow coloured nuts.
Swarnamangala	It is high yielding tall palm with medium thick stem and comparatively shorter internodes, partially drooping crown.

Hybrids

Hybrid	Parents	Salient features
VTLAH-1	Hirehalli dwarf x Sumangala	It is dwarf in nature. Sturdy stem with super imposed nodes, reduced canopy size, well spread leaves, medium sized oval to round shaped nuts and early stabilization and medium yielder.
VTLAH-2	Hirehalli dwarf x Mohitnagar	It is dwarf in nature. Sturdy stem with super imposed nodes, reduced canopy size, well spread leaves, medium sized oval to round shaped nuts and early stabilization are the salient features of this hybrid variety.

Coconut

Botanical name	*Cocus nucifera*
Family	Arecaceae
Origin	South East Asia
Chromosome number	2n=32
Ploidy level	Diploid
Type of inflorescence	Spadix
Bearing habit	Axile of leaf
Pollination system	Self pollination- Dwarf coconut and cross pollination- Tall coconut
Type of fruit	Drupe
Sex expression	Monoecious
Oil percent in copra	65-72%
Leading country in production	Indonesia
Leading state in production	Tamil Nadu
Leading state in area	Kerala

- The coconut palm, *Cocos nucifera* L. grows best under conditions of high humidity, at temperatures of 27–30°C and on moderately to well-aerated soils.
- Coconut is mainly an oil crop, particularly rich (48 %) in lauric acid.
- Copra normally has an oil content varying from 65 to 72 percant.
- The coconut industry offers a wide range of coconut products such as desiccated coconut, coconut cream, coconut milk powder, defatted coconut, coconut fibre and fibre products, shell charcoal and activated carbon, coconut vinegar and coconut arrack (liquor) etc. for the local and the international markets.
- Tender coconut as a natural beverage is presently gaining popularity and currently there is also a growing international market for fresh coconut water.
- Coconut oil is also the principal raw material used in the manufacture of soap, glycerine and margarine.
- The lauric acid from coconut oil is used to manufacture detergents, cosmetics and pharmaceuticals.
- The palm has adventitious roots continually produced from the base of the trunk, which is the swollen part or what is termed 'bole', in tall types and in some dwarf hybrids.
- It has no taproot or root hairs but has lots of primary roots which bear large quantities of rootlets.
- The main roots grow out somewhat horizontally from the bole and are mostly found within the topsoil. The main branches grow deeper and may extend laterally to as much as 10 m.
- The roots, having no cambium, are noticeably uniform – the main roots reaching a maximum diameter of about 1 cm. The root tip is the actively growing region and behind it is the absorbing area whose epidermis is a single layer of thin-walled cells that gradually thicken and become impervious with age.
- In old roots, the epidermis disintegrates and exposes the hard hypodermis which is generally red.
- The root centre has a stele surrounded by a single-celled pericycle sheath from which rootlets and aerenchymatous (respiratory exchange) protuberances or pneumatophores arise. The respiratory exchange occurs more abundantly nearer the soil surface to allow easy diffusion of root.

- The stem is called as trunk.
- It is unbranched, erect, stout and cylindrical.
- The leaf base encircles the stem as they have scars with pit mark on the trunk.
- The scars are helpful in determining the age of the tree. Based on these 12 to 14 scars, the age can be computed for a year.
- First few years there is increased growth in the thickness of the stem.
- The girth remains uniform for a number of years and finally tapers in growth when the tree grows very old.
- The trunk carries a big crown with large sized leaves and bunches of fruits.
- The leaves are crowded together at the top of the trunk in the form of a crown.
- The young leaf appears in the centre of the crown as a pointed structure having all the leaves closely held together.
- It takes five months for the complete development of a leaf from the time of initiation to its complete emergence.
- The crown has 15 open leaves, 15 younger leaves in different stages of development.
- The leaf consists of many leaflets arranged obliquely on the mid-rib or rachis.
- Each leaflet is long, linear, has a strong mid-rib with narrow lamina and parallel venation.
- The petiole is flat, very thick, broad and sheathe with fibre at the base and encircles the stem.
- It takes 34 months for the inflorescence to develop from the initiation to the opening of the spathe.
- The spadix is stout and erect, enclosed by the tough spathe which splits when inflorescence is mature.
- The axis of the inflorescence is branched, with the branches bearing sessile flowers, both staminate and pistillate.
- The palm is monoceious with male and female flowers on the same plant.

- Staminate flowers are small, numerous and are arranged closely on the secondary branches of the inflorescence.
- They have six perianth lobes in two whorls and the outer perianth are small, the inner perianth is bigger.
- Stamens are six and plenty of pollen is produced. An abortive pistillate or rudimentary ovary is present in the staminate flower. The pollen grains are spherical in shape and are smooth.
- Pistillate flowers are at about the base of the branches substened by one or two staminate flowers.
- They are small spherical structure known as 'button'.
- Six perianth lobes in two whorls arranged imbricate and more or less of the same size and shape.
- The perianth is rounded, concave and whitish, the stiga is found at the base of the ovary.
- Once pollination and fertilization occur, fruits set and develop to maturity in about 12 months, or less than 1 year for some dwarf cultivars. A count of bunch and fruit set can give a reasonable estimate of yield.
- Fruit is drupe, having three regions and one seeded .On the outside is the husk.
- The fruit is developed from a tricarpellary ovary.
- The outer layers of pericarp are fibrous called as husk which is initially green but turns brown after being picked and dried.
- The endocarp is very hard and forms the shell.
- Within the endocarp, soft white endosperm enclosing inside a big cavity filled with extra-cellular fluid called as "coconut milk." The seed coat, which is thin, and then the white flesh or copra. Both the copra and the milk are the endosperm of this seed.
- Initially the milk is fairly sweet and the copra is thin, but as the seed matures, the liquid is converted into solid endosperm rich in oils (triglycerides).
- The endosperm or the kernel also contains sucrose, fructose, galactose, glucose and raffinose.
- The moisture content in the kernel diminishes as the maximum quantity of oil is formed when the nuts are 9-10 months old.

- Liquid endosperm contains large quantities of cytokinin hormone which is useful to grow plants in test tubes from single cells and this develops into embryos

Breeding

- *Cocos nucifera* L. is a member of the monocotyledon family Arecaceae (Palmaceae) in the subfamily Cocoideae that includes 27 genera and 600 species and is the only species of the genus *Cocos*.
- Coconut possesses a diploid genome with 32 chromosomes (2n = 2x = 32).
- Crop improvement programmes in coconut led to development of hybrids, collection, conservation and cataloguing of germplasm.
- Collection of germplasm using different tissue culture techniques, DNA fingerprinting and MAS are the developments in biotechnology of coconut.
- Biotechnology can be used to improve coconut in several ways. By improving our knowledge of coconut diversity, molecular methods could help breeders effectively choose varieties to be used in a breeding programme to maximize heterosis and the chances of incorporating useful genes.
- A high level of genetic diversity in coconut has been observed using micro satellites in a collection of 130 coconut individuals sampled across the entire geographic range in the world.
- An recent study, found a reduced number of alleles in dwarf coconut group compared to tall coconut group, comparable with a reduction in the amount of genetic diversity in dwarfs.
- The first genome map for coconut was developed for an East African Tall x Laguna Tall F_1 population based on ISTR markers.
- Tall coconut palm, an allogamous and variable species, is highly heterozygous which is a major constraint in conventional breeding.
- Availability of homozygous lines would improve the efficiency of generating new cultivars through breeding programs.
- Gametic embryogenesis allows single-step development of completely homozygous lines from heterozygous parents.
- For the first time that there is a feasibility of developing homozygous lines in coconut via anther culture.

- Plant regeneration via anther culture of coconut (*Cocos nucifera* L.). (a) Embryos (em) derived from anther (an). Note the tiny connector (arrow head) between the embryo and the anther wall. (b) Mature embryo with germination point (gp). (c) Germinating embryo with a shoot (s) emerging by splitting of haustorium (h). (d) Calli (ca) bearing anther. (e) An anther bearing both callus (ca) and embryos (em). Note the embryo is opaque white whereas the callus has a translucent appearance. (f) Somatic embryos (se) developed from anther derived callus. (g) A complete plantlet derived from anther.

Production technology

Climate	Coconut grows at a altitude up to 900m above sea-level. Well-distributed rainfall of 800-2,500mm/ year is ideal. To get highest yield average annual temperature of 27°C with a diurnal variation 5°-7°C is optimum.
Soil	Laterite, lateritic red, sandy, alluvial sandy, alluvial coastal and black soils are good for coconut cultivation.
Propagation	By seed • Irrigate the nursery beds once in three days.· Keep the nursery free of weeds. To manage the weed problem in coconut nursery, growing sun hemp 2 times (each harvested at flowering stage) followed by one hand weeding at 6th month was found to be very effective besides yielding green manure for manuring the adult coconut palms. • Provide shade to the nursery by raising Sesbania or Leucaena on the sides of beds. • The seed nuts start germination 6 – 8 weeks after planting and germination continues upto six months. Select seedlings that germinate before 5 months after planting. Remove those nuts which do not germinate 5 months after sowing. • Regularly survey for pest and diseases.
Spacing	7.5 x 7.5 m with 175 plants/ha.
Time of planting	May-June.
Weed management	For the broad-leaved weeds, pre-emergence spraying of Atrazine @1.0 kg a.i./ ha for the control of grasses and sedges. Post emergence spraying of glyphosate @ 10 ml and 20 g ammonium sulphate/litre of water.
Harvesting	• Twelve months old nuts are harvested at the interval of 30-45 days for seed as well as copra making and culinary purposes. • For household use keep the nuts in vertical direction. However, for tender nut purposes 7 to 8 months old nuts are harvested. The nuts can be harvested using coconut climbers.

- Nuts which are 11 months old give fibre of good quality. This is suitable for coir fibre.
- In case of tall the nuts harvested for seed purpose can be stored for 2 to 3 months period before sowing, whereas in case of dwarfs and hybrids, nuts should be sown with in a period of 10 –15 days of harvest.· O n an average, we can have eight harvests, though the coconut palm produces inflorescence every month.

S .No	Variety	Nut yield (Nos / tree / year)	Earliness (year)
1.	Hybrid	100	3 – 5
2.	Tall	60 – 80	6 – 7
3.	Dwarf (tender coconut)	70 – 90	4 – 5

Improved varieties

Name of variety	Salient features
Andaman Ordinary	It is a tall variety suitable for production of copra and tender nut.
Chowghat Green Dwarf	It is an early bearing cultivar and takes about 3-4 years for initial flowering.
East Coast Tall	It is a tall variety suitable for production of copra and tender nut.
Gangabondam	It is a semi tall variety suitable for tender nut.
Malayan Green Dwarf	This is an early flowering cultivar and takes about 3-4 years for initial flowering.
Malayan Orange Dwarf	It is suitable for tender nut. This is an early flowering cultivar and takes about 3-4 years for initial flowering.
Tiptur Tall	It is a tall variety suitable for production of copra and tender nut.
West Coast Tall	It is a tall variety suitable for production of copra and tender nut.
Chandra Kalpa	It is relatively drought tolerant compared to the other varieties and hybrids.
Chowghat Orange Dwarf	It is a best cultivar for use as tender nut. This is an early flowering cultivar and takes about 3-4 years for initial flowering.
Kalpa Dhenu	It is a regular bearer, relatively tolerant to drought. The palms are tall, robust and bear large, green nuts.
Kalpa Mitra	It is a regular bearer, relatively tolerant to drought and produces large, yellowish green, oval shaped nuts.
Kalpa Pratibha	It is suitable for tender nut and copra. The variety is a regular bearer, relatively tolerant to drought.
Kalparaksha	A high yielding coconut variety with field resistance to coconut root (wilt) disease.
Kalpasree	It is mainly for tender nut. This variety was developed by selection from Chowghat Green Dwarf population in the root (wilt) disease hotspot regions of Kerala.

Name of variety	Salient features
Kalpatharu	It is a selection from Tiptur Tall accession, suitable for ball copra production.
Kera Chandra	A high yielding variety. This variety gives an estimated yield of 3.81 tons copra/ha, 2.52 tons oil/ha.
ALR(CN)-1	This is a selection from Arasampatti Tall. The average yield 125 nuts/tree.
VPM-3	This is a selection from Andaman Ordinary Tall.
Kerasagara	This is a selection from Seychelles (SE Asia).
Banawali	This is a pure line selection. The average yield 120 nuts/tree. Maharashtra.
Pratap	This is a selection from Banawali. The average yield 150 nuts/tree.
Indira Nariyal-1	The average yield 85 nuts/tree. The copra yield 2 t/ha.
Gauthami Ganga	The average yield is 90 nuts/tree. The average yield of copra 157g/nut having oil content 68%.
Kera Bastar	The average yield 110 nuts/tree. The average yield of copra 180 g/nut having oil content of 68.7%.
Kera Keralam	The average yield 147 nuts/tree. The average yield of copra 164 g/nut having oil content of 64%.
Kalyani Coconut-1	The average yield 180 nuts/tree. The yield of copra 154 g/nut having oil content of 66%.

Hybrids

Hybrid	Parents	Salient features
Kalpa Samrudhi	MYD x WCT	This hybrid has been found to be relatively drought tolerant compared to other released hybrids.
Kalpa Sankara	CGD x WCT	This coconut hybrid has tolerance to (wilt) disease.
VHC 2	East Coast Tall × Malayan Green Dwarf	The average yield 142 nuts/tree. The yield of copra 152 g/nut having 70% oil content.
VHC 3	East Coast Tall × Malayan Orange Dwarf	The average yield 156 nuts/tree. The average of copra 162 g/nut having 70% oil content.
VHC1	East Coast Tall × Malayan Green Dwarf	The average yield 115 nuts/tree. The yield of copra 142 g/nut having 69% oil content.
Godavari Ganga	East Coast Tall × Gangabondam	The average yield is 140nuts/tree. Average yield of copra 153 g/nut having oil content 67%.

Contd.

Hybrid	Parents	Salient features
Anandaganga	Andaman Ordinary × Gangabondam	The average yield 95 nuts/tree. The yield of copra 216 g/nut having oil content of 68%.
Keraganga	West Coast Tall × Gangabondam	The average yield 100 nuts/tree. The yield of copra 201 g/nut having oil content of 69%.
Kerasowbhagya	West Coast Tall × SSA	The average yield 116 nuts/tree. The yield of copra 196 g/nut having oil content of 65%.
Kerasree	West Coast Tall × Malayan Yellow Dwarf	The average yield 130 nuts/tree. The yield of copra 216 g/nut having oil content of 66%.

Oil palm

Botanical name	*Elaeis guineensis* Jacq.
Family	Arecaceae
Origin	Central and Western Africa
Chromosome number	2n=32
Ploidy level	Diploid
Type of inflorescence	Oil palms are monoecious, producing male and female inflorescences in leaf axils. The inflorescence of both sexes is a compound spadix with 100-200 branches, initially enclosed in a spathe or bract that splits 2 weeks prior to anthesis.
Pollination system	Cross pollination, Oil palms are primarily insect pollinated by various insects: in Africa, weevils *(Elaeidobius* spp), in Latin America, *Mystrops costaricensis* and *Elaeidobius* spp.
Bearing habit	Axile of leaf
Type of fruit	Drupe
Edible portion	Oil palm is the high oil content of the mesocarp and kernel of the fruits
Type of Sex expression	Monoecious
Leading country in production	Malaysia

- ***Elaeis guineensis*** is a species of palm commonly called **African oil palm** or **macaw-fat**.
- It is the principal source of palm oil. It is native to central and western Africa, specifically the area between Angola and the Gambia.
- Mature palms are single-stemmed and grow to 20 m tall.
- The leaves are pinnate and reach between 3-5 m long.

- The flowers are produced in dense clusters; each individual flower is small, with three sepals and three petals.
- The palm fruit takes five to six months to mature from pollination to maturity.
- It is reddish, about the size of a large plum, and grows in large bunches. Each fruit is made up of an oily, fleshy outer layer (the pericarp), with a single seed (the palm kernel), also rich in oil. When ripe, each bunch of fruit weighs 40–50 kg (88–110 lb).
- Oil is extracted from both the pulp of the fruit (palm oil, an edible oil) and the kernel (palm kernel oil, used in foods and for soap manufacture). For every 100 kg of fruit bunches, typically 22 kg of palm oil and 1.6 kg of palm kernel oil can be extracted.
- The oil palm originated in West Africa, but has since been planted successfully in tropical regions within 20 degrees of the equator. In the Republic of the Congo, or Congo Brazzaville, precisely in the Northern part, not far from Ouesso, local people produce this oil by hand.
- In 1995, Malaysia was the world's largest producer, with a 51% of world share, but since 2007, Indonesia has been the world's largest producer, supplying approximately 50% of world palm oil volume.
- Oil palm is a monocotyledonous tree belonging to the Family Arecacea, tribe *Cocoseae,* and subtribe *Elaeidinae*.
- This subtribe consists of the genus *Elaeis*. The genus name, *Elaeis*, is from the Greek 'elaia' for olive.
- The major interest in oil palm is the high oil content of the mesocarp and kernel of the fruits that are produced in heavy bunches.
- The plant is monoecious with out-crossing maintained through successive cycles of male and female inflorescence production on a single palm.
- The length of the male and female cycles varies widely according to genotype and environment.
- At maturity, the fruit is red-brown and consists of the pulp, the shell, and the kernel.
- The pulp yields edible, orange-red oil commonly known as palm oil.
- The endosperm or kernel when crushed produces clear yellowish oil that is known as palm kernel oil and is similar to coconut oil.
- Palm oil and its refined derivatives, palm olein and palm stearin, are the major commercial products of oil palm.

- The plant generation time is long, with seeds taking around 100–120 days to germinate, including a heat-treatment, followed by 10–12 months in the nursery before the young seedlings are ready for field planting.
- The oil palm starts to bear fruit after 2–3 years of field planting and approaches maturity at around 10 years.
- The economic life of plantings varies from 20–30 years, depending on local conditions.
- **National Research Centre for Oil Palm** was established by the Indian Council of Agricultural Research at Pedavegi in West Godavari district of Andhra Pradesh on February 19, 1995.
- In February, 2009 the institute was upgraded as **Directorate of Oil Palm Research (DOPR)** to cater to the location specific programmes across the country.

Biotechnology of Oil Palm

- The potential of oil palm biotechnology has been reviewed a number of times and one area where considerable progress has been made is clonal propagation.
- As oil palm has only a single vegetative meristem, it is not possible to replicate elite genotypes by cuttings or graftings.
- The absence of fully inbred lines means that it is also not possible to commercially propagate elite individual oil palm genotypes by seed.
- These two constraints led to the development of a clonal propagation system based on somatic embryogenesis.
- The development of suspension culture systems for the production of somatic embryos and of cryopreservation techniques for their storage also offers the possibility of mass production of artificial seed.

Production technology

Climate	Temperature is 210 C to 320C, annual rainfall - 200 cm and relative humidity - 75 – 100 %. Altitude – 450 -900 m above MSL.
Soil	Moist deep, loamy soils, rich in humus with good water permeability are suitable. Soil pH - 4 – 6.
Propagation	Through seeds.
Spacing	9m x 9m x 9m (Triangular method of planting).
Pit size	A pit size of 60cm x 60cm x 60cm is normally used for planting Oil Palm seedlings.

Method of planting	Take the pits prior to planting and allow to season. Transport quality seedlings to the planting site only at the time of planting. Apply 230 g single super phosphate or 250 g of rock phosphate and 50 g Phorate and mix with the soil at the base of the pit. Remove the bag by making a longitudinal cut and place the seedling in the pit along with the soil.
Planting time	Planting can be done in any season. Best period is June to December i.e., during monsoon. In case of planting during summer season adequate irrigation, mulching and growing cover crops like sunhemp in the basin will help to overcome the summer temperatures.
Plant Population	57 plants per acre or 143 plants per hectare.
Manuring and fertilization	Oil Palm is a gross feeder and demands a balanced and adequate supply macro, secondary and micronutrients for growth and yield. The following fertilizer schedule is considered satisfactory for oil palm under rainfed conditions the same is recommended in irrigated conditions also, but efforts must be made to modify this dose based on leaf nutrient analysis.
Irrigation	Oil Palm requires sufficient irrigation, as it is a fast growing crop with high productivity and biomass production. Do not grow Oil Palm if assured and adequate irrigation facility is not available. Insufficient irrigation lead to reduction in the rate of leaf production, affects the sex ratio (more male flower production) and results in inflorescence abortion and yield reduction.
Method of irrigation	Basin Method: When irrigation water is not a constraint, basin irrigation can be taken up. Required quantity of water can be given at weekly intervals. Prepare irrigation channels in such a way that the individual palms are connected separately by sub-channels. But don't form channels along palm rows.
Weeding	Take up regular manual weeding or use recommended herbicides. Use preferably contact herbicides. Among systemic herbicides Glyphosate @ 750 ml/ha/ year or 17.5 ml/basin is recommended for effective weed control. Mulching, growing cover crops and intercrops minimizes weed growth. Do not use 2, 4-D; 2, 4, 5-T; halogenated aliphatic acids and Dalapon. These compounds produce abnormalities in oil palm. See that the spray fluid does not spill on the oil palm plants.
Intercropping	Oil palm is a wide spaced perennial crop with a long juvenile period of 3 years. Hence lot of horizontal and vertical space both above and below ground is available. This inter space during the first three years can be used to generate income during the juvenile phase of the crop. Crop selected for inter-cropping should be compatible with the main crop and should not compete with oil palm for light, water and nutrients.
Flowering	Oil palm comes to flowering 14-18 months after planting. It produces both male and female inflorescence separately on the same palm (Monoecious). Male and female phases occur naturally in consequent cycles in a palm.
Pollination	Oil palm is a highly cross-pollinated crop. Wind and insects assist pollination, but wind pollination is not adequate. Effective pollinating insects like ***Elaeidobius kamerunicus*** helps in good pollination and fruit set. Release of this weevil after 2 and 1/2 year of planting is advisable. If the palms are not having good girth and vigour release the weevils after 3 years.

Mulching	Mulching of oil palm basins is essential to conserve moisture as well as to control weeds. Mulching can be done with dried oil palm leaves, male inflorescence, coconut husk, empty oil palm bunches brought from factory.
Cover crops	Common cover crops that can establish well in oil palm plantations are *Pueraria phaseoloieds*,*Calapagonium mucunoides*, *Centrosema pubescens, Mimosa invisa, Mucuna sp*.etc., These crops can be sown after taking intercrops for first three years. Cover crops are sown in the entire field leaving the basin. In pure plantations these cover crops can be sown at the time of planting of oil palm seedlings.
Ablation	Removal of all inflorescences during the initial three years to gain adequate stem girth, vigour and root system. Ablation is done at monthly intervals by pulling out the young inflorescence.
Harvesting	Harvesting should be aimed at recovering the whole harvestable produce without loss of oil or loss of fruit. Proper and timely harvesting of fruit bunches is an important operation which determines the quality and quantity of oil to a great extent.
Harvest index	When the bunch is matured and ready for harvesting a) fruits in the bunch turn yellowish orange, b) 5-10 fruits from each bunch drop on their own, c) when fruit is pressed hard with the fingers, orange yellow coloured oil exudates from the fruits.
Yield	Yield of FFB is 20-25tonnes/ha which yields 4-5 tonnes of oil/ha/annum.
By products	Palm oil can be used in preparation of various edible and non-edible products i.e. vanaspati, biscuits, ice creams, margarine, cooking fats, soaps, detergents, shampoo, etc., Theraputic drugs/products like carotenoides, vitamin E capsules are also produced from palm oil as a by product.

Hybrid

Tenera is the ruling hybrid grown all over the world. It is a cross between thick-shelled Dura and shell less Pisifera. Tenera has a thin shell, medium to high mesocarp content which in turn gives higher oil yield.

Palmyrah

Botanical name	*Borassus flabellifer* L.
Family	Arecaceae
Origin	Tropical Africa
Chromosome number	X=8 or X=9, n=18
Ploidy level	Allotetraploid
Type of inflorescence	Spadix
Bearing habit	The male and female flowers are held by two different trees, never in one tree. Both male and female trees produce spikes of flowers but only the female plant bears fruits.
Pollination system	Cross pollination

Type of fruit	Fleshy drupe
Type of Sex expression	Dioecious
Value added product	Toddy obtained by tapping the tip of the inflorescence

- The palmyrah palm belongs to the family Palmae, which is a very ancient family of trees, perhaps older than any other flowering plants.
- ***Borassus*** (**Palmyra Palm**) is a genus of six species of fan palms, native to tropical regions of Africa.
- They are tall palms, capable of growing up to 30 m high (98 ft).
- The leaves are long, fan-shaped, 2 to 3 m in length. The flowers are small, in densely clustered spikes, followed by large, brown, roundish fruits.
- *Borassus aethiopum* - African Palmyra Palm (and other names) (tropical Africa)
- *Borassus akeassii* - Ake Assi's Palmyra Palm (West Africa)
- *Borassus flabellifer* - Asian Palmyra palm/Lontar palm/Doub palm/Sea Coconut (southern Asia and southeast Asia)
- *Borassus heineanus* - New Guinea Palmyra Palm (New Guinea)
- *Borassus madagascariensis* - Madagascar Palmyra Palm (Madagascar)
- *Borassus sambiranensis* - Sambirano Palmyra Palm (Madagascar)
- There are two kinds of the palmyrah - the male and the female.
- The male and female flowers are held by two different trees, never in one tree.
- Both male and female trees produce spikes of flowers but only the female plant bears fruits.
- The flowers are small and appear in densely clustered spikes, developing into large, brown, roundish fruits.
- The male flowers are smaller than the female flowers.
- Each female palm may bear 6-12 bunches of about 50 fruits per year.
- The chief product of the palmyrah palm is the sweet sap (toddy) obtained by tapping the tip of the inflorescence.
- The palmyrah is believed to be a native of tropical Africa and grows extensively in the drier parts of India, Sri Lanka, Burma, Thailand, Vietnam, Malaysia and in some parts of Indonesia.

Production technology

Climate	Palmyrah is a tropical crop.
Soil	Arid, deep sandy & loamy soils are suitable. Dry areas with low rainfall are also ideal.
Propagation	Through seeds. Germination of 63.25 % is achieved 6 weeks after sowing.
Varieties	SVPR – 1
Spacing	3m x 3m (1110 palms/ha)
Pruning	Pruning of 30 per cent leaves @ 10 per cent in phased manner.
Growth	Palmyrah is a slow grower. First frond appears in about 5 months. First fan shaped tree leaves appears only in the 2nd year. When it attains the height of 12–18 m, comes to flowering (13–15 yrs) for padaneer (Sweet sap) purpose.
Yield	Average of 100 – 200 lit obtained for a period of 4 months from Feb – May. Padaneer and fruit yield are highly variable in individual palms. 10-12 lit Neer/plant.
Tapping	Extraction of sap (Neera/ Padaneer) from inflorescence is called tapping. According to sex of the palm & age of the inflorescence, different kinds of tappings are available.
Aripanai	In male palm sheath covering the young inflorescence is removed and dried for 3 weeks. End is cut every time & pot is tied (1- 1 ½ months).
Vallupanai	One month old spikes are selected. Each male spike bearing sessile flowers is pre-treated by pressing and stroking & 3 - 6 such spikes are brought together, wrapped with leaves and fitted to a pot.
Thattupalai	It has to be done in female palm to soften the tissue by hitting the inflorescence main axis with iron rod.
Kaivetty	Employed when the inflorescence is about 2 – 3 months old. Female palms are tapped for a longer period. i.e., April- Dec. and male for Dec – Feb. Sap is collected twice a day. Each time at the end of the collection of sap, a new cut surface is made by thin slicing. Tapped sap is called Neera or padaneer.

Rubber

Botanical name	*Hevea brasiliensis* Muell-Arg.
Family	Euphorbiaceae
Origin	Brazil
Chromosome number	X=18, 2n= 36
Type of inflorescence	Panicle
Sex expression	Monoecious
Pollination system	Allogamous (cross pollinated)
Pollinators	Pollination is primarily through insects, specifically midges and thrips.
Type of fruit	Capsule

Rubber Research Institute is situated at	Kottayam, Kerala
Leading country in production	Thailand
Highest productivity in the world	India
Leading state in production	Kerala

- ***Hevea brasiliensis***, the **rubber tree**, is a tree belonging to the family **Euphorbiaceae.**
- *Hevea brasiliensis* is the source of virtually all the world's rubber production.
- Many plant species produce natural rubber. Considerations of quality and economics, however, limit the source of natural rubber to one species, namely *Hevea brasiliensis*.
- It is the most economically important member of the genus *Hevea.*
- It is of major economic importance because the milky latex extracted from the tree is the primary source of natural rubber.
- The white or yellow latex occurs in latex vessels in the bark, mostly outside the phloem.
- These vessels spiral up the tree in a right-handed helix which forms an angle of about 30 degrees with the horizontal, and can grow as high as 45 ft.
- Harvesters make incisions across the latex vessels, just deep enough to tap the vessels without harming the tree's growth, and the latex is collected in small buckets.
- This process is known as rubber tapping. Latex production is highly variable from tree to tree and across clone types.
- *Hevea brasiliensis,* also known as the Para rubber tree after the Brazilian port of Para, is a quick growing, fairly sturdy, perennial tree of a height of 25 to 30 metres.
- It has a straight trunk and thick, somewhat soft, light brownish grey bark.
- The young plant shows characteristic growth pattern of alternating period of rapid elongation and consolidated development.
- The leaves are trifoliate with long stalks.
- The tree is deciduous in habit and winters from December to February in India. Refoliation is quick and copious flowering follows.

- The inflorescence is a **panicle** of separate staminate and pistillate flowers borne in the axils of basal leaves of new shoots that grow out after wintering.
- The flowers are small with no petals, bright or cream-yellow in colour and extremely pungent. They are either male or female but both are found in the same inflorescence (monoecious). Female flowers are apical, the more numerous male flowers lateral in the inflorescence.
- The fruit is an exploding 3-lobed, 3-seeded ellipsoidal capsule. Fruits burst open when ripe and the seeds are scattered up to 15 m from tree. Seeds are variable in size, 2.5-3 cm long, mottled brown, lustrous.
- The rubber tree may live for a hundred years or even more. But its economic life period in plantations, on general considerations is, only around 32 years – 7 years of immature phase and 25 years of productive phase.
- The Rubber Board is a statutory body constituted by the Government of India, under the Rubber Act 1947, for the overall development of the rubber industry in the country.
- The Rubber Research Institute of India (RRII) established in 1955 is located on a hillock in the eastern suburb of Kottayam, Kerala.
- Rubber plantations are predominated by small holdings of not more than 20 Ha of extent numbering over 11.6 lakhs and they share about 90% of total rubber area and 93% of total production during 2009-10. Average size of a small holding is around 0.53 ha.
- Rubber cultivation in India was started on a commercial scale in 1902.
- The area under rubber at the end of 2009-10 was 6.87 lakh ha., production 8.31 lakh tonnes and yield per ha 1775 kg.
- Production of rubber during 2010-11 was 861,950 tonnes. 90% of production is contributed by Kerala State.
- The average yield per ha of rubber plantation in our country was 1807 during 2010-11.
- India is a producer as well as consumer of natural rubber. The consumption of rubber during 2010-11 was 947,715 tonnes.
- The pattern of use of natural and synthetic rubber in India was in the ratio 70:30 during 2010-11, while the global pattern was 44:56.
- About 90% of rubber goods produced is for domestic use and the balance for export purpose. There are about 4528 rubber goods manufacturing units in India during 2009-10.

- The quantity of import of natural rubber during 2010-11 was 177,637 tonnes, whereas the export was 29,851 tonnes during the year.
- Rubber tree takes about 7 years to start giving yield. Although the crop is harvested throughout the year, the period September-January is peak production season and June-July and February-March are lean production season.

Production technology

Climate	Humid tropical climate prevails in the rubber-growing tract. Average annual rainfall in the tract varies from about 2000-4500 mm, temperature 21-35°C.
Soil	Soil in the rubber tract is generally highly weathered and consists mostly of laterite, lateritic types.
Tapping	It is the process for commercially collecting the latex from bark of trees. Tapping is done early hours of morning when the turgor pressure is maximum.
Propagation	By bud grafting.
Planting	4.5X4.5m.
Planting time	During starting of Monsoon.
Manure and fertilizers	For matured rubber trees under tapping apply NPK 10:10:10 grade mixtures at the rate of 900 g/tree (300 kg/ha) every year in two split doses. Add 10 kg commercial Magnesium sulphate for every 100 kg of the above mixture if there is magnesium deficiency.
Harvesting	• **Tapping:** Latex produced in the bark tissue is harvested through a process called tapping. • It usually takes 7 years for a rubber plant to attain tappable girth· Tappable girth- tree girth 50 cm at a height of 125 cm from the bud union.
Yield	In South India, the annual yield of rubber is 375 kg/ha from seedlings trees, whereas budded plants yield 800 - 1000 kg/ha.
Tapping	
S/2 d/2	(half spiral, alternate day for 6 months and rested for 3 months) 100% Intensity.
S /2 d/2 6m /9	(half spiral, alternate day for 6 months and rested for 3 months) 67 intensity.
S /2d/3	(half spiral, third day) 67 intensity.
S/2 d/31m/2	(half spiral, daily for one month and rested for next month) 100% intensity.
S /1 d/4	Full spiral, fourth day 100% intensity.
V /2 d/2 12m/16	Half circumstances and cut alternate day for 12 months and rested for the next 4 months 75% intensity.

Ethrel treatment

Ethrel is recommended to increase latex yield of trees. It is applied at 5% a.i. concentration with a brush below the tapping cut to a width of 5 cm after light scraping of the outer bark. The first application may be done after a drought period preferably after a few pre-monsoon showers and subsequent applications may be done in September and November. However, continuous application of Ethrel is not recommended for periods of more than 3 years at a stretch.

Clones

The planting materials approved by the Rubber Board are classified in to three categories I, II and III.

Multiclone planting recommendations - Traditional Region

Category I	RRII 105, PB 260, RRII 414 and RRII 430, RRII 417 and RRII 422
Category II	RRIM 600, GT 1, RRII 5, PB 28/59, PB 217, RRII 255, RRII 203, PB 280, PB 312 and PB 314
Category III	RRII 50, RRII 51, RRII 52, RRII 118, RRII 176, RRII 208, RRII 300, RRII 429, PR 107, PR 255, PR 261, PB 86, PB 5/51, PB 235, PB 311, PB 330, RRIM 605, RRIM 701, RRIM 703, RRIM 712, RRIC 100, RRIC 102, RRIC 130, KRS 163, IRCA 111, IRCA 130, SCATC 88-13, SCATC 93-114, Haiken 1, BPM 24 and Polyclonal seeds

RRII 105

- It is the highest yielding in the world.
- It has become very popular, occupying 80% of the area under rubber.
- Its average yield is 2400kg/ha/y.

Multiclone planting recommendations - North Eastern Region

Category I	RRIM 600
Category II	RRII 105, GT1, PB 235, RRII 203, RRII 208, RRII 429 RRII 417, and RRII 430
Category III	RRII 5, RRII 422, RRII 118, PB 260, PB 310, PB 311, RRIM 703, SCATC 88/13, SCATC 93/114 and Haiken I and Polyclonal seeds

Advisory on Clones suited to Karnataka and Konkan tract

Results of long term clone evaluations undertaken at HBSS, Nettana and RRS, Dapchari have helped in identifying clones/ planting material suited for cultivation in the South Kanara – Konkan belt along the western part of the Indian peninsula.

This region has seen a lot of rubber planting activity in recent years, by way of availability of land for expansion, it's soil suitability for the crop as well as the availability of the SW and NE monsoon rains.

Planting materials suitable for Karnataka and South Konkan

GT-1, RRII 203*, PB 260, PB 217*, RRII 414, RRII 430, RRII 422, RRII 429, PB 280, PB 312, PB 314, PB 235, RRII 5, RRII 300, PB 311* and RRII 105* have exhibited good growth and yield in the region. Polyclonal seedlings are advisable for planting in areas where proper management is not possible.

*RRII 105 is highly susceptible to Corynespora leaf disease. Hence recommended prophylactic and control measures are to be adopted. PB 217 and RRII 203 are also susceptible to the disease and require prophylactic protection. PB 311 is wind susceptible, hence planting of this clone in wind prone areas is to be avoided.

Planting materials suitable for North Konkan

RRIM 600, RRII 208, RRII 105, RRII 6, RRII 5, PB 260, PR 255 and RRIC 100 perform well in this region. The RRII 400 series clones have not been evaluated in North Konkan. However, the drought tolerant and stable clone RRII 430 shows good initial establishment and growth and could be planted in the region. Life saving irrigation is to be given to all clonal plantings in the first three years. For cultivation under rainfed/ unirrigated conditions, polyclonal seedlings are the best choice.

Clones Evolved and Released from RRII

Clone	Parentage	Status
RRII 105	Tjir 1 x Gl 1	Category I
RRII 414	RRII 105 X RRIC 100	Category I
RRII 417	RRII 105 x RRIC 100	Category II
RRII 422	RRII 105 X RRIC 100	Category II
RRII 430	RRII 105 x RRIC 100	Category I
RRII 5	Primary clone	Category II
RRII 203	PB 86 x Mil 3/2	Category II
RRII 50	Primary clone	Category III
RRII 51	Primary clone	Category III
RRII 52	Primary clone	Category III
RRII 118	Mil 3/2 x Hil 28	Category III
RRII 176	Mil 3/2 x PB 5/60	Category III
RRII 208	Mil 3/2 x AVROS 255	Category III
RRII 300	Tjir 1 x PR 107	Category III
RRII 429	RRII 105 X RRIC 100	Category III

Betelvine

Botanical name	*Piper betle* L.
Family	Piperaceae
Origin	Malaysia
Chromosome number	X=13, 2n=26 and 52
Ploidy level	Polyploid
Type of Sex expression	Dioecious crop
Nature of crop	Betelvine is a shade-loving perennial climber

- Betelvine (*Piper betle* L.) is a perennial, dioecious, evergreen creeper grown in many parts of the country.
- The plant belongs to the Piperaceae family, which consists about 10 genera and over 1,000 species of herbs, shrubs and climbers.
- The most probable place of the origin of betelvine is Eastern Malaysia.
- Basic chromosome number of the species is X=13. Samuel and Bavappa (1981) reported chromosome numbers as 2n=26 and 52.
- It is a dioecious crop and both male and female clones are cultivated based on local preference.
- The plant is a shade-loving perennial climber with dioecy where distinction between male and female plants in the vegetative state is not easy and gender distinction can be made only after flowering.
- The green, heart shaped leaves of this plant are masticatory and popularly known as *paan* in India.
- Fresh leaves are chewed with condiments such as arecanut, *kattha*, clove, cardamom, fennel and candied rose.
- The leaves are also widely used in the Indian system of medicine and health traditions.
- It was reported that fresh leaves contains: moisture 85.4, protein 3.1, fat 0.8, carbohydrate 6.1, fibre 2.3, calcium 230mg, phosphorous 40mg, iron 7mg, ionisable iron 3.5mg, iodine 3.4 μg, carotene (as vitamin A) 9600 IU, Thiamine 70 μg , riboflavin 30 μg , nicotininc acid 0.7mg and vitamin C 5mg/100g.
- Steam distillation of leaves gives essential oil (0.7-2.6%), the amount and nature of which is the important factor determining the aromatic value of the leaf. The oil consists of phenols and terpenes.
- A phenolic compound, hydroxyl-chevicol from the leaf was reported to possess anticarcinogenic property.

- Several attributes of the leaves such as 'digestive', 'carminative', 'stimulant', 'antiseptic' and 'antifungal' activities have been described.
- Roots are known for female contraceptive effects. They have a positive stimulatory influence on intestinal digestive enzymes, especially lipase, amylase and disaccharides.

Production technology

Climate	Betelvine require a cool shade, considerable humidity and regular supply of moisture in the soil.
Soil	Well drained fertile clay loams are suitable. It does not tolerate saline and alkaline conditions.
Varieties	Karpurakodi, Kallarkodi, Revesi, Karpuri, Vellai kodi, Patcha vethalai, SG 1 and SG (BV) – 2.
Propagation	The vines are propagated by terminal stem cuttings or setts about 30 – 45 cm long. Setts obtained from the top portions of the vines are easy to root and hence best for planting. Number of setts 1,00,000/ha. Setts with vigorous apical buds and nodal adventitious roots are selected and planted at the base of the live supports, which are to be planted 4 to 5 months earlier.
Field preparation	The field is prepared to a fine tilth and beds of 2 m wide are formed to a convenient length. Provide drainage trenches of 0.5 m width by 0.5 m depth in between two adjoining beds.
Planting	Plant the seeds of the live supports *i.e.* Agathi (*Sesbania grandiflora*) in long rows. About 750 banana suckers are planted at the edges of the beds, which are used, for tying the vines on the live support and for packing the betel leaf. When the Agathi supports grow to a height about 4 m they are topped. The crop is planted in two rows in beds of 180 cm width on Agathi plants with a spacing of 45 cm between plants in the row.
Season	November - December and January – February.
Irrigation	Irrigate the field immediately after planting and afterwards once in a week.
Manuring	Apply 150 kg N/ha/year through Neem cake (75 kg N) and Urea (75 kg N) and 100 kg P2O5 through Super phosphate and 30 kg Muriate of potash in three split doses first at 15 days after lifting the vines and second and third dose at 40 - 45 days intervals. Apply on beds shade dried Neem leaf or *Calotrophis* leaves at 2 t/ha and cover it with mud (2 t in 2 split doses).
Training of the vines	Training is done by fixing the vine at intervals of 15 to 20 cm along the standards loosely with the help of banana fibre. Training is done every 15 - 20 days depending upon the growth of vines.
Lowering of vines	Under normal cultivation, the vines grow to height of 3 m in one year period. When they reach this height their vigour to produce normal size leaf are reduced and they need rejuvenation by lowering during March - April. After the vine is lowered a number of tillers spring up

	from the nodes at the bends of the coiled vines at the ground level and produce many primary vines. After each lowering, irrigation should be given.
Harvesting	It depends upon the growth of the vines and market condition. Once harvesting starts it continues almost every day.
Yield	75 to 100 lakh leaves/ha/year.

Coffee

Particulars	Arabica	Robusta
Botanical name	*Coffea arabica*	*Coffea canephora*
Family	Rubiaceae	Rubiaceae
Origin	Ethiopia and central Africa	Western and central Subsaharan Africa, from Guinea to the Uganda and southern Sudan
Chromosome number	x=11, 2n=44 (Tetraploid)	x=11, 2n= 22 (diploid)
Ploidy level	Allo-tetraploid	Diploid
Pollination system	Self pollinated, self fertile	Cross pollinated, self sterile
Type of fruit	Pea Berry	Pea Berry
Caffeine content of beans	0.8-1.4%	1.7-4.0%
Soils	Deep, fertile, rich in organic matter, well drained and slightly acidic (pH 6.0-6.5)	Same as Arabica
Slopes	Gentle to moderate slopes	Gentle slopes to fairly level fields
Elevation	1000-1500m	500-1000m
Aspect	North, East and North- East aspects	Same as Arabica
Temperature	15 °C – 25 °C ; cool, equable	20 °C – 30 °C; hot, humid
Relative humidity	70-80%	80-90%
Annual rainfall	1600-2500 mm	1000-2000 mm
Blossom showers	March- April (25-40mm)	February – March (25-40 mm)
Backing showers	April-May (50-75 mm) well distributed	March-April (50-75 mm) well distributed
Propagation	By seeds	By cuttings

Area, production and productivity

Leading country in production	Brazil
Leading state in production	Karnataka
Leading state in area	Karnataka

- While the international coffee trade is concerned with only two coffee species - *Coffea arabica* and *C. canephora* - botanists regard as coffee trees all tropical plants of the Rubiaceae family, which produce seed resembling coffee beans.
- Coffee is a brewed beverage prepared from the roasted or baked seeds of several species of an evergreen shrub of the genus *Coffea*.
- The two most common sources of coffee beans are the highly regarded *Coffea arabica*, and the "robusta" form of the hardier *Coffea canephora*.
- The latter is resistant to the coffee leaf rust (*Hemileia vastatrix*), but has a more bitter taste.
- The basic chromosome number of the genus *Coffea* is x = 11.
- All the species of *Coffea* are diploid (2n = 22) except *Coffea arabica*, which is a tetraploid (2n = 44).
- All diploid species of coffee are self incompatible.
- *Coffea arabica* is the only self compatible species. In Coffea arabica the chromosomes present themselves as small rods, smaller than those of any of the diploid species (Mendes, 1959).
- Interspecific hybrids of *Coffea Arabica x Coffea canephora* have been reported to be diploids (Sybenga, 1960) or triploids (Mendes, 1959).
- Coffee is a short day plant. Flower buds are produced at the axils of mature green wood on short stalks, which are known as peduncles. In arabica, 4-5 inflorescence of 1-4 flowers each are produced per axil, while in robusta more number of flowers per inflorescence are commonly produced (Anonymous, 2000).
- Inflorescence is a condensed dichasial cymose fascicle subtended by bracts (Gamble, 1921).Coffee plants are cultivated in more than 70 countries, primarily in equatorial Latin America, Southeast Asia, and Africa.
- Several species of shrub of the genus *Coffea* produce the berries from which coffee is extracted. The two main species commercially cultivated are *Coffea canephora* (predominantly a form known as 'robusta') and *C. arabica*.
- *C. arabica*, the most highly regarded species, is native to the southwestern highlands of Ethiopia and the Boma Plateau in southeastern Sudan and possibly Mount Marsabit in northern Kenya.

- *C. canephora* is native to western and central Subsaharan Africa, from Guinea to the Uganda and southern Sudan.
- Less popular species are *C. liberica*, *C. stenophylla*, *C. mauritiana*, and *C. racemosa*.
- Arabica coffee is an allotetraploid inbreeder (2n = 44) and is a small tree with grey to light brown bark. It is a shrub or even bush under training.
- It branches profusely and the leaves are dark green.
- The flowers are white and generally pentamerous. The flowers blossom in 9-10 days after the receipt of summer showers known as blossom showers.
- The fertilized ovary grows into a fruit in about 8-9 months. The berries are 10-12 per node and are oblong to round in shape. Removal of silver skin imparts bluish green tint to the seed.
- Arabica trees come into bearing 3-4 years after planting and are in full bearing by 6-8 years. Good yields are expected for 15-18 years, after which it declines, becoming unprofitable by 20-30 years (Wrigley, 1988).
- Caffeine content in arabica coffee is 0.5% to 1.8%, which is less when compared to robusta coffee (Charrier, 1983).
- *Coffea canephora* is a low land species, cultivated up to 1000 m above MSL in a belt of 10° N and 10° S of equator (Willson, 1999).
- Though it tolerates warmer temperature and high humidity it is more sensitive to cold (Illy and Viani, 1995).
- It is a diploid (2n = 22) species, bigger tree when compared to Arabica with broader and larger pale green leaves.
- Flowers are white, fragrant and generally pentamerous. They are borne in clusters larger than that of arabica.
- Berries are small but higher in number per node varying from 40-60 or more.
- The flowers open on 7th or 8th day after the receipt of blossom showers.
- It is self-sterile and hence cross pollination is the rule.
- The fruits mature in 10-11 months. They are generally ready for harvest two months later than Arabica (Shanmugavelu et al., 2002).
- The species contain more caffeine, ranging from 1.5% to 3.8%, but more neutral in cup quality although known for body (Wellman, 1961).

- *Coffea canephora* var. *robusta* is the cultivated variety of the species. Besides this there are two other varieties of some importance, var. kouillon and var. niaouli.
- *Coffea canephora* var. *robusta* is generally called robusta coffee (Wrigley, 1988).
- Robusta coffee possesses several useful characters like, high tolerance to leaf rust pathogen, white stem borer, nematode invasion and potentiality to give consistent yield.
- Because of these reasons, cost of robusta cultivation is relatively less compared to that of arabica.
- The other hand, inability to endure long drought, late cropping, as well as later stabilization of yields and inferior quality compared to arabica, are some of the negative aspects of robusta coffee (Anonymous, 1996).
- *Coffea arabica* is predominantly self-pollinating, and as a result the seedlings are generally uniform and vary little from their parents.
- In contrast, *Coffea canephora*, and *C. liberica* are self-incompatible and require outcrossing. This means that useful forms and hybrids must be propagated vegetatively.
- Of the two main species grown, arabica coffee (from *C. arabica*) is generally more highly regarded than robusta coffee (from *C. canephora*); robusta tends to be bitter and have less flavour but better body than arabica.
- For these reasons, about three-quarters of coffee cultivated worldwide is *C. arabica*.
- Robusta strains also contain about 40–50% more caffeine than arabica.
- For this reason, it is used as an inexpensive substitute for arabica in many commercial coffee blends. Good quality robusta seeds are used in traditional Italian espresso blends to provide a full-bodied taste and a better foam head (known as *crema*).
- However, *Coffea canephora* is less susceptible to disease than *C. arabica* and can be cultivated in lower altitudes and warmer climates where *C. arabica* will not thrive.
- In 2014 Brazil was the world leader in production of green coffee, followed by Vietnam, Indonesia and Colombia.
- Originally, coffee farming was done in the shade of trees that provided a habitat for many animals and insects. Remnant forest trees were used for this purpose, but many species have been planted as well.

- These include leguminous trees of the genera *Acacia*, *Albizia*, *Cassia*, *Erythrina*, *Gliricidia*, *Inga*, and *Leucaena*, as well as the nitrogen-fixing non-legume sheoaks of the genus *Casuarina*, and the silky oak *Grevillea robusta*.
- Instant coffee is dried into soluble powder or freeze-dried into granules that can be quickly dissolved in hot water.
- Canned coffee has been popular in Asian countries for many years, particularly in China, Japan, South Korea, and Taiwan.
- Brazil remains the largest coffee exporting nation, but Vietnam tripled its exports between 1995 and 1999 and became a major producer of robusta seeds.
- The primary psychoactive chemical in coffee is caffeine, an adenosine antagonist that is known for its stimulant effects.
- Coffee also contains themonoamine oxidase inhibitors â-carboline and harmane, which may contribute to its psychoactivity.
- Central Coffee Research Institute is situated at Balahanur, Karnataka.
- Coffee Board was established by the Government of India through a constitutional act "Coffee Act VII of 1942" under the administrative control of Ministry of Commerce and Industry.
- India is the only country that grows all of its coffee under shade.
- The two main varieties of coffee viz., Arabica and Robusta are grown in India. Arabica is mild coffee, but the beans being more aromatic, it has higher market value compared to Robusta beans. On the other hand Robusta has more strength and is, therefore, used in making various blends.
- Arabica requires more care & nurture and is more suitable for large holdings whereas Robusta is suitable irrespective of size of the farm.
- The harvest of Arabica takes place between November to January, while for Robusta it is December to February.
- Arabica is susceptible to pests & diseases such as White Stem Borer, leaf rust etc., and requires more shade than Robusta.
- Black bean/Black Joloo/ Normal Joloo is a physiological disorder due to depletion of CHOs reserve during bean development.

Plant Breeding

Coffea arabica

- *C. arabica* is a tetraploid (44 chromosomes) and is self-pollinating.
- There are two distinct botanical varieties: *arabica* (typica) and *bourbon.*
- Historically, typica was cultivated in Latin America and Asia, whereas bourbon arrived in South America and, later, East Africa via the French colony of Bourbon (Reunion).
- Because *C. arabica* is self-pollinating, these varieties tended to remain genetically stable.
- However, spontaneous mutations showing desirable characteristics have been cultivated in their own right, as well as being exploited for cross-breeding purposes.
- Some of these mutants and cultivars are described below.

Mutants: Caturra - a compact form of bourbon; Maragogipe - a mutant typica with large beans; San Ramon - a dwarf typica; and Purpurascens - purple leaved forms.

Coffea canephora

- *C. canephora* is diploid and self-sterile, producing many different forms and varieties in the wild.
- The identification of cultivars is confused, but two main forms are recognised:
- 'Robusta' - upright forms
- 'Nganda' - spreading forms

Arabica / Robusta hybrids

- Coffee has been selectively bred to improve characteristics of: growth and flowering, yield, bean size and shape, cup quality, caffeine content, disease resistance and drought resistance.
- Crosses between *Arabica* and *Robusta* aim to improve Arabica by conferring disease resistance and vigour or to improve the cup quality of Robusta.
- Hibrido de Timor is a natural hybrid of *Arabica* x *Robusta* which resembles Arabica coffee and has 44 chromosomes.
- Catimor is a cross between Caturra and Hibrido de Timor and is resistant to coffee leaf rust (*Hemileia vastatrix*).

- A new dwarf hybrid called Ruiru Eleven, developed at the Coffee Research Station at Ruiru in Kenya, was launched in 1985. Ruiru 11 is resistant to coffee berry disease and to coffee leaf rust. It is also high yielding and suitable for planting at twice the normal density.
- Icatu hybrids are the result of repeated backcrossing of interspecific Arabica x Robusta hybrids to Arabica cultivars Mundo Novo and Caturra.
- Arabusta hybrids are fertile interspecific Fl hybrids from crosses between Arabica and induced auto-tetraploid Robusta coffee.

Techniques used in coffee breeding

1. Controlled pollination and multiplication by seed
2. Vegetative (clonal) propagation
    - Traditional methods: grafting, taking cuttings
    - New methods (tissue culture): micropropagation, somatic embryogenesis
    - In recent years the potential of genetic manipulation of Coffea using recombinant DNA technology and tissue culture techniques has been investigated.

By introducing new genes for characteristics such as resistance to pests or to herbicides, or genes coding for desirable cup quality attributes, it may be possible to produce plants with any combination of features required.

Production technology

Spacing	Arabica Coffee: 1.5 to 2.0 m either way Dwarf varieties: Sanraman- 1 x 1 m Robusta coffee: 2.5 m either way
Planting time	Planting spreads from June - December
Irrigation	It is generally grown as a rainfed crop. But irrigation with sprinkler during March - April increases blossoming and results in higher yields.
Harvest	Harvest starts during November and harvesting extends up to February. Coffee fruits should be harvested as and when they become ripe. Coffee is just ripe when on gently squeezing the fruits the beans inside come out easily.
Fly picking	Small scale picking of ripe berries during October to February.
Main picking	Well formed and ripened berries are harvested during December. Bulks of the yields are obtained from this picking.
Stripping	Picking of all the berries left irrespective of ripening.
Cleanings	This is collection of fruits that have been dropped during harvesting. Unripe fruits should be scrupulously sorted out before using the fruits for pulping. They may be dried separately as cherry.
Yield	750 - 1000 kg dry parchment /ha.

Manuring

Species	Pre- Blossom March N:P2O5:K2O	Post – blossom May N:P2O5:K2O	Mid monsoon August N:P2O5:K2O	Post- monsoon October N:P2O5:K2O	Total
Arabica					
Young coffee 1st year after planting	15:10:15	15:10:15	—	15:10:15	45:30:45
2nd and 3rd year	20:10:20	20:10:20	—	20:15:20	60:45:60
4th year	30:20:30	20:20:20	—	30:20:30	80:60:80
Bearing coffee 5 years and above for less than one tonne/ha crop	40:30:40	40:30:40	—	40:30:40	140:90:120
For one tonne / ha and above	40:30:40	40:30:40	40:30:40	40:30:40	160:120:160
Robusta					
For less than one tonne/ ha crop	40:30:40	—	—	40:30:40	80:60:80
For 1 tonne /ha and above	40:30:40	40:30:40	—	40:30:40	120:90:120

Important varieties

Arabica varieties

Sln 795, Sln 7, Sln 9, Sln 10, Cauvery and its selections and HRC (Hawaian Red Cuturra), Chandragiri and san Roman

Robusta varieties

Sln 274, Sln 270, Sln 3

Improved varieties

Varieties	Parentage	Special characters
Selection 1 (S288)	Tetraploid hybrid from S26 (natural cross – *C. liberica* x *C. arabica*)	Resistance to leaf rust race 1 & 2, high yielding. Wider adaptability
Selection 3 (S795)	Cross between S.288 x Kent	Resistance to leaf rust race 1 and 2, 700 -1200kg/haBold fruits 75% A grade
Selection 5	Devamachy x S -881	Small, oblong fruits 900 – 1100kg/ ha
Selection 6 z	S-274 (Robusta) x Kents	High A grade beans, 900 – 1000kg/ ha
Selection 7	San Ramon –a short internode arabica	Dwarf in stature, segregates to tall by 30%
Selection 8	*Pure line selection of Hibrido –de -Timor*	*Highest resistance to leaf rust, drooping branches*
Selection 9	Sln 8x Tafarikela	Drought hardy, suitable to different coffee zones
Selection 10	Double cross hybrid, Cattura x (S795 XHDT)	Resistant to leaf rust
Selection 11	Progeny of *C.liberica* X *C..eugenoides*	Field resistance to rust and drought hardiness
Cauvery	Catimor x Hibrido de Timor	Plants are dwarf, suitable for high density plantingYield – 3000kg/ haMore a grade beans with superior cup quality
Selection 12	Cross Caturra x HDT, followed by selfing	Precocious, suitable for close planting, resistant to leaf rust

Improved varieties

Variety	Salient features
Kents	Kents is the earliest variety of Arabica, selected by an English planter of the same name during the 1920s
S.795	This is by far the most popular Arabica selection released during the 1940s with high yields, bold beans, superior quality and relative tolerance to leaf rust. This selection was developed using 'Kents' Arabica, known for its high quality
Cauvery	Popularly known as Catimor, Cauvery is a descendant of a cross between 'Caturra' and 'Hybrido-de-Timor'. Caturra is a natural mutant of the famous Bourbon variety. Thus, Cauvery inherited the high yielding and superior quality attributes of Caturra and the resistance of 'Hybrido-de-Timor'
Sln.9	Selection 9 is a derivative of a cross between an Ethiopian Arabica collection, 'Tafarikela', and 'Hybrido-de-Timor'. Sln.9 has inherited all the superior cup quality traits of Tafarikela. This variety has won the Fine Cup Award for best Arabica at the 'Flavour of India - Cupping Competition 2002' organised by Coffee Board of India

Cocoa

Botanical name	*Theobroma cacao* L.
Family	Sterculiaceae
Origin	South America
Chromosome number	2n=20
Ploidy level	Diploid
Type of inflorescence	Cauliflory
Type of Sex expression	Dioecious
Bearing habit	Cauliflorous
Pollination system	Cross pollination
Type of fruit	Drupe
Nutritive value	Fat (37%) and protein (7%)
Leading country in production	Côte d'Ivoire
Leading state in production	Kerala
Leading state in area	Tamil Nadu

- The **cocoa bean**, also **cacao bean** or simply **cocoa** is the dried and fully fermented fatty seed of *Theobroma cacao*, from which cocoa solids and cocoa butter are extracted.
- The cacao tree is native to the South America.
- Cacao trees will grow in a limited geographical zone, of approximately 20 degrees to the north and south of the Equator.
- Nearly 70% of the world crop is grown in West Africa.
- The cacao plant was first given its botanical name by Swedish natural scientist Carl Linnaeus in his original classification of the plant kingdom, who called it *Theobroma* ("food of the gods") *cacao*.

- There are three main varieties of cocoa plant: Forastero, Criollo, and Trinitario.
- The first is the most widely used, comprising 95% of the world production of cocoa.
- Overall, the highest quality cocoa beans come from the **Criollo** variety, which is considered a delicacy.
- Criollo plantations have lower yields than those of **Forastero,** and also tend to be less resistant to several diseases that attack the cocoa plant, hence very few countries still produce it. One of the largest producers of Criollo beans is Venezuela (Chuao and Porcelana).
- Trinitario is a hybrid between Criollo and Forastero varieties. It is considered to be of much higher quality than the latter, but has higher yields and is more resistant to disease than the former.
- In general chocolate and cocoa is considered to be a rich source of antioxidants such as procyanidins and flavanoids, which may impart anti aging properties.
- Chocolate and cocoa also contain a high level of flavonoids, specifically epicatechin, which may have beneficial cardiovascular effects on health.
- Cocoa beans is the primary raw material for confectioneries, beverages, chocolates and other edible products.

Production technology

Climate	The optimum range of temperature of cocoa growing regions is 15°-32°C. Rainfall above 300mm favour incidence of black pod and vascular streak dieback.
Soil	Virgin, freshly cleared forest soils are used for cultivation of cocoa. A depth of up to 1.5m is necessary for cocoa.
Propagation	By seeds, soft wood grafting.
Planting	2.7 X 2.7 m
Pit size	50 cm x 50 cm x 50 cm
Time and method of planting	Cocoa is a shade loving plant. During its seedling period it requires about 50% shade and later the shade requirement is about 40%. The plant is grown as a mixed crop with other plants like spices and rubber mainly under rainfed conditions. Cocoa is planted as an intercrop in coconut and arecanut gardens.
Planting time	Planting should coincide with the onset of monsoon, but in places where irrigation is resorted to, flexibility in the time of planting is possible.

Harvesting and Processing	Pods take about 140 – 160 days to ripen. Generally, cocoa gives two main crops in a year, *i.e.* September – January and June to August. The stage of maturity of the pod is best judged by change of colour of pods. Pods that are green when immature turn yellow when mature and reddish pods turn yellow or orange.
Yield	Cocoa produces flowers from the 3rd year of planting onwards and economic yield starts from 5th year. Properly well maintained (irrigation/ nutrient management) tree yielded 1-2 kg of pods / tree/ year.

Drip (Fertigation)

Fertilizers can be applied through drip irrigation system (fertigation). Weighed quantity of fertilizers as per schedule was dissolved in water and then injected to sub-main through venturi and then to lateral lines as per treatment. Drip irrigation is done once in 2 days. The fertilizers are applied through drip irrigation at weekly intervals. 100:40:140 kg of N, P and K are applied through drip and fertigation in different stages of crop growth as detailed below.

Phases of crop	Months	Weeks	No. of weeks	*Water soluble fertilizers (NPK (%))
1st season flower initiation	January & February	1-8	8	12.5
Fruit set	March	9-13	5	10.0
Fruit development	April and May	14-22	9	12.5
Fruit maturity, harvest and 2nd season flower initiation	June and July	23-30	8	20.0
Fruit set	August	31-35	5	12.5
Fruit development	September & October	36-43	8	15.0
Fruit maturity and harvest	November & December	44-52	9	17.5
Total				100

***Water soluble fertilizers :** All 19(19% N: 19 % P: 19 % K), Mono – Ammonium Phosphate (MAP – 12:61:00 % NPK), Proprietary water Soluble form of N and K fertilizers containing 13% N and 45 % K (Multi 'K') and Urea (46% N)

Improved varieties

Variety	Salient features
Forstero	It is survive under adverse condition. It is a recommended variety by CPCRI, Kasargod, Kerala
Criollo	The best quality cocoa
CCRP-1	It is a clonal selection
CCRP-2	It is a clonal selection

Contd.

Variety	Salient features
CCRP-3	It is a selection from open pollinated seedlings of T76/1224/1201
CCRP-4	It is a clonal selection
CCRP-5	It is a clonal selection
CCRP-6	It is a clonal selection
CCRP-7	It is a clonal selection
VTLCC-1	It is a clonal selection

Hybrids

Hybrid	Salient features
Hybrid-7	It is a recommended by Kerala Agricultural University.
CCRP-8	It is a hybrid between CCRP-1 and CCRP-7. It is high yielding, producing medium sized smooth green pods turning yellow on ripening.
CCRP-9	It is a hybrid between CCRP-3 and CCRP-4. It produces medium sized smooth green pods turning yellow on ripening.
CCRP-10	It is a hybrid between CCRP-3 and GVI 68. It is high yielding, producing medium sized smooth green pods turning yellow on ripening.
VTLCH-1	It is early, heavy bearer, tolerant to water stress. The colour of the pod is red to pink. The average yield 9 q/ha (dry beans).
VTLCH-2	It is early, heavy bearer, black pod disease tolerant. The colour of the pod is yellow. The average yield 8 q/ha (dry beans).
VTLCH-3	It is early, heavy bearer, drought tolerant. The colour of the pod is yellow. The average yield 10 q/ha (dry beans).
VTLCH-4	It is early, heavy bearer, drought tolerant. The colour of the pod is yellow. The average yield 9 q/ha (dry beans).

Tea

Botanical name	*Camellia sinensis* (L.) Kuntze
Family	Theaceae
Origin	China (South East Asia)
Chromosome number	2n=30
Type of inflorescence	Terminal or axillary
Flowers	Flowers are solitary in the axils of the leaves. The flowers are white , middle yellow & 1 ½ inches in diameter and are fragrant
Bearing habit	Flowers are bracteate, pedicellate, hermaphrodite and actinomorphic and born on inflorescences that are terminal or axillary.
Pollination system	Cross pollination
Type of fruit	Capsule
Nature of crop	Calcifiuge crop and can tolerate soils with less amount of calcium but high quantities of potassium and silicon
Leading country in production	China
Leading state in production	Assam
UPASI	United Planters' Association for South India

- ***Camellia sinensis* (L.) Kuntze** is probably one of the most investigated plant for its medicinal and food applications, and its main use is to prepare tea, a very popular beverage made by brewing **leaves and buds.**
- It is of the genus *Camellia*, a genus of flowering plants in the family **Theaceae.**
- White tea, yellow tea, green tea, and black tea are all harvested from this species, but are processed differently to attain different levels of oxidation.
- Kukicha (twig tea) is also harvested from *Camellia sinensis*, but uses twigs and stems rather than leaves.
- There are two major varieties used for tea: **Chinese tea**, *Camellia sinensis* var. *sinensis*, and **Assam tea,** *Camellia sinensis* var. *assamica.*
- *Camellia Senensis* is an evergreen shrub or small tree that is usually trimmed to below 2 m (6.6 ft) when cultivated for its leaves. It has a strong taproot.
- Tea flowers are bracteate, pedicellate, **hermaphrodite** and actinomorphic and born on inflorescences that are terminal or axillary,
- The leaves are 4–15 cm (1.6–5.9 in) long and 2–5 cm (0.79–1.97 in) broad.
- The young, light green leaves are preferably harvested for tea production; they have short white hairs on the underside.
- Older leaves are deeper green. Different leaf ages produce differing tea qualities, since their chemical compositions are different.
- Usually, the tip (bud) and the first two leaves are harvested for processing. This hand picking is repeated every one to two weeks.
- Assam tea comes from the northeastern section of the country. This heavily forested region is home to much wildlife, including the rhinoceros. Tea from here is rich and full-bodied. It was in Assam that the first tea estate was established, in 1837.
- Darjeeling – the Darjeeling region is cool and wet, and tucked in the foothills of the Himalayas. Tea plantations reach 2200 metres. The tea is delicately flavoured, and considered to be one of the finest teas in the world.
- The Darjeeling plantations have 3 distinct harvests, termed 'flushes', and the tea produced from each flush has a unique flavour.

- First (spring) flush teas are light and aromatic, while the second (summer) flush produces tea with a bit more bite. The third, or autumn flush gives a tea that is lesser in quality.
- Nilgiris tea comes from a part of India that is almost as high as Darjeeling. This southern Indian region has elevations between 1,000 and 2,500 metres. The flavours of Nilgiris teas are subtle and rather gentle. They are frequently blended with other, more robust teas.
- Tea industry provides direct employment to more than one million workers directly, nearly 50% are women.
- The present Tea Board set up under section 4 of the Tea Act 1953 was constituted on 1st April 1954.
- *Camellia sinensis* is the source of tea of commerce. The young processed leaves yield tea – the world's most important caffeine beverage.
- The most important bioactive compounds present in *Camellia sinensis* leaves are alkaloid, flavonoids, steroids and terpenoids which serve as valuable starting material for the medicine development.
- In general, fresh green tea leaves contain 36 % polyphenols, among which catechins prevail.
- Pharmacological properties of tea are due primarily to methylxanthines and catechins of which EGCG is the predominant in green tea leaves (48–55 % of total polyphenols).
- The most abundant components in green tea are polyphenols, in particular flavonoids such as the catechins, catechingallates and proanthocyanidins.
- Tea polyphenols are well-known for their antioxidant properties
- The beneficial effects of the tea have been attributed to the strong antioxidant activity due to the phenoloic compounds.
- The carotenoids, flavonoids, benzonic acid, ascorbic acid, tocotrienols, cinnamic acid, folic acid, tocopherols are some antioxidents produced by the plants for their substance.
- Tea is called the queen of all beverages.

Production technology

Climate	The performance of tea is excellent at elevations ranging from 1000 – 2500 mm. optimum temperature: 20 – 27° C.
Soil	Tea requires well drained soil with high amount of organic matter and pH 4.5 to 5.5
Propagation	By semi hard-wood cuttings.
Preparation of cuttings	Cuttings are taken on April – May and August – September. Semi hard-wood cuttings are prepared with one leaf and an internode with a slanting cut at the bottom.
Methods of planting	**Single Hedge System:** In this method, the spacing adopted is 1.20 x 0.75 m accommodating 10,800 plants/ha.
	Double Hedge System: In this method, the spacing adopted is 1.35 x 0.75 x 0.75 m accommodating 13,200 plants/ha.
Season and planting	May – June or September – October. Sleeves should be opened lengthwise without injuring the roots and planted in the pit and the soil is gently pressed.
Irrigation	Subsoil irrigation may be given for young tea seedlings during summer months.
Manuring and fertilization	Manuring should be done 2 months after planting. Phosphorous should be applied at 80 – 100 kg/ha as Rock phosphate once in a year by placement at 15 – 25 cm depth up to the first pruning and thereafter once in two years. N : K ratio 2 : 3 should be adapted for the first 3 years and a ratio 1 : 1 thereafter.
Time of application	Application of fertilizers should be done before the onset of monsoon. Fertilizers should be broadcast around the drip circle avoiding contact with the collar. Calcium Ammonium Nitrate is the best N fertilizer.
Centering	To induce more laterals, centering should be done 3 – 5 months after planting. The main leader stem should be cut, leaving 8 – 10 matured leaves.
Tipping	Tipping is the practice of removal of terminal portion of shoot. It is done at a height of 35 cm from the second tipping at 60 cm from ground level.
Pruning	Pruning is done to maintain convenient height of bush and to remove dead and diseased branches
	Area to be pruned every year = Total extent of the garden Pruning cycle
	Pruning interval = (Elevation in feet / 1000) + 1
	Pruning should be done in April – May or August – September.
Skiffing	This is the lightest of all pruning methods. A removal of only the top 5 – 8 cm new growth is done so as to obtain a uniform level of pruning surface (more than 65 cm).
Shade regulation	Pollarding of shade trees should be done prior to heavy rains at a height of 8 – 10 m from the ground level.

Annual lopping	Cutting the erect type branches on the laterals in shade trees before monsoon season.
Crop duration and harvest	Plucking commences when the tea bush is 3 years old. The plucking of extreme tip of the growing branch consists of an unopened bud together with two leaves is popularly known as "Two leaves and a bud", while fine plucking is anything less than this. In South India plucking continues throughout the year at weekly intervals during March – May and at intervals of 10 -14 days during the other months.
Maturity	Tea takes 18-20 months from plantation to reach plucking.
Yield	The yield of green leaves is 10 t/ha.
Manufacturing process	CTC= Crush, Tear and Curl, Orthodox
Fermentation	• In tea processing, fermentation is the term employed to denote enzymic oxidation, by which the polyphenols in the leaf get oxidized with the help of PPO (Poly Phenol Oxidase). • The end product of fermentation are the flavins and the arubigins which are responsible for the briskness, brightness and colour of tea liquor.
Moisture %	Fermented dhool- 55-60%, Made Tea- 2.5-3%

Tea cultivars with special characters

S. No	Special characters	Clone	Originators
1.	Wind Tolerance	UPASI-2 UPASI-10	UPASI-TRF India
2.	Drought resistance	UPASI-9	UPASI-TRF India
3.	Frost resistance	B-26	HPKV-TES India
4.	Smallest leaf	CH-1	IHBT India
5.	Biggest Leaf	Betjan	BETJAN T.E, India
6.	Blister blight tolerance	TRI-2043,	TRI, Sri Lanka
7.	High pubescence	TRI-2043	TRI, Sri Lanka
8.	High anthocyanin pigentation	TRI-2025	TRI Sri Lanka
9.	High tolerance to pH	TN-14-3	TRF, Kenya
10.	Poor fermenter	12/2	TRF, Kenya
11.	Mite tolerance	7/9	TRF, Kenya
12.	Scale insect tolerance	TN-14-3	TRF, Kenya
13.	High polyphenol content (53.7%)	Luxi white tea	TRI, China
14.	High amino acid content (6.5%)	Anji white tea	TRI, China
15.	Low caffeine content (0.14%)	Guangdong tea	TRI, China
16.	High caffeine content (6.96%)	Wild tea at Yunnan	TRI, China
17.	Water logging tolerance	TV-9	TES, India

(Mondal *et al.,* 2004)

Improved clones

S.No	Clone	Important features
1.	UPASI 1 (Ever green)	Hardy, Quality-Above average
2.	UPASI 2 (Jayaram)	Hardy, Quality-Above average, tolerant to drought and wind
3.	UPASI 3 (Sundaram)	Natural triploid quality clones and very high yielding
4.	UPASI 6 (Brooklands)	Suited to mid and higher elevations
5.	UPASI 8 (Golconda)	Suited to all elevations, high yielding
6.	UPASI 9 (Arthrey)	Fairly tolerant to drought and withstand slightly high pH, , high yielding
7.	UPASI 10 (Pandian)	Hardy, Quality-Above average, tolerant to drought and wind
8.	UPASI 14 (Singara)	Suited to higher elevations , High yield
9.	UPASI 15 (Spring field)	Flushes throughout the year
10.	UPASI 17 (Swarna)	Flourishing well at mid and high elevations
11.	UPASI 24	Hardy
12.	UPASI 25	High yielding
13.	UPASI 16	High yielding
14.	UPASI 27	Drought tolerant
15.	UPASI 28 (UPASI 10 * TRI2025)	Biclonal, Good strength and high quality

□□□□

13

Recent Initiatives in Crop Improvement

High-yielding Plantation crops varieties and hybrids with tolerance to pests / diseases and to various farming constraints such as soil and water salinity, soil acidity, drought, flood and such other factors have been developed.

Three coconut varieties (IND 045S, IND 048S and IND 058S), two arecanut varieties (VTL 62-Shriwardhan selection and VTL7-Nalbari) and two selections of cocoa (VTLC 1 and VTLC 57) were identified for commercial cultivation. Coconut variety, Kalpa Samrudhi, was recommended for commercial cultivation in Kerala and Asom, and Kalpa Sankara and Kalpasree for coconut root(wilt)-affected tracts in Kerala. Molecular-markers associated with hybridity in coconut, dwarfness and resistance to root (wilt) and yellow leaf diseases were utilized. Homology-basedmodelling of somatic embryogenesis receptor-like kinase (SERK) protein in coconut, development of a computational tool for detection of microsatellites in whole genome sequences, development of algorithm for gene annotation and genome-wide analysis of microsatellites in strains of plant growth promoting (*P. fluorescens*) are achievements in bioinformatics.

To develop dwarf and compact oil palm, Dwarf Tenera 1 from Palode (Kerala) was selfed and its progeny was evaluated. Of the 58 palms evaluated, 23 recorded lesser than 40 cm and of which 4 palms recorded increased height of lesser than 30 cm per annum. One palm which recorded lesser than 25 cm height increment had a Bunch Index of 0.43 against the standard value of 0.3 for normal D × P material.

Source: DARE/ICAR Annual Report 2012–13

Coconut hybrid, IND 058S × IND 042S, with a yield of 140 nuts/palm/year and copra yield of 4.66 tonnes/ha has been identified for release. A superior selection, IND 045, with uniformly green coloured fruits, higher endosperm content of 300 g and copra content of180-200 g has been identified. This population also performs well under rainfed condition with an annual yield of 110 nuts/palm.

Of the 16 tall **arecanut** hybrids, Shrivardhan × Sumangala, Shriwardhan × Mangala and Mohitnagar × Sumangala exhibited higher yield potential. Among 21**cocoa** hybrids grown under arecanut in high-plantingdensities, hybrid SCA 6 × ICS 6 appears promising and recorded >2 kg dry bean yield/tree/year. Three hybrids, VTLCH 2, VTLCH 3 and VTLCC 1, showed high performance and adaptability both under arecanut and coconut shades.

In **cashew**, hybrids H 43, H 66, H 68, H 125 and H 126 gave annual yield of 5.20, 6.25, 6.55, 5.95 and 5.70 kg/tree, respectively, with a cumulative yield of 35.33, 34.61, 35.55, 37.60 and 34.39 kg/tree for seven harvests.

Source: DARE/ICAR Annual Report 2011–12

Microsatellite analysis of DNA isolated from 308 samples of 58 coconut accessions exhibited gene diversity of 0.215 and heterozygosity of 0.617. Clustering with Neis' genetic distance indicated two distinct clusters of coconut accessions. First web-basedutility for phytoplasma information containing 776,165 RNA sequences, 1,514 genes and 2,102 proteins entries was developed. Coconut varieties, Kalpasree and Kalpa Sankara, were recommended for release in root (wilt) prevalent tracts of Kerala. Kalpasree was recommended for release as a variety suitable for cultivation in homesteads in root (wilt)-diseased tract owing to its superior quality of oil and sweet tender nut water (240 ml/nut) containing total sugars 4.80 g/ml, potassium 2150 ppm, sodium 22.40 ppm with a TSS of 4.8° Brix. Kalpa Samrudhi hybrid of coconut was recommended for release in Kerala and Asom based on high copra yield under rainfed condition and good tender nut quality. The hybrid is high-yielding (117 nuts/palm) and has high copra yield (25 kg/palm/year) and oil yield of 3.04 tonnes/ha under rainfed condition. Kalpatharu coconut was recommended for release in Kerala and Karnataka. High recovery of ball copra without much spoilage of kernel (3.92%) is another beneficial trait of Kalpatharu variety.

A high-yielding arecanut variety, Kahikuchi (VTL- 64), has been recommended for release in Asom and north-eastern region. The maximum yield of 607 nuts/palm (4.50 kg of chali/palm) was observed in Nalbari accession. Based on high yield record, Shriwardhan and Nalbari varieties were recommended for release to farmers. In cocoa, 274 accessions were maintained and evaluated and during the year under report, 44 new accessions were also added.

In oilpalm, 3 promising dwarf interspecific hybrids, namely 47 (361*Eg* × 11*Eo*), 48 (16*Eo* × 18*Eg*) and 6 (12*Eo*× 82*Eg*) were identified, which can be used for backcrossing progarmme. Screening of Duras collected from four African countries (Cameroon, Guinea Bissau, Tanzania and Zambia) for drought tolerance and/or high water-use efficiency is being done at different locations in India. Among African germplasm, Zambian accessions, in general, performed better under both irrigated and stressed conditions. Bunch index was

highest in Zambian material. Fatty acids composition of 18 palms from 5 African accessions (GB22/311, GB5 10, ZS 01, ZS 02, TS 04) and 1 indigenous accessions (204D × 281D) was also carried out.

In cashew, 29 accessions were evaluated while 67 accessions were genotyped using ISSR and SSR markers. The Jaccard's similarity coefficient values(0.38–0.89) indicated low-diversity existing among the accessions. High shelling (30%) was observed in accessions, JG 147 and JG 151. Accessions, RFRS 173 and RFRS 177, had higher number of panicles/m2. The genotype, T.No 10/19, produced highest cumulative nut yield (78.69 kg/tree), followed by T No 30/1 (66.49 kg/tree) in 14 harvesting. Under hybridization trials, at Jhargram centre, the highest annual yield of 14.7kg/ tree was obtained in H 41, followed by H 57 (13.60 kg/tree) and H-23 (13.2 kg/tree).

Source: DARE/ICAR Annual Report 2010–11

Coconut: Of the 20 accessions evaluated, higher nut yield was recorded in IND 071S, IND 027S,

IND 148, while higher copra yield was obtained in IND 027S, IND 026S, IND 071S, IND 148, IND 002S, respectively. Of the 71 coconut accessions, including 58 tall accessions and 13 dwarf accessions, copra content ranged from 70.4 to 349 g. The accessions, IND 085S, IND 023S and IND 034S, were found promising for copra content; with 300 g copra/ nut. Among dwarfs, coconut germplasm accession, IND 092S, recorded copra content of 219.80 g.

Based on the storability and firmness of copra, IND 030S, WCT, Kalpa Mitra and IND 082, respectively, were found better suited for ball copra production. The hybrid IND 058S × IND 069S, was found superior for nut, copra yield and tender nut traits.

Analysis of dwarfs using 14 microsatellite markers showed that, MYD and COD were monomorphic, whereas heterozygosity ranged from 0.3 to 0.4 in talls. The UV treatment was found to induce more growth in coconut embryos under tissue culture conditions.

The nucleotide and protein sequences pertaining to genes induced during somatic embryogenesis, viz. *SERK* (somatic embryogenesis receptor kinase) and *BBM* (*BABY BOOM*) were retrieved from the NCBI. Nucleotide sequences coding conserved domain amino acid was selected for oligomer designing. These degenerate primers were used to amplify, clone and sequence *SERK* and *BBM* genes in coconut.

For molecular tagging of coconut root (wilt) disease, 18 degenerate primer pairs were used to amplify Resistance Gene Analogues (RGAs) from coconut. Initially, the annealing temperature for each primer pair was standardized using gradient PCR. After PCR, electrophoresis was carried out and amplicons of expected sizes were eluted, cloned and sequenced. Two of the clones showed homology to RGAs from other plant species.

Kalpatharu: Considering higher coconut yield (15,750 nuts/ha), better copra outturn of 2.7 tonnes/ ha @ 15.48 kg/palm/year with oil content of 67.2%, drought tolerance attributes and adaptability to water deficit regions, accession IND 125 S is recommended for cultivation in Karnataka, Tamil Nadu and Kerala.

Kalpa Samrudhi: Considering the superiority of coconut hybrid IND 376 (a hybrid between IND 058 S × IND 069 S) for higher nut yield, (117 nuts/palm), high copra out turn (25.72 kg/ palm/year), oil yield of 3.04 tonnes/ha under rainfed conditions, semi-tall habit, good tender nut water quality (TSS 6% Brix), and yield (346 ml/ tender nut) and drought tolerance nature, this hybrid is recommended for cultivation in Assam and Kerala.

Arecanut: Performance of arecanut germplasm accessions collected from NE region was found to be better than others and, accessions, VTL-29 II,VTL-12, VTL 18 III, Nalbari and K & J hills, recorded higher yield ranging from 2.81 to 3.58 kg chali/palm/ year.

Forty-nine tissue culture derived plantlets from YLD resistant arecanut palms were field planted for evaluation against YLD. Clonal fidelity test using RAPD indicated that 98% of progenies are similar to mother palms.

The DNA of arecanut sample from YLD resistant and susceptible palms was screened with ISSR primers (UBC820-840) and 40 RAPD primers (OPAF and OPAB). Reproducible primers are selected (UBC828, UBC835, UBC822, OPAF11, OPAF13, OPAF18, OPAB3 AND OPAB7) for further testing of resistant and susceptible palms. The CNZ 01, CNZ 03, CNZ05, CNZ 18, CnCirB3 and CnCirC09 primers showed amplification of arecanut DNA. Resistant genes were collected from NCBI database and analyzed for conserved regions based on multiple sequence alignment. The primers were synthesized for NBC-LRR regions. The primers were used to amplify the resistant genes from YLD resistant areca palm. Amplicons at expected size were eluted from the gels and cloned, and are being sequenced. Resistant genes from different categories from various crops were analyzed for consensus region and motifs.

Under bioinformatics, databases on plant growth promoting rhizobacteria (PGPR) and sequences of all the available phytoplasma have been developed. Protein kinase gene family in palms was analyzed for disease resistance, OBP sequences were analyzed for red palm weevil of coconut.

Oil palm: Under characterization of interspecific hybrids of *Elaeis oleifera* or the American oil palm produces more liquid oil due to higher unsaturated fatty acid content compared to commercial species (*Elaeis guineensis*). However, due to erratic and poor yield, *E. oleifera* is not cultivated commercially. Interspecific hybrids have potential of combining yield and quality in oil. Interspecific crosses were made involving *E. guineensis*and *E. oleifera* palms. Bunch component analysis of hybrids showed intermediate values for bunch weight as well as bunch related parameters including proportion of parthenocarpic fruits. However, oil/mesocarp oil/bunch were lower than those of the parents. Fatty acid composition showed intermediate value between the two parental species for all fatty acids. Wide variability in fatty acid composition was found in progenies of two specific interspecific crosses. Since performance of each palm is different, individual interspecific hybrid palms were assessed based on total unsaturated fatty acids and oleic acid content and 20 superior palms were selected, which could be employed for further back crossing to combine quality of palm oil and yield.

After several extensive experiments with spear leaves collected from mature palms, callus induction, somatic embryo induction and maturation including plantlet regeneration have been obtained from spear leaves. Standardization of media for callus induction and somatic embryo induction was also possible from immature inflorescence collected from mature palms. Refinement and confirmation of the above protocols are under progress.

Cashew: Three hybrids, H-66, H-68 and H- 43, yielded 4.0, 5.2, and 5.6 kg/ tree in fifth harvesting with a cumulative yield of 23.8, 24.0, and 25.1 kg/tree, respectively. Hybrids, H-125 and H-126, yielded 7.0 and 6.0 kg/tree in fifth harvesting with a cumulative yield of 26.4 and 23.7 kg/tree, respectively. At Jhargram, maximum nuts were recorded with Kanaka (14.75), followed by Vengurla 6 (13.75) and Dhana (11.75).

Source: DARE/ICAR Annual Report 2009–10

Recent Initiative by Central Plantation Crops Research Institute, Kasargod

Evaluation of coconut hybrids

In HET VI planted at Kasaragod with eight hybrids, MYD x WCT, COD x WAT, COD x LCT and COD x ADOT recorded higher annual yield of over 100 nuts palm-1 year-1. Fruit component studies undertaken in the hybrids with WCT as control indicated significant differences among the hybrids for fruit weight, fruit polar and equatorial circumference, husk and copra weight. Higher copra content was recorded in COD x ADOT (232g) followed by MYD x LCT (219g).

In HET VII planted at Kasaragod, the released hybrid, LCT x COD recorded 189 nuts palm–1 followed by LCT x GBGD with 149 nuts palm-1. The percentage of spoilage during ball copra production ranged from 20 to 57 among the four hybrids and WCT. Among the hybrids, LCT x GBGD recorded lower spoilage making it suitable for ball copra production.

In HET IX, and X at Kidu, the hybrids, PHOT x GBGD, CGD x PHOT, CGD x LCT, LCT x CGD and CRD x WCT palms continued to show better performance for growth, flowering and bunch production.

Morphological characteristics of seedlings and flowering in young palms were recorded in mutant and elite palm progeny trials planted at Kidu. Observations on bunch production and tender nut traits in Dwarf x Dwarf trials revealed that the hybrids MYD x CGD, COD x GBGD, MYD x NLAD, MYD x GBGD are better for yield and tender nut traits. The MYD x NLAD palms are robust, having thick stem resembling the male parent, closely arranged leaf scars indicating dwarfness, large bunches of over 1.3m length, with an average of 35 female flowers bunch-1 with good setting percentage. The nuts are bright green, round with about 400ml of tender nut water and free from mite infestation. The identified hybrids are proposed for testing under more locations. Tender nut analysis revealed higher quantity water in hybrids GBGD x NLAD, MYD x NLAD and GBGD x HPOD with 400 to 500 ml nut^{-1}.

Among the seven T x T hybrids and their four parents and planted for drought tolerance evaluation, significant differences were observed for the nut yield and related traits. The hybrid, FJT x WCT recorded higher nut yield and was on par with WCT.

Production of new experimental hybrids

Twenty eight new Dwarf x Tall (D x T) cross combinations involving seven dwarf(COD, GBGD, MOD, MYD, CRD, CGD and MGD) and four tall parents (CCNT, SNRT, KWGT, PHLT) were developed for evaluation for dwarfness, higher nut yield with higher copra out turn, aromatic tender nuts and soft endosperm. A total of 1053 hybrid nuts of different D x T and D x D cross combinations were sown during the year for new hybrid evaluation trials and also for gap filling in the existing trials. Crossing programme is in progress for production of required number of seedlings in the identified cross combinations. One batch of pollen of SNRT and CCNT was provided to CARI, Andaman for crossing work involving the four dwarf accessions of Pacific origin being maintained at CARI. Based on the morphological traits, dwarf mother palms were identified in the Pacific Ocean collections for crossing works. Pollination programme for production of identified/released hybrids viz., MYD x Kenya Tall, MYD x Tiptur Tall and Kalpa Samrudhi was initiated for multi location evaluation under AICRPP centres. Nursery with over 1200 experimental hybrid seedlings at various stages was maintained and their morphological traits and vigour observed for seedling selection.

Population improvement in WCT

Towards developing a uniform population of West Coast Tall, 55 WCT mother palms were selected based on the morphological grouping from the 280 palms studied at the Institute. The selected WCT palms were characterized using 16 SSR primer pairs and palms showing lesser heterozygosity have been identified.

Trait-specific genetical investigations

About 220 inter se mated progenies of two LMT mother palms were produced and evaluated at seedling stage for use as a mapping population. A selection from this population based on morphological traits showed a highly uniform population of LMT with compact seedling traits. A total of 40 seedlings of this homogenous selection will be evaluated further for its uniform performance.

A comprehensive programme was prepared for the utilization of Thairu Thengai palms existing in the farmers' fields for crop improvement and their utilization for product diversification. Pollen of Thairu Thengai was collected from in situ palms of Andaman and used for production of D x T hybrids involving MYD as female parent for further evaluation.

A rare bulbiferous twin coconut palm, producing a number of bulbil shoots in the leaf axils in both the vigorously growing main palm and its weaker twin, was identified in Karnataka. The bulbil shoots, on examination, revealed the production of secondary bulbils in their leaf axils indicating the complete vegetative state of the palm. Microsatellite analysis using 10 polymorphic SSR primer pairs, revealed monomorphic bands for the main palm, its twin and all the bulbils, indicating their genetic uniformity.

Investigations on this rare phenomenon will be helpful in understanding the flowering and shooting process in palms manipulation of flowering and vegetative growth in palms and standardizing viable vegetative propagation protocols for development of uniform populations for crop improvement. The production of early axillary buds and their conversion as floral/ vegetative buds will throw light on the breeding for earliness in coconut as the bulbil shoots exhibited axillary growth in 6th to 9th axil while it is exhibited at 18th to 36th axil in normal coconut seedlings.

Selfed progenies for development of inbred lines in coconut

Seedling observations were recorded in the trial on performance of inbred progenies of coconut planted at Kidu with S1 and open pollinated progenies of 5 coconut accessions. Nuts of WCT (S3) progenies and MYD x WCT (S3) hybrid produced at RARS (KAU), Pilicode were sown at Kidu for evaluation. Morphological observations were recorded on about 400 seedlings produced through selfing of S2 palms and progenies of MYD x S2. Twenty eight S2 palms in RARS Pilicode which were used for selfing and generating S3 population have been characterized for morphological characters. The highest variation among palms used for selfing was observed for number of nuts produced annually, followed by number of female flowers. Variation of 35% was observed for number of leaf scars metre^{-1} indicating high level of segregation for the trait over selfing. Reproductive traits such as number of spikelets inflorescence^{-1} and length of spikelet showed more than 25% variation.

Breeding for resistance/tolerance to coconut root (wilt) disease

A survey was carried out in hotspots of root (wilt) disease and 60 WCT and 32 CGD mother palms were identified. Morphological characterization carried out in disease-free WCT mother palms (located in disease hotspots) and its inter se mated progenies (planted at CPCRI, Kayamkulam) revealed that inter se mated progenies were more vigorous compared to its parents for vegetative, reproductive and nut characters. Crossing programme was continued in 13 disease-free S1 progenies of WCT to generate self/ inter se progenies. A total of 170 S2 progenies of WCT were planted at CPCRI, Kayamkulam. Hybridization was carried out on 21 CGD palms involving 130 inflorescence and 3,180 female flowers. A total of 764 CGD x WCT hybrid seed nuts were harvested and sown in nursery. Nucleus planting material of Kalpasree (variety released for cultivation in the root (wilt) disease prevalent tract) has been generated. Besides, seed nuts of different green dwarfs (ANGD, LGD, MGD, CGD) were collected for evaluating its performance in the root (wilt) disease prevalent tract.

The incidence of root (wilt) disease (RWD) was recorded from the new evaluation trial (planted with MGD, MOD, MYD, MGD x WCT, MYD x WCT, CGD x WCT, three and a half years after planting. Observations indicated that MOD is more susceptible to RWD as compared to MGD and MYD. Disease incidence in hybrids was lesser in CGD x WCT (Kalpa Sankara) and CGD x MGD as compared to MYD x WCT and MGD x WCT. All reported symptoms of RWD like flaccidity, necrosis, yellowing and inflorescence die back appeared in the susceptible varieties even at the young stage.

Evaluation of arecanut hybrids

Yield component traits were studied in the dwarf hybrids evaluation trial, and hybrids viz., Hirehalli Dwarf x Sumangala, Hirehalli Dwarf x Mohithnagar and Hirehalli Dwarf x Mangala continued to show superiority for yield with 2.48, 2.60 and 2.71 kg dry kernel palm^{-1}. Data on morphological characters and production of inflorescences were recorded in the MLTs of dwarf hybrids both at Kahikuchi and Mohitnagar.

Among the 16 TxT hybrids under evaluation for high yield and economic traits, Shriwardhan x Sumangala, Shriwardhan x Mangala and Mohitnagar x Sumangala exhibited higher yield potential and recorded 36.67, 33.15 and 29.86% heterosis over mid parental value, respectively.

Production of promising hybrids involving Hirehalli Dwarf and released tall varieties was continued and a total of 26,780 female flowers were pollinated in four combinations. Morphological data viz., seedling height, girth and production of leaves were recorded in eight hybrid combinations planted during October 2011 at Vittal.

Disease indexing and recording of growth traits have been carried out in YLD screening trial comprising of dwarf hybrids, three species and one related genera. Normal growth of seedlings was observed and so far no symptoms of YLD are noticed.

A replicated field trial with eight arecanut hybrids (Mangala x HD, HD x Mangala, Sumangala X HD, HD x Sumangala, Sreemangala x HD, HD x Sreemangala, Mohitnagar x HD and HD x Mohitnagar) along with the 5 parents (Mangala, Sumangala, Sreemangala, Mohitnagar and Kahikuchi variety) Hirahath Dwarf, planted during 2007, showed significant differences for annual growth characters. Higher number of functional leaves (9.46) was recorded in Kahikuchi, whereas, Hirehalli Dwarf recorded the highest collar circumference (57.5 cm). Among the hybrids, higher number of functional leaves (9.7) was recorded in Mohitnagar x HD and highest collar circumference (60.5 cm) in HD x Sumangala.

Evaluation of cocoa selections/hybrids for intercropping in coconut/arecanut gardens

Under arecanut canopy at high density planting, the cocoa progenies viz., VTLCP-20, VTLCP-6 and VTLCP-17 yielded an average of 92, 83 and 81 pods tree^{-1}, respectively. Under coconut, at Vittal, VTLCH-3, VTLCP-9 and VTLCH-4 recorded 70, 62 and 61 pods tree^{-1} and at Kidu the yield of hybrids was considerably lesser. Comparative performance of grafts and seedlings of 12 selected clones showed considerable difference in their bearing behaviour. Among the grafted clones, the yield ranged from 21.6 to 44.2 pods tree^{-1} and among the seedlings, it ranged from 15.1 to 30.5 pods tree^{-1}, at 7 years of growth.

Studies on insect pollinators of coconut

The insect fauna associated with coconut inflorescence and putative pollinators were recorded from Malayan Dwarf varieties (MGD, MOD and MYD) at two hourly intervals during 8 AM to 4 PM. Insect diversity was found to be maximum during male phase with bees (*Apis cerana indica, Apis dorsata and Trigona iridipennis)* and ants (*Camponotus* sp., *Anoplolepis gracilipes*) being the predominant floral visitors. The diversity and population of insects declined during the female phase and *Apis dorsata* was not recorded during the female phase. This indicates that the insects depend on coconut inflorescence mainly as a pollen source and only to a limited extent as a nectar source. Microscopic observation on pollen load on the body parts of bees and ants revealed that only bees carried substantial quantity of pollen and ants carried very less pollen, suggesting that bees (*Apis cerana indica* and *Trigona iridipennis*) are the major pollinators among the floral visitors.

Effect of novel growth regulators on regeneration from plumular explants in coconut

The effective concentration of glutathione for callus induction was found to be 100 µM. For healthy plantlet formation from plumule-derived callus, supplementation of medium with glutamine at a concentration of 178.5 µM was found to be effective. Rhizogenesis in tall varieties was enhanced by treating the meristemoid-derived plantlets in medium supplemented with NAA for four weeks, followed by sub-culturing in a hormone-free media.

Palm to palm variation was observed for callus induction and somatic embryogenesis in GBGD, CGD, MGD, COD and PHOT cultivars. The palms responding well were identified for multiplication based on experiments conducted during the previous two years. The embryogenic potential was found to be 16.4% in CGD, 50% in COD, 42% in PHOT and about 50% in MGD. The subsequent plantlet regeneration ranged between 2-6% from embryogenic calli.

The inter-se mating carried out in MGD palms at CDB Farm, Neriamangalam during the year 2011-12 resulted in the production of 1538 nuts. These nuts were utilized for the plumule culture experiments. The clonal fidelity of plantlet derived from plumule culture was confirmed with SSR primers.

Cell suspension culture studies were initiated from coconut embryogenic calli and cell multiplication was achieved on sub culturing in medium supplemented with decreasing levels of 2,4-D.

In vitro multiplication of dwarf hybrids of arecanut

Inflorescences from adult palms of Mangala, Hirehalli Dwarf x Sumangala and Hirehalli Dwarf x Mohitnagar were sampled and inoculated into Y3 media with 200ìM of picloram. The inoculated plates were sub cultured once in 15 days to lower the levels of picloram. Callus induction was noticed in 53.34% of cultures from inflorescence length ranging from 10- 38 cm and somatic embryos were developed in callus. These embryos were cultured into regeneration media and started shoot induction. The six dwarf hybrids planted in YLD endemic area have not shown symptoms of YLD for the past two and half years.

Boinformatics

Resistance gene analogue database and tool for prediction of RGA domains

RGA sequences of coconut, arecanut, oil palm and date palm were collected and sorted based on domains. All the sequences were searched in the PRINTS database and among 150 sequences, four domains and 261 motifs were obtained. A tool for prediction of the RGA motifs was developed using support vector machines (SVM), standalone BLAST and Hidden Markov Models (HMM).

Gene prediction tool for gibberellin biosynthetic genes in palms

A computational method based on SVM was developed for prediction of gibberellic acid biosynthetic enzymes in palms. The domains of these enzymes were collected using Pfam and grouped into clusters. Biosynthetic enzymes domain classifiers were developed for each of these enzymes using sequence pair wise alignment based SVM.

Development of a DNA barcode database of palms

DNA barcode data of three genes of indigenous palms were organized into a database, implemented in MySQL. The front-end of the database was developed using HTML and the client server connectivity was established by PHP program. This database contains palm- specific similarity search (BLAST) program.

Algorithms for gene annotation and isolation of genes from plant growth promoting rhizobacteria

LTTRPred tool was developed using SVM based on the composition of amino acid and amino acid pairs, for identifying and predicting LTTR genes, responsible for the activation of Plt transcription regulators, in Pseudomonas spp. The performance of the tool was tested using various datasets of LTTR genes from different species of Pseudomonas and compared with other gene predicting tools. This tool can be used for the functional annotation of the microbial data providing insights into genome evolution and finding different PGPR strains containing LTTR genes.

A gene prediction tool using a machine learning algorithm called WEKA was developed to predict Non- Ribosomal Peptide Synthetases (NRPS) genes, which are involved in the pyoverdine biosynthetic pathway of Pseudomonas fluorescens.

BACPred is a tool developed, using WEKA, for the identification of bacilysin synthesizing genes in whole genome of Bacillus spp. The tool performs individual prediction of six genes based on its function contributing to bacilysin biosynthesis.

Elucidation of pyoverdine gene cluster in *Pseudomonas fluorescens*

Pyoverdine gene cluster in Pseudomonas fluorescens Pf0-1 was elucidated using comparative genome analysis. Three strains of Pseudomonas fluorescens (Pf0-1, SBW 25 and Pf- 5) and Pseudomonas aeruginosa (PA01) were compared using bioinformatics tools. The analysis revealed the presence of 33 genes involved in PVD biosynthesis, transport or regulation metabolism in the Pseudomonas fluorescens Pf0-1 genome, which are distributed across five different regions of the genome.

PMU Finder for identification of potential mobile units in phytoplasma

The phytoplasma genome contains many multicopy gene clusters called potential mobile units (PMU). A functional PMU resemble composite replicative transposons which can integrate anywhere in a genome. PMU Finder tool is a heuristic nucleotide local alignment software, employed for the identification of PMUs from whole genome sequences. The tool is written in Perl scripts in CGI environment. Output of the identified PMUs is exported to an HTML report, which provides the results in a tabular format.

Homology modelling of adenylation domain of non-ribosomal peptide synthetases (NRPS)

The 3-D structure of adenylation domain of *Pseudomonas fluorescens* Pf-5 NRPS was predicted using MODELLER 9.10 based on templates showing similar identity to the target protein. Loop refinement was performed to increase the quality of the model with the best discrete optimized potential energy (DOPE) score. The secondary structure of the protein was determined using PORTER, SOPMA, GOR, NetsurfP. STRIDE was used to calculate the secondary structure of NRPS protein. Evaluation and verification of the model was done using Procheck via Ramachandran plot which showed that 86.4% of the residues were in most favourable region and around 9.4% of the residues were in additional allowed regions. Based on the PROMOTIF program, it was found that the percentage of β-strand was 14.7%, percentage of 3,10 helix was 2.6% and percentage of α-helix was 31.9%.

Elucidation of ABC transporter genes in coconut root (wilt) phytoplasma

The 3-D structure of the PotC protein of coconut RWD and sugarcane grassy shoot (SCGS) phytoplasma was predicted by I-TASSER server. Six transmembrane helices were identified for coconut and sugarcane phytoplasma PotC protein. Both the phytoplasma PotC proteins contain ABC transporter intergral membrane type-1 domain and bindingprotein- dependent transport systems inner membrane component motif region.

Development of an analysis pipeline for the palm sequences

An integrated, configurable and time effective microsatellite mining and annotation pipeline was developed and made available at the public domain (http://210.212.229.56/front_end/). The usage of this tool can be extended to the SSR detection and annotation in genome/EST sequences of other crops as well. The primary analysis output generated along with the results can be used for other general sequence analysis.

Analysis of partial genome of coconut

The partial coconut genomic sequences, 22,76,863 sequences of size 917Mb, obtained by 454 sequencing were analysed and 57,547 SSR containing sequences were detected. All the 233 homologues regions were annotated, with 72 sequences showing homology with putative uncharacterized protein and the remaining sequences had similarity with many important housekeeping proteins such as retrotransposon protein, ribosomal protein, transcription factor, cytochrome 450 and F-box family proteins.

Understanding the biosynthetic pathways in cocoa

Eight libraries of ESTs derived from Phytophthora palmivora infected cocoa leaf and pod tissues were analysed and a total of 14,831 redundant sequences were retrieved from ESTtik database. Clustering and assembling using CAP3 resulted in 6844 nonredundant (1743 contigs and 5101 singletons) sequences with GC percentage of 42.66 and the sequence length of 101-3610 nucleotides. Functional annotation revealed that most of the sequences are related to molecular function, stress response and other biological processes. An analytical tool SEMAT (Standalone EST Microsatellite Mining and Analysis Tool) for cocoa EST- SSR analysis was also developed.

For cocoa genome analysis, the reference cocoa chloroplast genome (NC_014676) was retrieved from Genbank database and analysed with SRA software Pathway tool version 16.0. Metabolic pathways were created using the program PathoLogic and cocoa specific pathway database was developed. Totally four metabolic pathways (Calvin-Benson-Bassham cycle, Oxygenic Photosynthesis, Photosynthesis light reaction, Rubisco Shunt) were constructed from cocoa chloroplast genome. A total of 19 genes involved in the stress related pathways (ornithine, citruline, polyamine and ethylene biosynthesis) were annotated from the cocoa nuclear genome, which can be used for in silico reconstruction of stress related pathways in cocoa.

Source: Annual Report 2012-13, Central Plantation Crops Research Institute, Kasargod

Salient Achievements of Central Plantation Crops Research Institute, Kasargod

Genetics and Plant Breeding

- World's largest germplasm collection of coconut (comprising 401accessions 269 indigenous and 132 exotic) is being maintained in the Institute. The exotic collections are from 28 countries of South Asia, South-East Asia, Africa, Caribbean Islands, Indian Ocean Islands and Pacific Ocean Islands.
- Designated as National Active Germplasm Site for Plantation crops and host for National Gene Banks for coconut, arecanut and cocoa.
- Host to International Coconut Gene Bank for South Asia under Coconut Genetic Resources Network of Bioversity.

- Released twelve improved varieties of coconut (Chandra Kalpa, Kera Chandra, Chowghat Orange Dwarf, Kalpa Pratibha, Kalpa Dhenu, Kalpa Mitra, Kalparaksha, Kalpasree, Kalpatharu, Kalpa Jyothi, Kalpa Surya and Kalpa Haritha.
- Released five high yielding hybrids of coconut - Chandra Sankara, Kera Sankara, Chandra Laksha, Kalpa Samrudhi and Kalpa Sankara.
- Dwarf varieties viz., Chowghat Orange Dwarf, Kalpa Jyothi, Kalpa Surya were released exclusively for tender coconut.
- Kalpa Pratibha, Kalpa Haritha, Kalpa Samrudhi, Chgandra Sankara, Kalparaksha and Kalpasree were released as dual purpose varieties suitable for copra and tender nuts.
- Chandra Kalpa, Kalpa Mitra, Kalpa Dhenu, Kalpa Pratibha Kalpatharu and Kalpa Samrudhi are relatively tolerant to drought.
- Kalpa Haritha is comparatively free from Eriophyid mite infestation amidst heavily infested palms of other varieties.
- Dwarf varieties viz., Kalparaksha and Kalpasree are recommended for root (wilt) affected areas with higher degree of root (wilt) disease resistance.
- The hybrid Kalpa Sankara is released for root (wilt) disease tracts for its tolerance to the disease.
- Kalpatharu is released for ball copra production.
- Coconut descriptors prepared for 74 accessions and important accessions registered.
- In arecanut,164 germplasm collections of which 23 exotic and 141 indigenous are conserved at CPCRI,RS, Vittal.
- Released six high yielding varieties (Mangala, Sumangala, Sreemangala, Mohitnagar, Swarnamangala and Kahikuchi) and two hybrids of arecanut (VTLAH1 and VTLAH2).
- In cocoa, 321 germplasm collections including 281 exotic and 40 indigenous clones are being conserved under Arecanut and Coconut at CPCRI, Regional Station, Vittal, Karnataka.
- Seven varieties of cocoa are released including four hybrids (VTLCH-1, VTLCH-2, VTLCH-3 and VTLCH-4), one clone VTLCC-1 and two selections VTLCS-1 and VTLCS-2. They exhibited high pod and dry bean yields, bigger and bold beans with less shell contents, high nib recovery,

richness in fat, tolerance to water stress and suitability to be grown both in arecanut and coconut gardens.

- Softwood grafting technology has been standardized in cocoa and utilised for mass multiplication.
- Scion bank and compact block on varieties are established with mother trees.
- Six bi-clonal and one poly-clonal orchards are established with 1100 trees for F1 seed production at CPCRI, Research Centre, Kidu, Karnataka.
- Quality planting material of mandate crops are being produced to the extent of more than 1.2 lakh coconut seed nuts, 3 lakh arecanut seed nuts and 1.1 lakh cocoa grafts/ seedlings/ seed pods.
- CPCRI nurseries at Kasaragod, Kidu and Vittal are recognized and accredited under 4* category by National Horticulture Board.

Biotechnology

- Protocol for regeneration of coconut plantlets from plumule explants through somatic embryogenesis developed.
- Cryopreservation techniques have been standardized for mature coconut zygotic embryos and coconut pollen.
- DNA finger printing using molecular markers *viz*. RAPD, AFLP, DAF and micro satellites have been carried out in coconut and cocoa accessions to document the genetic integrity and diversity. About 139 coconut accessions have been characterized using SSR markers. Mother palms used for seed production have been genotyped to develop markers for hybridity. Diverse germplasm accessions including dwarfs and talls are being used for developing markers for dwarf plant habit.
- Tissue culture protocol has been standardised for mass multiplication of arecanut from inflorescence explants and genetic fidelity of *in vitro* derived plantlets has been assessed using molecular markers.
- Transcripts induced during water stress have been identified using differential display RT-PCR technique (DDRT-PCR). The clones were characterized into different functional groups based on the sequence similarities.

Cloning and identification of transcripts induce during water-stress in coconut

- The nucleotide and protein sequences pertaining to known genes induced during water stress *viz. DREB*, 14-3-3, aquaporins,9-cis-epoxycarotenoid dioxygenase, *WRKY* , *NAC* and epicuticular wax (*CER, GL and Wax*) were retrieved from the NCBI. Nucleotide sequences coding for conserved domain amino acids were selected for oligomer designing. The degenerate primers were used to amplify putative water stress responsive genes in coconut via RT-PCR. Amplicons of expected sizes were eluted, cloned and sequenced. Positive sequences were deposited in Genbank.

Isolation of genes induced during somatic embryogenesis in coconut

- The nucleotide and protein sequences pertaining to genes induced during somatic embryogenesis *viz. SERK* (somatic embryogenesis receptor kinase) and BBM (*BABY BOOM*) were retrieved from the NCBI. Nucleotide sequences coding conserved domain amino acid was selected for oligomer designing. These degenerate primers were used to amplify, clone and sequence *SERK* and *BBM* genes in coconut

Cross transferability of SSR makers from coconut to related palms

- To study the cross transferability of SSR markers from coconut in other palms, 86 micro satellite markers specific to coconut were screened out of which 55 primers gave strong, clear bands of expected size range (100-300bp). These primers were tested for their cross-taxa amplification in oil palm, arecanut, palmyrah and date palm. The percentage of cross-amplification of coconut SSR loci in other palms were 36.36% in oil palm, 29.09% in arecanut, 18.18% in palmyrah and 12.70% in date palm.

Cloning of MAP kinase induced during water stress in coconut

- Coconut plantlets *in vitro* were subjected to water stress with PEG treatment and RNA was isolated from the leaves. Degenerate primers, designed for amplification of *MAP kinase*, were used in single-step RT-PCR reactions with the isolated RNA as template. Bands of expected size were eluted, cloned and sequenced. One fragent showed homology to known *MAP kinase* from other plants.

Bioinformatics

- Under the Bioinformatics initiative, several comprehensive databases for the mandate crops have been developed.

Source: Central Plantation Crops Research Institute, Kasargod.

□□□□

14

Recent Initiatives in Crop Production

Cost-effective farming systems for plantation crops cultivation for increasing production and productivity in different agro ecologies have been developed.

In coconut-based cropping system, elephant-foot yam cv. Gajendra yielded high amount of corm. Guinea grass (var. GGCo 3) produced higher green fodder (89.2 tonnes/ ha/year) under husk application. There was lower specific volume of soil and soil porosity (higher soil compaction), and soil dehydrogenase activity (an index of biological activity) as important abiotic pre-disposing factors in incidence of root wilt disease. Growing *Gomphrena globosa* as intercrop in root wilt- affected coconut gardens has been shown to be cost-effective and sustainable. In areca-based mixed farming system, total cash inflows and outflows from arecanut + dairy (3–4 milch cows) amounted to '4.41 and '5.96 lakh, with a net profit of' 1.55 lakh. The system productivity of arecanut + cocoa system (3,127 kg/ha) was 30% higher per unit area than arecanut alone (2,405 kg/ha). Integrated nutrient management on four improved arecanut varieties recorded highest chali yield (4.19 kg).

Studies on spore associated bacteria (SAB) in coconut/ arecanut based cropping systems under organic management practices revealed an increase in AMF spore load with the number of intercrops. Identification based on BIOLOG revealed the presence of bacteria, viz. *Citrobacter amalonaticus, Staphylococcus arlettae*, *Bacillus subtilis* and *B. amyloliquefaciens,* in association with spores of *Glomus* spp. The potent phosphate solubilizes isolated from coconut rhizosphere were identified as *Enterobacter cloacae* RNF 267, *Pseudomonas plecoglossicida* KnSF 227 and *P. putida* Biotype B HSF 132 by conventional biochemical tests and confirmed by BIOLOG Microbial Identification System and 16S rRNA gene analysis.

Of the 90 genotypes of walnut, 38 found protogynous and rest were protandrous. Forty-two genotypes showed7–14 days of male and female bloom overlapping, indicating synchronized flowering while the rest were non-synchronous. The genotype, CITH-W 40, showed maximum synchronizing phase of 14 days, followed by CITH-W 37 and Nugget of 13 days while genotypes CITH-W 10, CITH-W 19, CITH-W 24 and CITH-W 48 showed non-synchronization with respect to male and female blooming period, indicating need for pollinizer varieties. The most suitable rootstock and scion girth for grafting in walnut was found to be rootstocks with the girth of 25–30 mm and scion of 15–20 mm size, which are fully developed and plumpy.

Foliar sprays on cashew at three different stages, viz. flushing, flowering and nut development, with urea (3%) + H3PO4 (0.5%) + K2SO4 (1%) and foliar spray of secondary and micronutrients, viz. ZnSO4 (0.5%) + MgSO4 (0.5%) + Solubor (0.1%) recorded highest number of bisexual flowers and nut yield/tree. However, application of paclobutrazol (0.5, 1 and 1.5 g a.i./plant) showed reduction in growth of plants with respect to plant height, canopy spread and internodal length. High-density planting showed significantly higher (3.23 times) nut yield (2,841 kg/ha) as compared to the normal density (880 kg/ha). Intercropping in cashew with Amorphophallus gave highest net returns (139,639/ha), followed by tapioca (129,992/ha).

Source: DARE/ICAR Annual Report 2012–13

In cropping of Gajendra variety of elephant-foot yam in coconut plantation under coastal littoral sandy soil, yielded high corm yield (17.6 tonnes/ha) when grown with soil moisture conservation measures like husk and coir pith application. Guinea grass (var. GGCo 3) gave higher green fodder yield under husk application (ranging from 8.0 to 8.5 tonnes/cutting with a total yield of 82.2 tonnes/ha/year). Thevam variety of black pepper recorded significantly higher dry berry yield (1.71 kg/vine) when grown as a mixed crop in coconut garden. Recycling of biomass along with application of organic manures (cow dung, cow urine, biogas slurry) in cropping system is found to be self-sustainable in terms of nutritional requirement of coconut and other component crops in a mixed farming system for root (wilt)-affected coconut gardens.

The maximum annual nut yield (6.23 kg/tree) of cashew was recorded in moderate dose (0:250:125 g NPK/tree). The increase in yield under high-planting density (600 plants/ha) was 36.90% higher as compared to normal density

(200 plants/ha). In high-density cashew plantations, number of nuts/m2 was highest (24.70) in 200 plants/ha at 75:25:25 NPK (kg/ha) at Jhargram. Under intercropping trials, highest net returns were obtained with intercropping *Amorphophallus* (139,639/ha), followed by tapioca (129,992).

Results of a study on stress responses of the selected PGPR (22 coconut and 21 cocoa isolates) indicated that *Bacillus cereus* (ESB 15), *Bacillus* sp. (RSB 14),*Serratia marcescens* (KiSII) isolated from the rhizosphere of coconut and 5 *Bacillus subtilis* isolates (CSB 8, KGEB 10, PEB 2, PEB 4 and VEB 17) from cocoa rhizosphere could tolerate a maximum temperature of 60°C and were also able to grow on TSA medium amended with 12% NaCl.

Source: DARE/ICAR Annual Report 2011–12

In coconut, intercropping of ash gourd, pumpkin or amaranth in orchard gave higher yield of coconut (124 nuts/palm/year). Perennial *Stylosanthes* was found to be a suitable fodder legume under coconut in root(wilt)-affected areas for higher fodder production. Similarly, yield of cocoa in oil palm plantations ranged from 0.25 to 1.5 kg/plant/year, whereas cocoa yield under coconut gardens varied from 0.5 to 2.7 kg.

In oil palm, sap flux in 9th and 17th fronds was close to 0 in morning and reached peak fluxes of 45 cm/hour during noon. The peak sap flux in 25th frond was lesser than that of 17th frond. While in the trunk, peak sap flux was only 8 cm/hour. Thus, sap flux is more in younger fronds of palm, indicating its importance in maintaining growing parts. The amount of carbon sequestered by oil palm hybrids ranged from 17.98 to 38.10 tonnes carbon/ha with Papua New Guinea and Ivory Coast hybrids sequestering the highest and lowest carbon contents, respectively. Direct embryogenesis without callus phase was obtained from cotyledonary nodes of germinated immature zygotic embryo. The globular embryos with clear suspensor region appeared directly on the explants and multiplied. Other stages such as to pedo and heart shaped embryo were seen on sub-culturing. On transfer to light in Y 3 media, they matured into complete plants.

The cashewnut yield was found to be 63% more at a spacing of 5 m × 4 m and 45% more at 6 m × 4 m compared to normal spacing, i.e. 10 m × 5 m. On top-worked trees, nut production was higher in trees receiving paclobutrazol compared to trees receiving just pruning treatments and no further pruning after limb pruning. The highest cumulative nut yield (75.97 kg) was recorded in cashew in the treatment 500 : 125 : 125 g NPK/tree, followed by 500 : 250 : 125 g NPK/tree (70.96 kg).

Source: DARE/ICAR Annual Report 2010–11

Coconut: Intercropping banana, hybrid Napier grassCO-3 and ash gourd in coconut garden under littoral sandy soil with husk and coir pith application resulted in higher yield. Higher coconut yield was recorded under coconut + vegetable intercropping system (130 nuts/palm/year) compared to monocropping of coconut (118 nuts/ palm/year).

Economics of coconut based farming system involving coconut with integration of grass, pepper (trailed on coconut), banana (on border of garden), dairy and poultry resulted in the net return of Rs 1,29,070 /ha. Sale of coconut, milk and broiler was found to give 91% of the revenue generated from the system.

Cow dung slurry application on vanilla grown as mixed crop in coconut garden was found to increase the number of inflorescences (21), number of beans (208) and fresh bean yield (2.0 kg/vine), followed by vermiwash application and vermicompost + biofertiliser application.

Allelopathic studies in coconut-based cropping system indicated that the micro organisms are capable to catabolise the allelochemicals in leachates and nullify their plant suppressive properties.

The coconut yield was higher with the treatment, organic + inorganic manuring + tillage practice (158 nuts/palm/year) under rainfed situation. Application of 50 % VC +50% NPK recorded significantly higher nut yield (68.8 nuts/ palm/ year) and was on a par with application of 25% VC + 75% NPK (T2) treatment (66.3 nuts per palm per year.

Arecanut: In arecanut garden, mixed cropping with pepper, banana and citrus was found to be economically advantageous in North-East Region. In Sub-HimalayanTerai Region, aster, marigold and gladiolus performed better under arecanut garden.

Cashew: The yield in high-density plots (416 and 500 trees/ha) was significantly higher (1.13 and 1.12 tonnes/ha, respectively) than in normal planting system (200 trees/ha) (0.72 tonnes/ha) 7 years after planting. Cashew yield was appreciably higher in coconut husk burial (2.04 tonnes/ha) and modified crescent bund treatments (2.03 tonnes/ ha) than in reverse terrace (1.94 tonnes/ha), catch pit (1.39 tonnes/ha) and the control (1.48 tonnes/ ha) treatments.

Source: DARE/ICAR Annual Report 2009–10

Recent Initiatives by Central Plantation Crops Research Institute, Kasargod

Coconut

Management of coconut based farming systems for sustainable productivity under coastal ecosystems

The sustainability and profitability of coconut based farming system comprising coconut, pepper trailing on the coconut trunk, banana in the border of the plots, fodder grass (hybrid Bajra Napier Co 3) in the interspaces of coconut, dairy unit (7 cows of Holstein Friesian and one Jersey cross breed), poultry (100 broiler birds batch^{-1}), Japanese quails (100 layers) and aquaculture (1,000 fingerlings) was assessed. From one ha of coconut based integrated farming system 23,100 coconuts, 14,311 litres of milk, 238 kg broiler chicken, 6,095 quail bird eggs, 4,387 hen eggs, 2,009 kg of banana and 362 kg of pepper were obtained. The net return (Rs. 2,67,141) was the highest in the CBIFS receiving combined application of 50 percent organics (produced from the system) and 50 percent inorganic. Fodder grass with the application of 50 per cent NPK through vermicompost and farm yard manure + 50 percent NPK through chemical fertilizers recorded yield of 114 t ha^{-1} and was comparable with fully organic treatments (102 t ha^{-1}) and significantly differed from chemical fertilizers alone (100 t ha^{-1}). The coconut palms maintained under CBIFS receiving integrated nutrient management practices (combined application of organic and chemical fertilizer) recorded significantly higher yield (132 nuts palm^{-1}), which was comparable with other nutrient treatments and also higher than coconut monocropping (106 nuts palm^{-1}). During the year, a new component of goat rearing with 2 female and one male was integrated with the system.

Management strategies for poor yielding coconut palms under littoral sandy soil

The average yield of last five years indicated that the treatment of growing vegetables as intercrops has shown to increase the yield of coconut in all the categories. The highest yield recorded was 63 nuts in low yielders, 72 nuts in medium yielders and 86 nuts in high yielders against the initial yield of <15 nuts in low yielders, 15-30 nuts in medium yielders and >30 in high yielders. The percent increase in yield was higher in low yielders compared to high yielders. The nutrient status of the soil at 0-50 cm soil depth was not significant among the different yielders. But the treatments significantly influenced nitrogen and potassium contents. The nitrogen content was significantly higher with vegetable intercropping followed by cow dung slurry application. Potassium content was significantly higher with poultry manure applications. Similar trend was noticed at 50-100 cm soil depth also. With regard to leaf nutrient status, both nitrogen and potassium were significantly higher with intercropping of vegetables. Difference between different yield categories was not significant.

Impact of drip fertigation on productivity of coconut under coastal sandy soil

Effect of fertigation on coconut yield under coastal sandy soil with treatments viz., 25, 50, 75 and 100 % NPK through drip irrigation and 100 % NPK through soil application was evaluated. The fertilizer applied through drip irrigation was given in six equal splits avoiding the monsoon period. Irrigation was given @ 66 % of open pan evaporation through drip @ 32 litres palm^{-1} day^{-1}. The conventional fertilizer application was followed as per the recommended practice. Six years mean coconut yield was the highest (76 nuts palm^{-1} year^{-1}) with application of 100 % NPK through drip irrigation. It was at par with 50 % NPK through drip irrigation (73 nuts palm^{-1} year^{-1}) and 75 % NPK through drip irrigation (74 nuts palm^{-1} year^{-1}). All these treatments recorded significantly higher yield over 25 % NPK through drip irrigation and 100 % NPK through soil application.

Coconut based cropping system for better resource use efficiency, productivity and profitability

Mean yield of coconut for five years (2007-12) showed no significant difference among the treatments. The average yield was 155 nuts palm^{-1} with fully organic treatment, 152 nuts palm^{-1} with 1/3rd recommended chemical fertilizers and recycling biomass through vermicompost + biofertilizer + green manuring + vermiwash and 145 nuts palm^{-1} with 2/3rd recommended chemical fertilizers and recycling biomass through vermicompost. The cost of production was higher in fully organic treatment (Rs.1,12,450 ha^{-1}). The net return was also the highest under fully organic treatment (Rs.1,77,825 ha^{-1} during 2011-12) followed by other treatments. The annual carbon sequestration from coconut was 0.53 t C ha^{-1}.

Precision farming in coconut: Evolving site specific management practices for increasing the coconut productivity

Soil samples (1,500 nos.) at 0-25 cm, 25-50 cm and 50-100 cm depths were analysed for the macro and micro nutrient status and the yield data of coconut were collected. The soil fertility status was assessed based on soil depth weighted average (0-100 cm soil depth) value. Organic carbon content of 36 per cent of the samples was found to be less than 0.5 per cent. Available K was found to be less than 100 ppm in 14 per cent of the samples and less than 200 ppm in 75 percent of the samples. Available Zn status was less than 0.5ppm in 27 per cent of the samples. Available Cu status was less than 0.5 ppm in 4 per cent of the samples. Available Fe and Mn status were found to be more than 2.5 ppm in 99 per cent of the samples.

In red sandy loam, soil organic carbon, available N, K, Ca, Mg and Zn were found to be deficient. Yield relationship with the soil nutrients was studied using correlation and significant positive relationships between yield and soil pH, EC, available P, K, Na, Ca and Mg was observed in red sandy loam soil. In coastal sandy soil, available K, Ca, Mg, Fe, Mn, Cu and Zn were found to be deficient and showed positive correlation with yield. Yield showed positive correlation with leaf N, Mg, Fe and Mn content. The sandy soils of Onattukara region showed that 90 percent of the samples tested were having less than 200 ppm of Ca and 120 ppm of Mg. Available K status of 65 percent of the samples tested were less than 60 ppm.

Evaluation of organic cultivation of coconut in coastal agro-ecosystem

The effect of organic cultivation practices on coconut productivity with four organic cultivation practices (T1: Vermicomposting in the basin, application of biofertilizers and cover cropping in the inter space, T2: Vermicomposting in the trenches, application of biofertilizers and cover cropping in the basin, T3: Vermicomposting in the basin, application of biofertilizers and growing vegetables in the inter space, T4: Vermicomposting in the trenches, application of biofertilizers and raising vanilla and pepper) and a control (T5) was studied. Biomass of cover crop which was raised in basin as well as in the inter space was weighed and ploughed back into the system as per the treatment requirement. All the fallen leaves from coconut palms of respective treatments were vermicomposted in situ using *Eudrilus* sp. of earth worm and applied to coconut palms. *Pueraria phaseoloides* was raised as the green manure cover crop and the biomass was incorporated into coconut basins at the appropriate time. Weeding was done and the biomass yield was estimated. All the organic cultivation treatments recorded significantly higher nut yield when compared to the control in both WCT and COD x WCT. However, the organic cultivation treatment, vermicomposting in the trenches and cover crop in the basin (T2) recorded higher yield of 102 nuts and 133 nuts palm-1 for WCT and COD x WCT, respectively. The population of function-specific and beneficial microorganisms was also found to be more in the rhizosphere of coconut palms receiving organic treatments.

Integrated approaches for root (wilt) disease management in Coconut

Development of cropping system model with dwarf variety of coconut for RWD affected areas The treatment with coconut husk burial in planting trenches recorded significantly highest yield of pineapple (13.52 t ha^{-1} of intercropped area), followed by an yield of 9.08 t ha^{-1} with application of dried leaves in planting trenches. Ninety two percentage of the young coconut palms planted during June 2009 have flowered. On an average, 7.8 nuts $bunch^{-1}$ was recorded in the palms during 2012-13.

Nutritional requirement of dwarf varieties of coconut in root (wilt) disease affected area

Observations recorded during the month of December indicated that so far around 36, 43 and 22 percentages of plants under Kalparaksha (MGD), Kalpasree (CGD) and COD varieties of coconut, respectively have flowered. The general trend continued to indicate that application of inorganic manures favoured better growth in Kalparaksha (MGD), while organic manures was found to enhance growth in COD and Kalpasree (CGD) varieties. Application of chemical fertilizers favoured more flowering in Kalparaksha (39 %), while application of organic manures or integrated application resulted in more flowering in Kalpasree variety of coconut (67 and 44%, respectively). Microbial population for both general and function specific communities were found to be higher with either organic manures or an integrated application along with inorganic manures.

Soil and plant health management through organic recycling in coconut based production system

Low cost vermicomposting units with provision for collection of vermiwash were constructed in 12 coconut gardens (8 in Kerala and 4 in Tamil Nadu covering a total area of 10 ha) during 2009 and the bio resources generated were recycled by vermicomposting and applied to various crops. Farmers adopted inter/mixed cropping system in their coconut garden as per their choice. Application of organic manure as vermicompost/ vermiwash influenced the rhizosphere microbial community. Inclusion of more intercrops increased the population of function- specific soil microflora, which in turn has increased the soil enzyme activities, which are key indicators of soil health.

Nutrient management practices for intercropping selected flower crops in root (wilt) disease affected coconut garden

Heliconia stricta cv. Iris Red is found to be a potential intercrop in root (wilt) disease affected coconut gardens. Both organic and integrated nutrient management were found to be economical. The plants had a superior vegetative growth and inflorescence production in sandy soils by supplying cow dung @

1kg plant^{-1} + bonemeal @250g plant^{-1} basally and topdressing equal doses of Vermicompost (VC) @ 200g plant^{-1} + Neemcake (NC) @ 100g plant^{-1} at quarterly intervals from three months after planting followed by the treatment supplying cow dung @ 1kg plant^{-1} + bonemeal @250g plant^{-1} basally and topdressing equal doses of VC @ 100g plant^{-1} + NC @ 50g plant^{-1} + 13:5:13 NPK@2.5 g plant-1 with a BCR of 3.9 and 3.4 respectively. Heliconia recorded a Land Equivalent Ratio of 1.4.

Intercropping Marigold (October to April) followed by Gomphrena (May to September) as catch crops in coconut garden of 25 % shade intensity can ensure continuous supplies for market oriented flower production with a mean BCR of 2.62.

The basal application of FYM and VC @ 20t ha^{-1} + 6.25t ha^{-1} respectively + 56.3:60:60kg NPK ha^{-1} at 15 days after planting (DAP) + Vermicompost extract spraying at 30 DAP and 45 DAP produced highest number of marketable flowers in marigold with a BCR of 3.5. This treatment combination was demonstrated with marigold in an area of 100 cents by farmer participation at Kadakkarapally Panchayath during 2012, which recorded a BCR of 2.82.

Intercropping of flower crops substantially reduced the number of barren bunch produced per palm (2.6 to 1.0) and increased the average number of buttons (12.3 to 23.3) and nuts (5.99 to 7.96) produced per palm over a period of three years. The nutrient status of coconut leaf and basin soil recorded were also higher in the intercropped area.

Predisposing factors of root (wilt) disease of coconut

Field surveys and monitoring of symptoms along with soil and plant sample collection was conducted in areas of Cumbum, Colachel, Kozhinjampara and Thrissur. Site-specific variation in soil parameters were observed in these localities and variation in symptom expression of the disease were also observed. Available P content in the soil of the diseased site was very low (ranging from traces to up 2.0 ppm), whereas, in the healthy area, it was up to the extent of 30 ppm. The clay content was apparently more in the diseased area as compared to the healthy area.

A marked variation on the particle size distribution at the two layers in both the sites was observed. In the diseased area, there was more clay and silt content in the sub surface layer as against the condition in the healthy area. Predominant difference was also noticed with respect to the content of organic matter between the two sites. In the diseased area, it was 0.736% and 0.449% at 0-25 cm and 25-50 cm depths respectively, whereas, in the healthy area, it was 1.06% and 1.41% respectively at 0-25 cm and 25-50 cm depths.

Arecanut

Cropping/farming system

The project on arecanut based mixed farming system was initiated in 2007 to study the feasibility and efficiency of resource management. The analysis of five years data clearly showed that the net income gradually increased and on an average, dairy unit with three - four milching cows generates around 5,000/- as net income per month. The fisheries component with 250 fingerlings (rohu, katla and common carp in the ratio of 4:3:3) generated gross profit of 14,500/-

The project on drip fertigation in arecanut + cocoa mixed cropping system was initiated in 2007 with the objectives to study the feasibility of supplying organic and inorganic fertilizers through drip irrigation, to observe the effect of drip fertigation on growth and yield of arecanut and cocoa and to evaluate the effect of fertigation on soil quality indicators. The experiment was laid out in split plot design with three replications. The main plots included two systems: arecanut and arecanut + cocoa mixed cropping system. The sub plots included six treatments: control + OMR through vermicompost, 75% recommended NPK, VCE (vermicompost extract) 10% of N, VCE 20% of N, VCE 10% of N + 25% NPK and VCE 20% of N + 25% NPK. Pooled analysis of 2008-2011 indicated that arecanut registered similar yield levels in sole and arecanut-cocoa cropping situations (3,022 -3,117 kg ha^{-1}). Among fertigation treatments, yield levels for 75% NPK, VCE 20% N and application of VCE (10 and 20% N) + 25% NPK were at par (3,029-3,375 kg $ha-^{1}$). Dry bean yield of cocoa was at par with fertigation of 75% NPK and 20% N VCE + 25% NPK (291-335 kg $ha-^{1}$). Fertigation @ 75% NPK increased the yield of cocoa by 52% over VCE alone. The productivity per unit area (kg $ha-^{1}$) was significantly higher in arecanut-cocoa system (3,450) than arecanut sole (3,090) accounting for 12% increase despite initial bearing stage of cocoa. Fertigation of 75% NPK, 20% N VCE and VCE (10 and 20% N) + 25% NPK registered similar productivity levels (3,316-3,665 kg ha^{-1}). Leaf nutrient status of arecanut and cocoa indicated lower levels of N and K and above normal levels of Ca, Mg and micronutrients.

Nutrient management

A field trial was laid out at Vittal in 0.29 ha area to standardize the methods to manage the nutrient disorders. The palms were classified based on symptoms like crown choking, bending and cross nodes. Scoring was given on 0-5 scale based on severity of symptoms. Spraying of 1% ZnSO4 resulted in normal growth of emerging leaves in palms with score of 1-3.

An experiment was laid out in two farmers' gardens to demonstrate the Yellow Leaf Disease (YLD) management strategies, including nutrient aspect of the YLD management. The leaf N, P and K concentrations were below optimum and the concentrations of secondary and micronutrients were above optimum range in all three gardens. Deviation from optimum (DOP) percentage was reduced for both major and micro nutrients during 2007-2010 with the interventions. In farmer's plots affected by yellowing, the yields were below 2000 kg ha^{-1}. The yield levels obtained with different treatments (averaged over all three locations) indicated that treatments with recommended nutrient inputs managed the YLD at low index values up to 20. However, for index level above 20, precision application of organic and inorganic inputs is required to avoid antagonistic nutrient interactions in soil and plant.

Coconut/arecanut based cropping system

Medicinal plants such as *Swertia sp., Asparagus officinalis, Mentha arvensis, Piper longum, Rauwolfia serpentina* and *Aloe vera* were planted under Arecanut and coconut gardens and also in the open field. The yield of *Swertia* sp. was good under arecanut garden (153.6g) compared to open (147.9g) and coconut plantation (107.2g).

Yield performance of *Rauwolfia serpentina* and *Asparagus officinalis* was better in the open condition with 124g dry root $plant^{-1}$ and 1.57 kg dry root plant-^{1}, respectively, compared to that under arecanut and coconut plantations. The performance of *Mentha arvensis* was almost equal in three conditions. Jute, introduced as inter crop under arecanut, yielded 116 g sq m^{-1} fibre.

Integrated nutrient management of released arecanut varieties

An experiment was started with four released arecanut varieties (Mangala, Sumangala, Sreemangala and Mohitnagar) and five nutrient doses (no fertilizer, recommended fertilizer, 100% N applied through vermicompost (VC), 50% N through VC and 50% N by inorganic fertilizer, and 75% N through VC and 25% N with inorganic fertilizer) of organic, inorganic and their combination to study the effect of organic and inorganic fertilizers on yield and yield attributing characters.

Maximum number of inflorescence $palm^{-1}$ was produced by the varieties Mangala (3.7) at T5 (75% N substitution by VC), Sumangala (3.5) at T2 (recommended fertilizer dose), Sreemangala (3.2) at T5 (75 % N through VC) and Mohitnagar (3.7) at T4 (50% N through VC), respectively, whereas, the maximum number of nuts $palm^{-1}$ was produced by Mangala (376) at T3 (100% N through VC), Sumangala (343.7) at T2 (recommended fertilizer doses),

Sreemangala (352.7) at T5 (75% N through VC) and Mohitnagar (414) at T4 (50% N through VC). The maximum chali yield (4.3kg palm^{-1}) was recorded in Mohitnagar, than the other varieties at T4 level (50% N through VC). But in Sumangala and Sreemangala, maximum chali yield was recorded at recommended fertilizer doses i.e. fully inorganic form of fertilizers. In the case of Mangala, the maximum chali yield (3.315 kg palm^{-1}) was recorded in 100% N substitution by vermicompost.

Micro irrigation and water management practices

A field trial on the water requirement and comparative efficiency of different systems of irrigation (drip, sprinkler, basin) at different levels of fertilizer (1/3rd, 2/3rd and full dose of recommended fertilizers) in arecanut is in progress. Drip irrigation at 66% open pan evaporation was found effective for growth, yield (3.20 kg chali palm^{-1}) and moisture distribution/utilization at 2/3rd dose of fertilizer level.

Development of in situ water conservation measures

A trial to study the effect of different in situ conservation measures in the hilly slopes of Assam was initiated in the farmer's field with imposition of different soil and water conservation measures like (i) catch pit filled with coconut husk, (ii) catch pit with pineapple downstream, (iii) contour staggered trenches, (iv) half moon trenches, and (v) without any conservation measures. Among the different treatments, catch pit filled with coconut husks produced 3.22 kg chali of arecanut palm^{-1}.

Development of an arecanut based cropping system model

An experiment involving two models of Arecanut based HDMSCS is in progress to study the economics and biomass generation in the models under different fertilizer levels. Model I included arecanut (Kahikuchi selection) + pepper (Karimunda) + banana (Chenichampa) + citrus (Assam lemon) + clove and Model II with arecanut (Kahikuchi selection) + pepper (Panniyur1) + banana (Chenichampa) + citrus (Gandharaj lemon) + nutmeg (grafts). Maximum biomass was obtained from the plots applied with 2/3rd dose of recommended chemical fertilizer, irrespective of models. The highest B:C ratios (3.61 and 3.38, respectively) were observed under 2/3rd dose of recommended fertilizer for both the models.

Integrated nutrient management for arecanut

A field experiment on substitution of chemical fertilizers with vermicompost was laid out with five released varieties of arecanut (Mangala, Sumangala,

Sreemangala, Mohitnagar and Kahikuchi) and seven nutrient combinations (T1: Control (no fertilizer), T2: 100% Vermicompost (to substitute only N requirement), T3: 200% Vermicompost (as per the N requirement), T4: 100% chemical fertilizers, T5: 50% chemical fertilizers + 50% Vermicompost, T6: 2/3rd chemical fertilizers + 1/3rd Vermicompost and T7: 1/3rd chemical fertilizers + 2/3rd Vermicompost. Palms fertilized with 1/3rd chemical fertilizers + 2/3rd vermicompost resulted in maximum yield (3.17 kg chali palm-^{1}) of arecanut palm. Among the varieties, Kahikuchi performed well in terms of yield and it gave maximum yield (3.71 kg chali palm^{-1}) under 1/3rd chemical fertilizers + 2/3rd vermicompost (T7). Maximum leaf nutrient content of N, P and K was recorded under the treatment 1/3rd chemical fertilizers + 2/3rd vermicompost (T7).

Application of Microorganisms in Agriculture

Plant Growth Promoting Rhizobacteria (PGPR) for coconut seedlings

Two plant growth promoting bacteria designated as KiSII and RNF 267 isolated from the rhizosphere of coconut palms were identified as *Serratia marcescens* and *Enterobacte*r sp. based on their phenotypic features, BIOLOG studies and 16S rRNA gene sequence analysis. Biolog provided the identification of KiSII as *Serratia marcescens* sp. *marcescens* with similarity index 0.617 and distance 5.087 (99.3% probability), validating the conventional identification. BIOLOG assay of RNF 267 gave similarity index (0.046) lower than the acceptable value (0.5), though it showed maximum similarity with Enterobacter cloacae in the database. The molecular analysis based on 16S rRNA gene homology identified the PGPR KiSII as Serratia marcescens and RNF 267 as Enterobacter sp. With 99% and 97% similarity with the respective strains in NCBI GenBank database. The nucleotide sequences were deposited in NCBI GenBank database under accession numbers JX498910 (*S. marcescens*) and JX498911 (*Enterobacter* sp.).

Both bacteria exhibited phosphate solubilization, ammonification, and production of indole acetic acid, β-1, 3 glucanase activities and 1-aminocyclopropane- 1-carboxylate-deaminase activity. They could also tolerate a range of pH conditions, low temperature and salinity (NaCl). In addition, S. marcescens KiSII exhibited N- fixation potential, chitinase activity, siderophore production and antibiotics production. Seed bacterization with these bacteria increased the growth parameters of test plants such as paddy and cowpea over uninoculated control in green house assay. In coconut seedlings, significant increase in growth and nutrient uptake accompanied with higher populations of plant beneficial microorganisms in their rhizospheres were recorded on inoculation with both the PGPRs.

Effect of co-inoculation of PGPRs on the growth parameters of coconut and cocoa seedlings

Experiments conducted for evaluating the effect of co-inoculation of potent PGPRs, on growth parameters of coconut (WCT) and cocoa (Bulk Forastero) seedlings in polybags under field condition showed that co-inoculation of *B. megaterium* TSB 16 with B. coagulans RSB 14 had significant effect in the dry weight of coconut seedlings (30 % increase over *B. megaterium* TSB 16) when compared to its single inoculation. Dual inoculation of *B. subtilis* VEB 4 and *B. licheniformis* KGEB 16 on cocoa seedlings recorded significantly more enhancement in total dry weight (28 % over *B. subtilis* VEB 4 and 14 % increase over B. *licheniformis* KGEB 16) than individual inoculations. However, when *Pseudomonas* spp. were inoculated in combination with *Bacillus* spp., they had no further stimulatory effect on the growth parameters of both coconut and cocoa seedlings compared to their single inoculation. In coconut, P. putida KnSF208 and *Bacillus* sp. RSB14 exhibited higher growth when inoculated individually than their combinations. Significant difference was observed among the treatments with regard to the presence of beneficial microorganisms and the treatments which enhanced the growth of seedlings also maintained higher microbial population such as *Bacillus* spp., P-solubilizers and nitrogen fixers.

Species diversity of AMF in arecanut based cropping systems under organic management practices

Arbuscular mycorrhizal status and species diversity of AM fungi was studied in farmers' fields under organic management practices (only cow dung and husk as the inputs) of arecanut based cropping systems at three different locations (Maneikkara, Cheruvanjeri and Nedumpoyil) in Kannur district of Kerala. Arecanut based intercropping system in Maneikkara had banana as the intercrop, Cheruvanjeri had banana and pepper as intercrops and Nedumpoyil had banana, pepper and cardamom as component crops. The highest AMF colonization was observed at Maneikkara with a spore load of 90 spores 10g^{-1} soil and 80% frequency of root colonization and least at Cheruvanjeri plot with an average spore load of 28 spores 10g^{-1} soil and 49 % frequency of root colonization. However, it was observed that species diversity increased with the number of intercrops. *Funneliformis mosseae, Rhizophagus fasiculatus, Funneliformis geosporum, Glomus macrocarpum, Glomus aggregatum, Glomus multicaule, Glomus glomerulatum* and *Acaulospora bireticulata* were the AMF species associated with different crops in arecanut cropping systems. The relative distribution of AMF species varied between cropping systems as well as among intercrops. Studies indicated that *Funneliformis geosporum* was the most abundant (29-50 %) followed by *Glomus macrocarpum* (9.4 -

29%) and *Glomus multicaule* (11.9 - 29.38 %) and the least relative abundance (0-2.93%) was observed in *Acaulospora* sp.

Diversity analysis of Bacilli based on BIOLOG and 16S rDNA sequencing

PCR targeting the 16S rDNA has been used extensively to study prokaryote diversity and it allows identification of prokaryotes as well as the prediction of phylogenetic relationships. It is the basic molecular tool for amplified rDNA fragents (ARDRA) which detects interspecies and interstrainal variability and provides accurate genetic diversity information on microbes. Diversity of aerobic endospore-forming bacteria, *Bacillus* spp., one of the major types of soil bacteria, isolated from the rhizosphere of coconut, growing in acidic soils of Kerala and Karnataka (soil pH varying from 2.18- 4.99), was studied.

Carbon source utilization pattern based on BIOLOG and ARDRA analysis of these isolates using twelve Fast digest® restriction enzymes viz., HaeIII, HinfI, AluI, TaqI, MspI, RsaI, HhaI, EcoRI, HpaII, HindIII, BgaIII and XbaI was employed for diversity studies. The identification of acid tolerant bacterial isolates through Microbial Identification System (BIOLOG) and 16S rDNA sequencing revealed *B. subtilis, B. amyloliquefaciens, B. megaterium, B. cereus, Lysinibacillus sphaericus* and *B. tequilensis* as the major species. Among the 12 fast digest enzymes, TaqI, AluI and HhaI endonucleases gave higher discrimination of the isolates. TaqI was found to be the most discriminating marker for *Bacillus* species and, hence, can be used for distinguishing between closely related species of *Bacillus*. The results suggested presence of genotypically and phenotypically diverse population of acid-tolerant Bacillus spp. in rhizosphere region of coconut palms. However, ARDRA was a better tool to distinguish taxonomic groups than BIOLOG phenotypic fingerprints.

Source: Annual Report 2012-13, Central Plantation Crops Research Institute, Kasargod

Salient Achievements of Central Plantation Crops Research Institute, Kasargod

Coconut

- Nursery management techniques comprising selection of garden, mother palm and seed nuts, planting and maintaining the nursery, and the technique for raising polybag nursery were standardized.
- Square system of planting at a spacing of 7.5 x 7.5 m with a plant density of 175 palms/ ha is recommended for monocrop and coconut based cropping system.

- Recommended dose of fertilizer for coconut palm is application of Organic Manure @50kg/palm or 30 kg green manure, 500 g N, 320 g P_2O_5 and 1200 g K_2O/palm/year in two split doses during September and May. Application of magnesium @500 g MgO per palm was found to be advantageous in areas where palms show yellowing of leaves (Coconut root (wilt) affected gardens) through soil application.
- Application of 200 litres of water once in four days was recommended for irrigating coconut palms by basin irrigation method.
- Drip irrigation @ 66 % of the open pan evaporation (27 to 32 litres of water per palm per day under Kasaragod conditions) from December to May is ideally suited for coconut resulting in 34 per cent saving of water.
- Sprinkler irrigation or perfo irrigation with 20 mm (IW / CPE = 1) water was found to be the best suited to inter or mixed cropping systems where the entire surface requires wetting.
- Fertilizer saving through drip fertigation i.e., 50% of the recommended dose of fertilizer when applied through drip fertigation is sufficient to give a yield equivalent to 100 % of the recommended dose of fertilizer. Fertilizers viz., 70 g Urea, 60 g DAP and 170 g Muriate of Potash are recommended for a single dose for one palm. A total of 7 times the fertigation can be applied through drip irrigation. For phosphorus application, commercial phosphoric acid also can be used.
- Moisture conservation methods such as mulching with coconut husk, coir dust, coconut leaves, etc. in addition to application organic manures or green manures, intercrop cultivation, bunding, terracing, etc. are recommended.
- In sloppy terrains, trench filled with coconut husk, half moon bund, staggered catch pit reinforced with pineapple and growing CO3 grass across the slope, were proved successful in soil and water conservation.
- Surface runoff water harvesting and Roof water harvesting were successfully implemented and harvested water was utilised for irrigation.
- The technique for utilization of leguminous cover crops such as *Pueraria phaseoloides, Mimosa invisa and Calopogonium* species as green manures to supply biologically fixed nitrogen and easily decomposable biomass to coconut, to substitute 50 % nitrogen fertilizer, was standardized.
- Growing Glyricidia as green manure crop and using the biomass as green manure was found to be ideal for management of littoral sandy soils.

- The technology for vermicomposting of coconut palm wastes by using a local earthworm, *Eudrilus* sp., closely related to the African night crawler, was standardized.
- Multiplication technique for the local Eudrilus sp. of earthworm using 1:1 cow dung-decayed leaves mixture was standardized and the earthworms are being distributed to the farmers to initiate vermicomposting.
- Production of vermiwash from coconut leaf vermicompost has been standardized and its beneficial impact on crop growth is being assessed.
- Utilization of coconut wastes for oyster mushroom cultivation (*P. florida, P. sajor caju, P. flabellatus, P. opuntia* and *P. eous)* was found to be economically feasible.
- *Beijerinckia indica, Azospirillum spp., Burkholderia sp., Azoarcus* sp., etc. were effective bioinoculants for better establishment of nursery seedlings.
- Four coconut PGPRS - *P. putida* KnSF 208, B. *coagulans* RSB 14, B. *megaterium* TSB 16 and B. *megaterium* TEB 2 and four cocoa PGPRs - *P. putida* KDSF 23, B. *licheniformis* KGEB 16, B. *cereus* ASB 3 and B. *subtilis* VEB 4 gave the highest increase in planr growth parameters. All the eight isolates have been identified using biochemical, BIOLOG and molecular based procedures.
- Developed organic farming practices for coconut to achieve coconut yield of over 100 nuts per palm per year with recycling of farm wastes, application of nitrogen fixing legumes and biofertilizers.
- Coconut based cropping systems involving cultivation of compatible crops like tubers, flowering, medicinal and aromatic crops, fruits, vegetables, spice crops, in the interspaces of coconut was economically superior to coconut monocropping.
- Coconut-based high-density multi species cropping systems (HDMSCS) involving many crops like banana, pineapple, clove, and pepper was established.
- Application of 2/3rd of recommended NPK along with recycling available biomass in the farm of vermicompost resulted in sustainable yield in the system. The system provided a net income of Rs.85000/- per ha/annum.
- Mixed farming system including coconut, dairy, poultry, rabbitry, sericulture and pisciculture has been successfully demonstrated in coconut, under kasaragod condition. System provided organic manures for recycling, additional man power and additional income to a coconut farmer.

- Mixed cropping with CO_3 hybrid Nopier fodder grass provided sufficient green grass to feed 5 to 7 cows from one ha coconut garden.
- Banana variety, Grand Naine, Elephant foot yarn(gajendra), pineapple, vegetables were found to be a suitable intercrop for coconut gardens in littoral sandy soil with coir pith and husk as amendments.
- Using Remote Sensing and GIS map, coconut plantation area and coconut root (wilt) disease affected coconut palms were identified with > 98% accuracy.
- Carbon sequestration potential of coconut, arecanut and cocoa were determined and the values were 8 to 32 t CO_2/ ha/ year in coconut, 5.14 to 10.94 t CO2/ ha/ year in arecanut and 2.02 to 3.89 t CO_2/ ha/ year in cocoa depending on cultivar, agro-climatic zone, soil type and management practices.

Arecanut

- A spacing of 2.7 X 2.7 m is found to be optimum for the growth of arecanut crop.
- Irrigation with 200 litres of water per palm once in 6 days through hose is recommended.
- Recommended fertilizer dose for arecanut is 100g N, 40g P_2O_5 and 140gK_2O/ palm/ year.
- Studies in arecanut-cocoa cropping system revealed that it is possible for carbon sequestration to the tune of 7.16 t/ha to 14.83 t/ha annually.

Cocoa

- Softwood grafting is the most successful propagation technology.
- Canopy architecture by pruning and planned cutting of branches have been standardized for canopy management under intercropping and in sole crop.
- Cocoa is planted at a spacing of 2.7m x 5.4m in pits of 60 cm^3 filled with compost in areca garden planted at a spacing of 2.7m x 2.7m with provision of shade.
- When cocoa is to be raised as a mixed crop with coconut, cocoa can be planted at 2.7 m apart in single hedge system and 2.5mapart in paired rows in double hedge system between two rows of coconut palms.

- The crop is to be irrigated once in a week during November-December, once in 6 days during January-March and once in 4-5 days during April-May with about 175 litters of water.
- Systematic study for 10 years on drip irrigation and fertilizer requirement of cocoa mixed cropped in arecanut revealed that drip irrigation at E_0 of unity (20 litters of water per day per tree) and a fertilizer dose of 100: 40: 140 g of N, P_2O_5 and K_2O per tree per year would be optimum for cocoa.

Source: Central Plantation Crops Research Institute, Kasargod

□□□□

15

Recent Initiatives in Crop Protection

Crop protection research has brought out tangible and effective methods to suppress various pests in major agro-ecosystems. Indigenous knowledge in this regard has been tested and validated for its suitability.

Among the 137 *Phytophthora* isolates collected from bud rot/fruit rot affected coconut gardens in Tamil Nadu, Kerala and Karnataka, four isolates from bud rot affected gardens were identified as *P. nicotianae* and one as *P. capsici*. Since, *P. nicotianae* is found associated with bud rot disease of coconut in three districts representing three states, it is expected that this may emerge as a major pathogen of coconut. Further, placing *Trichoderma* coir pith cake in the innermost leaf axils of coconut palm in disease endemic areas just before the onset of southwest monsoon (May-end) and thereafter at two months interval was found to be very effective in the management of bud rot disease of coconut.

An organic and biodegradable pesticide slow release product, using coir pith was developed, for slow release of Mancozeb to coconut crown and preliminary field trials show that it could effectively replace the polythene sachets. Copper oxychloride (Blitox 0.5%), Metalaxyl + Mancozeb (Ridomil Gold 0.5%) and potassium phosphonate (Akomin 0.5%) were observed to be effective in the management of black pod disease. Isolation and characterization of ABC transporter system in coconut root (wilt) phytoplasma has been accomplished. In phylogenetic tree based on *sec*A gene, the YLD phytoplasma clustered with sugarcane grassy shoot, coconut root wilt and napier grass stunt phytoplasma, all members of 16SrXI group. A novel nanomatrix with ordered pore channels was developed for loading the pheromone and kairomone for controlling red palm weevil. Similarly, a nanomatrix and polymer composite was also developed to load the rhinoceros beetle pheromone (ethyl 4 methyl octonate), which could extend the life-span to 6–8 months.

Surveillance for Eriophyid mite (*Aceria guerreronis* Keifer) infestation was undertaken in five districts of Asom. The highest incidence of mite (35.7%) was recorded in Morigaon, followed by Nagaon (17.67%). The hymenopteran solitary ecto-larval parasitoid recovered from parasitized (*Leucopholis coneophora*) grub was identified as *Campsomeriella collaris* (Fabricius) and the solitary endo-larval parasitoid as *Prosena* sp. nr. *siberita* (Fabricius).

Chemical control of pest complex in cashew increased nut yield from 28.34 to 41.68 % and maximum (41.68%) being with L-Cyhalothrin (0.003%). Similarly, in post- extraction prophylaxis trial for management of cashew- stem- and root-borer (CSRB), with Chlorphyriphos (0.2%) treatment, the treated trees without reinfestation varied from 80 to 93.3%. Spiders were observed as one of the most abundant natural enemies of tea mosquito bug (TMB) in cashew agro-ecosystem. A total of 104 species spread over 56 genera and 11 families were identified. The virulence study of entomopathogenic nematodes (EPN), viz. *Heterorhabditis indica, Steinernema abbasi* and *Steinernema bicornutum,* against grubs of cashew stem- and root-borers (CSRB), *Plocaederus* spp. revealed mortality of grubs at a mean duration of 14.1, 12.9 and 12.4 days respectively. *Steinernema bicornutum* induced more than 50% of mean mortality of bait species (wax moth) even after 150 days under simulated conditions.

Source: DARE/ICAR Annual Report 2012–13

In **coconut**, control of bud rot disease was obtained by pouring Mancozeb 5 g in 300 ml water coupled with placing of two perforated sachets containing 5 g Mancozeb to innermost leaf axils at bimonthly intervals. Assessment of trap density using the refined PVC traps of 1.5 m height and size of windows of 10 cm × 15 cm indicated maximum catch of beetles in traps placed @ 1 trap/ha area. Laboratory screening of new molecules against red palm weevil revealed 100% mortality of grubs with Imidacloprid (Confidor 200% SL).

In cashew, stem- and root-borer (CSRB) grubs, reared entirely on semi-synthetic diet (SSD) from nascent stage (immediately after hatching), could successfully complete their life-cycle leading to adult emergence for the first time. The incidence of flower drying disease was lower in Mancozeb (8.9%)-

treated trees, followed by Copper oxychloride (11.9%) and Propiconazole (14.6%). The damage rating in untreated control was 22.8%. For control of tea mosquito bug (TMB) and other insect pests, Lambda- cyhalothrin (0.85%) was the best compared to the control (2.48%). Under curative control trial for management of CSRB, Chlorpyriphos 0.2% resulted in 83.33% trees without re-infestation at Bapatla and 90% at Bhubaneshwar.

Source: DARE/ICAR Annual Report 2010–11

Coconut: Fourty fluorescent pseudomonad isolates were obtained from coconut leaf vermicompost, of which 34 produced IAA in varying quantities. Nine endophytic bacilli from coconut roots and 12 coconut rhizospheric bacilli were found to inhibit both *Thielaviopsis paradoxa* and *Ganoderma* sp.

The Nonnegative Constrained Least Square (NCLS) algorithm, sub pixel classification of root (wilt) affected palm in IRS P-6 satellite data map showed that, among 7 southern districts of Kerala, nearly 30% of the coconut land cover is affected by coconut root (wilt) disease.

Combined application of *Bacillus subtilis* + *Pseudomonas fluorescens* was found effective in the management treatment of leaf rot disease in coconut. Placement of Mancozeb sachets in leaf axils around spindle leaf has been found to be an effective prophylactic measure for bud rot on coconut. For managing stem bleeding disease of coconut, soil application of *Trichoderma harzianum* +*T. virens* and root feeding of Tridemorph were effective.

By releasing parasitoids, infestation on coconut leaf by leaf-eating caterpillar could be controlled effectively. Evaluation of talc-based formulation of two virulent strains of *Hirsutella thompsonii,* viz. CPCRI-19 and CPCRI-51(II) resulted in 74- 82% mortality of coconut mite in treated buttons. The release of egg/larvae predator, *Cardiastethus exiguus,* in the crown of 10 % of coconut palms in garden, @ 50 bugs / palm at 5 days interval for six times consecutively at egg-early larval stages of coconut black headed caterpillar (*Opisina arenosella)*gave effective control of the pest in Tamil Nadu and Andhra Pradesh.

Cashew: The *Tribolium castaneum* and *Ephestia cautella* were found to be the common storage insect pests noticed in cashew processing units surveyed in Tamil Nadu, Goa and Kerala. At Bapatla, Triazophos (0.1%) was significantly superior against inflorescence thrips followed by Profenofos (0.05%) which recorded a damage score of 0.46 and 0.64, respectively. Profenophos could effectively check the damage by leaf caterpillar, leaf miner and thrips at Jagdalpur. The ë-cyhalothrinwas effective in managing tea mosquito bug, thrips and apple and nut-borer at Vengurla and Vridhachalam. Chloropyriphos (0.2%)

was the best treatment resulting in 100% of the treated trees without reinfestation by cashew stem and root- borer (CSRB) at Vengurle and Jhargram and 86% at Bhubaneswar, 78.38% at Chintamani and 72.73% at Jagdalpur.

Source:: DARE/ICAR Annual Report 2009–10

Recent Initiatives by Central Plantation Crops Research Institute, Kasargod

Screening coconut and cocoa accessions against Phytophthora disease

The rate of spread of infection (lesion size) caused by the six selected Phytophthora isolates of coconut viz., KL/CO-8, KL/CO-1 and KL/CO-78 (*P. palmivora* isolates), KL/CO-16 (*P. meadii*), TN/CO- 127 (*P. nicotianae*) and KA/CO-110 (*P. capsici*) was determined using the detached leaf technique in 10 commonly cultivated accessions/hybrids viz., West Coast Tall (WCT), Chowghat Orange Dwarf (COD), Chowghat Green Dwarf (CGD), COD x WCT, WCT x COD, Laccadive Orange Dwarf (LCOD), Malayan Green Dwarf (MGD), Malayan Yellow Dwarf (MYD), Ganga bondam Green Dwarf (GBGD) and Benaulim Tall (BENT). Phytophthora infection varied with the varieties and the Phytophthora species used for inoculation. Though all the accessions were susceptible to the infection caused by *P. palmivora* and *P. nicotianae*, the rate of infection/susceptibility varied. In general, LCOD was found to be highly susceptible to Phytophthora infection and MGD was least susceptible to all three isolates of *P. palmivora*. The lesion sizes caused by all 4 Phytophthora species were the smallest in MGD and *P. capsici* could not cause infection in this cultivar, indicating the high level of resistance of MGD to Phytophthora infection.

A preliminary study was conducted to assess the susceptibility of cocoa varieties and hybrids to Phytophthora species isolated from cocoa by artificially inoculating on detached cocoa pods. The study clearly revealed the variability in virulence of the five species as well as the two isolates of *P. palmivora*. Lesion size on pods of different accessions varied with the isolate and accession. *P. palmivora* was found to be more virulent than other four species. Among the two isolates of *P. palmivora*, the canker isolate, AP/CA-334 was more pathogenic than the black pod isolate. *P. meadii* was the least pathogenic species. The lesion size caused by canker isolate of *P. palmivora* (AP/CA-334) was more than that caused by KL/CA-216 on all the seven varieties. The accessions, ICS-6 x SCA-6, II-67 x NC-29/66, NC x 45/53 and II-67 were more susceptible to *P. citrophthora* than the *P. palmivora* isolate, KL/CA- 216 (black pod isolate). *P. tropicalis* was more pathogenic on pods of II-67 x NC-29/66 than both *P. palmivora* isolates and *P. citrophthora*. II-67 was more susceptible to *P. tropicalis* than *P. citrophthora* and KL/CA-216 isolate of *P. palmivora*.

P. capsici could not cause infection on pods of I-56 x II-67, ICS-6 x SCA-6, II-67 and ICS-6. Similarly, I-56 x II-67, II- 67 x NC-29/66, ICS-6 and NC-29/66 were resistant to *P. meadii* infection. Among the seven varieties tested, only NC x 45/53 was susceptible to the infection of all six isolates.

Fruit rot and crown rot of arecanut and their management

Phytophthora infected arecanut samples were collected from the endemic areas of Kerala and Karnataka state and eight isolates of *Phytophthora* spp. were obtained. All the isolates were identified as *Phytophthora meadii* based on morphological characters and among the isolates, P8 was found to be highly virulent, which produced the highest lesion size of 11.7 cm in detached arecanut assay.

Eight new fungicides (Akomin, Biofight, Biopot, Companion, Curzate, Melody duo, Mixol and Samarth) were screened against highly virulent isolate of *P. meadii* (P8) by poisoned food technique. Melody duo and Mixol showed 100 % inhibition of mycelial growth at 125 ppm. Curzate showed 100 % inhibition of mycelial growth at 500 ppm and companion at 1000 ppm.

Among the three fungal biocontrol agents viz., *Trichoderma harzianum, T. viride* and *T. virens* screened in vitro against virulent isolate (P8) of *P. meadii, T. virens* exhibited the highest mycelial growth inhibition up to 62.5% in simultaneous inoculation of *Phytophthora* and *Trichoderma* and 35.2% inhibition in inoculation of *Trichoderma* 48 hours after inoculation of *P.meadii. T. virens* was also found compatible with the tested fungicides (Akomin, Biofight, Biopot, Curzate, Melody duo, and Mixol) except with Companion.

Field trial conducted at Vittal and Kidu on management of fruit rot of arecanut with the presently available fungicides among the arecanut farmers did not indicate any significant difference in control of fruit rot over 1 % Bordeaux mixture considering the less overall disease (10-15 %) incidence during the current season.

***In vitro* studies on major diseases of coconut and arecanut**

Aqueous extract of thirty botanicals were tested in vitro at 5%, 10% and 20% concentrations, against *Ganoderma lucidum* (causal organism of Ganoderma wilt of arecanut) and *Thielaviopsis paradoxa* (causal organism of stem bleeding of coconut). *Allium sativum* at 10% conc. was found to have 100% inhibition over control against both *Ganoderma lucidum* and *Thielaviopsis paradoxa* and *Solanum nigrum* at 20% showed 100% inhibition over control against *Ganoderma lucidum* after 144 hr of incubation.

Twenty two isolates of *Trichoderma* spp. and 12 isolates of *Pseudomonas fluorescens* were isolated from healthy arecanut palms from different locations. Among the isolates, T16 showed maximum inhibition of *Ganoderma lucidum* with 88.55 per cent inhibition over control. Among the Pseudomonas isolates, Pf6 showed maximum inhibition of *Ganoderma lucidum* with 77.55 per cent inhibition over control.

Management of Ganoderma wilt/basal stem rot of arecanut in NEH region

Under the NEH Mini Mission project, a field trial on management of Ganoderma wilt/basal stem rot of arecanut was initiated during 2009 to study the efficacy of different treatments viz., T1: Soil drenching with Captan 0.3% (30 g 10 lit-^{1} of water) + Neem cake (2 kg palm-^{1}); T2:T1+ root feeding with Calixin @1.5 % (125 ml palm^{-1}); T3: T1 + root feeding with 100% culture filtrate of Trichoderma viride (25 ml palm^{-1}); T4:T1+ root feeding with 100% culture filtrate of Pseudomonas fluorescens (25 ml palm^{-1}), T5: T1 + root feeding with Neem leaf extract (10%), and T6: Untreated control. Treatment T3 showed the lowest mortality percentage (17) with higher annual leaf production (8.5) and fewer drooping leaves (0.92) than the rest of the treatments.

Subsequently, a field trial was initiated in the farmer's field during 2010 to study the efficacy of different treatments, viz., T1: Recommended dose of fertilizer, T2-T1+ Neem cake (2 kg palm^{-1} year^{-1}); T3- T2 + Trichoderma viride (100 g palm^{-1}); T4-T2 + Calixin (0.3%) @ 10 lit palm^{-1}, T5-T2 + Bavistin (0.3%) @ 10 lit palm^{-1}, T6- T2 + *Clerodendron infortunatum* @1kg palm^{-1}; T7-T2 + green leaves (Glyricidia) @ 10 kg palm^{-1}; T8- T1+green leaves (Glyricidia) @ 10 kgpalm^{-1} and T9-control. Among the treatments, T3, T4 and T6 were found to be promising with low mortality percentage (33.33).

Biochemical and serological detection of root wilt disease (RWD) of coconut

Determination of isozyme and phenolic compounds

Superoxide dismutase isozyme pattern from spindle leaf extracts were done in 5 different sets, with one each of healthy, RWD early, RWD middle and RWD advanced adult palm of CGD. Gels were run with both equal volume and equal protein load basis of soluble protein extract. Four to six bands of superoxide dismutase isozyme were observed in all the samples analyzed. The number of SOD isozyme bands was found to be the same in both healthy and RWD palms. Consistent differences in the intensity of these bands between healthy and RWD palms were also not observed across the five sets.

To determine the HPLC profile of phenolic compounds in healthy and root (wilt) diseased palms, the HPLC procedure was standardized. A combination of two separate solvents (Sovent A -1% acetic acid and Solvent B -100% methanol) in the proportion of 80:20 with total flow rate of 0.9 ml/minute and a run time period of 15 minutes were found to be optimum for obtaining dose response peaks to different concentrations of each of eleven phenolic compound standards. The R2 value of standard curve of each phenolic compound was above 0.97 when averages were taken for two consecutive days determinations.

Serological testing of samples through ELISA

Analysis of spear leaf samples of coconut collected from various sources were subjected to serological testing through ELISA. Of 48 samples subjected to ELISA test, 24 palms reacted positively and 24 negatively to the disease.

Preparation of polyclonal/monoclonal antisera for ELISA For the production of fresh polyclonal antisera against RWD phytoplasma, fresh antigen was purified from spear leaf samples of RWD affected coconut palms. Four intramuscular injections with purified antigens were administered to two Newzealand white rabbits at weekly intervals. Two batches of leaf samples of RWD affected palms were supplied to Chromous Biotech, Bangalore for the production of monoclonal antibodies specific to coconut root (wilt) disease phytoplasma and the identification and generation of recombinant surface antigens of phytoplasma of RWD is in progress.

Inducing systemic resistance

The experimental palms under various treatments (Digitonin, Triazole, Salicylic acid, *Pseudomonas fluorescens* and *Bacillus amyloliquefaciens* in comparison with control) were monitored for appearance of symptoms of root (wilt), leaf rot, major pests etc. The root (wilt) symptoms were confirmed in 10 palms. Data on growth characters such as number of leaves, flowering/inflorescence emergence, girth of bole, height and general outlook of the experimental palms were recorded twice (July 2012 and January 2013). Seed nuts of WCT and MGD were sown for the new induced systemic resistance trial with *P. fluorescens*, salicylic acid and Triazole treatments.

Molecular detection of phytoplasma of coconut RWD

Spindle leaf samples from 116 coconut palms were collected from Cumbam/ Theni regions of Tamil Nadu. DNA was extracted through phytoplasma enrichment protocol and PCR was performed using primer pairs 1F7/7R3 and nested primer pair 1F7/7R2. Out of 116 samples tested, 66 samples showed amplification in nested PCR. Embryos were also collected from 50 coconuts from root (wilt) affected coconut palms from Cumbam/Theni regions. DNA was extracted using Qiagen DNEASY plant mini kit. PCR was done by using primer pair 1F7/7R3 nested with 1F7/7R2. Out of 50 embryos tested, 19 embryos gave positive amplification at 480 bp.

Quick detection of root (wilt) disease of coconut using molecular kit

TaqMan assay was performed for different samples collected from Theni and Cumbam areas. Out of 58 samples tested, positive amplification was obtained for 47 samples (81%). Loop mediated isothermal amplification (LAMP) primers were designed using the software "Primer Explorer" and coconut root (wilt) phytoplasma sequence (GenBank Acc.No.FJ794816) as the template. All the designed primers were validated and finally one set of primers and three sets of LB primers were synthesized. The reaction conditions were optimized for the LAMP assay and positive amplification was obtained for the root (wilt) diseased samples.

Standardization of chromosome preparation for *fluorescent in situ* hybridization

Chromosome spreads were prepared from actively growing root tips of coconut. The coconut specific probe was prepared and hybridized with chromosome spread. Strong signal was obtained from the chromosome spread.

Testing of seedlings raised from nuts collected from diseased palm

Mature nuts from diseased coconut palms were collected from Kayankulam and sown in net house and observed for germination. Leaf samples were collected after germination and total DNA was extracted using DNeasy Plant mini kit (Qiagen). The samples (14 seedlings) were tested for phytoplasma using seminested primers 1F7-7R3-1F7-7R2. All the 14 seedling samples tested gave negative result for phytoplasma in the nested PCR assay. In vitro embryo culture of eight embryo samples collected from diseased palms was performed as per the protocol standardized at CPCRI. The plantlets were sampled after four months of inoculation for DNA extraction and PCR. These samples also tested negative for phytoplasma in the nested PCR assay with primers 1F7-7R3-1F7-7R2.

Characterization of arecanut YLD phytoplasma

Arecanut YLD phytoplasma partial secA gene (420 bp) was isolated and characterized. Primers for secA gene amplification was designed and nested PCR was done for 35 cycles with primer pair cocsnf/cocsr. The PCR fragents of expected sizes were excised from the gel, eluted, cloned and sequenced with M13 forward and reverse primers. Blastn search showed that the sequence had 99% identity to sugarcane grassy shoot phytoplasma secA gene.

Identifying the pheromone synergists of red palm weevil

Placing of food baits (banana) with pheromone lure had synergistic effect in trapping the weevils. The compounds released from the fermenting banana head space volatiles were ethyl acetate, iso amyl alcohol, acetoin and iso amyl acetate. When yeast was added to fermenting banana, the level of ethyl acetate, isoamyl alcohol, iso butyl acetate, ethyl hexanoate, iso amyl methyl butyrate, ethyl octonoate and ethyl decanoate increased.

The physiologically relevant odour to red palm weevil antennae released by fermenting banana was identified by Gas chromatograph coupled with electroantennography. Three blends A,B and C were prepared based on the proportion in the host volatiles and electrophysiological response using synthetic chemicals. Behavioural assay by wind tunnel revealed that blend C that had the major and minor compounds from host volatiles caused increased orientation of the weevils.

Blends prepared were loaded into the nanomatrix along with 4 methyl 5 nonanol and 4 methyl 5 nonanone (9:1). A good combining ability was observed between all the three blends and pheromone. In field trial, blend C (having combination of major and minor compounds released from the fermenting food baits) along with pheromone was effective in trapping in weevils.

Refinement of IPM of palms and cocoa

Oryctes rhinoceros

Prophylactic leaf axil filling experiment conducted in a farmer's field at Thiruvalla indicated significant reduction in the leaf damage ranging from 18.3% to 45.2%. The highest reduction in leaf damage by black beetle was observed in palms treated with Pongamia cake (45.2%) followed by Neem granules (41.4%) admixed with sand. A botanical cake comprising of hexane and methanolic extracts of *Clerodendron infortunatum* and *Euphatorium* sp. was developed for prophylactic leaf axil filling against black beetle infestation in coconut. The botanical soap @ 5-10% induced moulting aberrations (66.6-77.7%) during larval-pupal transformation of black beetle in laboratory bioassay.

In vitro study was carried out on the efficacy of botanicals viz., leaves of *Curcuma longa* (5 and 10% conc.), *Manihot esculenta* (5 and 10% conc.), *Vinca rosea* (5 and 10% conc.), *Carica papaya* (5 and 10%) and control (cow dung), against the grubs of rhinoceros beetle of coconut. Results revealed that 10% concentration of turmeric leaves showed 90% mortality of third instar rhinoceros larvae after 10 days of treatment.

Terpenoids were screened for their repellency to rhinoceros beetle. In the olfactometer assay, terpenoids repelled over 80 per cent of the beetles. The terpenoids caused antennal response when assessed by EAG. A delivery device for slow release of terpenoids was developed.

A field trial on development of management practices against rhinoceros beetle of coconut has been initiated at CPCRI (RC) Kahikuchi under Mini mission project, with five treatments viz.,T1: powdered Neem cake (200g) + sand (200g), T2: powdered mustard oil cake (200g) + sand (200g), T3: chlory dust (chlorpyriphos 1.5% D) 50g + 2 kg sand for 8 palms, T4: *clerodendron* dry leaves 200g + sand 200 g and T5: control. Pre-treatment observations on damaged leaves, spindle damage and spathe damage per palm were taken.

Nanomatrix loaded with pheromone trapped the maximum (105) beetles/trap. Polymer membrane dispenser containing 800 mg of pheromone trapped 72 beetles per trap. The phermone loaded in nanomatrix had a slower release rate as compared to the commercial lure in polymer membrane. Studies on optimum trap density revealed that 1 trap ha^{-1} is ideal, but if the population load of the pest is high, then 2 traps ha^{-1} could trap the beetles better.

The entomopathogenic nematode, *Steinernema abbasi* was pathogenic to third instar grubs of rhinoceros beetle at 350 IJS/cc of vermicompost in laboratory bioassay. *Steinernema carpocapsae* was oriented to over 7.3 cm in 72 hours of inoculation using volatile cues in vermicompost to find O. rhinoceros grubs.

Rhynchophorus ferrugineus

Laboratory evaluation of Cartap hydrochloride (Boregan 4% G) and Fipronil (Sargent 0.3% G) @ 0.01g grub-1 induced 65% and 95% mortality of red palm weevil grubs (n=20) in a period of 24 h. Precocious pupation, unfastened fibrous cocoons and malformed adults were formed in *Rhynchophorus ferrugineus* grubs treated with Cigna 0.01% (Lufenuron 5.4% EC).

Three isolates of Bacillus species from coconut ecosystem were tested on red palm weevil grubs. RPW grubs treated with Bacillus isolate (RC KD B2) induced 50% grub mortality. Further, 35% of the treated grubs transformed to pupae, but failed to metamorphose to adults, while the remaining15% of the grubs transformed to sub sized adults (n=30). Grubs treated with isolate EKA2-3 showed normal feeding and grubs treated with isolate TEC1-3 showed precocious pupation without feeding.

Aprotinin (Bovine pancreatic trypsin inhibitor) inhibited all the three endopeptidases of red palm weevil grubs viz., trypsin (77.4%), elastase-like chymotrypsin (31.9%) and chymotrypsin (8.6%) at the highest level (50 ìg) followed by soybean trypsin inhibitor and phenyl methyl sulphonyl fluoride.

An entomopathogenic nematode, *Heterorhabditis indica* (1500 IJ) with Imidacloprid (0.002%) was highly effective against red palm weevil grubs, inducing 95% mortality.

A prototype of red palm weevil (RPW) detector based on acoustic system was developed by CDAC, Trivandrum and the electronic gadget is being standardized for effective field detection of RPW infestation on coconut.

Aceria guerreronis

A significant reduction (67.4 to 69.8%) of mite incidence was observed in palms treated with palm oil (200 ml) and sulphur (5g) in field trials at Kasaragod and Minicoy. Under the Mini mission project, two field trials have been initiated for development of an integrated management system against eriophyid mite of coconut in the NEH region.

Field evaluation of talc based formulation of *Hirsutella thompsonii* in combination with botanicals was continued at farmer's plot, Nileswaram, Kasaragod District. There were 5 treatments including *H. thompsonii* talc formulation @20g palm-^{1} alone and in combination with botanicals (Neem oil/ garlicsoap emulsion/Azadirachtin 10000ppm) and control, each treatment replicated in 25 palms. Posttreatment observations indicated 51-61% reduction in mite incidence and 59-78% suppression in mite population. H. *thompsonii* talc formulation was supplied to 3 AICRP palms centers viz., HRA Ambajipeta (Andhra Pradesh), RCRS, Ratnagiri (Maharashtra) and ARS, Aliyarnagar (Tamil Nadu) and CPCRI (RC) Kahikuchi for multi-location field evaluation of the biocontrol agent against coconut mite. Coconut water was found as an ideal medium for mass production of *H. thompsonii*, as evidenced by comparable growth rate (1.91 cm in 20 days), spore production (12.9 x 104 cm3) and yield of dry mycelium mat (1.017 g 100 ml^{-1}) to that of standard fungal growth media.

Opisina arenosella

Application of cigna 0.01% (Lufenuron 5.4% EC) induced larval-pupal intermediates (80-90%) of black headed caterpillar.

Medium level of *Opisina arenosella* was noticed in Puthiyavila (Trivandrum) with leaf infestation of 59.6% and population of 141/100 leaflets. Awareness campaign was conducted in the area with collaboration of Parasite Breeding Station, Trivandrum and Dept. of Agriculture, Kerala. Regular monitoring and release of stage specific parasitoids resulted in 55.7% reduction of leaf damage and 94% reduction in pest population over a period of 8 months. Outbreak of *O. arenosella* was also noticed in Kallara (Kottayam) region during August 2012 with leaf infestation of 83.4% and pest population of 288/100 leaflets. Systematic monitoring and release of larval parasitoids viz., *Goniozus nephantidis* and *Bracon brevicornis* could considerably reduce leaf damage (42%) and pest population (93%) in a period of 7 months.

The farmers from Pollachi were trained on mass multiplication of *Goniozus nephantidis* and they are carrying out the multiplication in their gardens and releasing them in outbreak areas.

Leucopholis spp.

Metarhizium anisopliae spore suspension tested at higher dose treated on second and third instar grubs of *L. coneophora* could not infect the grubs. Natural infection of Cordyceps on *Leucopholis* spp. was recorded from Sringeri area where *L. lepidophora* is predominant. In laboratory assay, *S. carpocpasae* at 3850 infective juveniles (IJs) in combination of 0.25 ppm Imidacloprid was found promising against coconut root grub, *Leucopholis coneophora* (late II instar) grubs causing 48 % mortality.

S. carpocapsae @ 1.2 crore IJs/palm basin effected 30.2 and 52 % reduction in coconut root grub (*L. coneophora*) infested coconut garden on 40 and 80 days after application. Application of 60 cadavers of *G. melonella* per palm basin caused 26.72 and 57.94 % reduction of grubs after 40 and 80 days of treatment, respectively. Application of *Steinernema carpocapsae*@1.20 crore IJs with 1 ml of Imidacloprid resulted in 47.04 % reduction of *Leucopholis burmeisterie* (late second-instar) grubs in infested arecanut garden, whereas, when used in isolation caused 24.7% reduction after 90 days of treatment.

Helopeltis spp.

Cocoa is damaged by three species of tea mosquito bugs viz., *Helopeltis bradyi, H. theivora and H. antonii,* with *H. bradyi* being the most predominant species. Population dynamics of tea mosquito bug in cocoa was studied to understand the recent trends in seasonal abundance. Damage of tea mosquito bugs in cocoa was noticed immediately after monsoon (October - November), with the peak incidence during November and February to May, with the population declining during December to January.

Field screening of cocoa genotypes for tea mosquito bug resistance showed percentage damage ranging from 0-42.9. The Malaysian clones VTLC-1 and VTLC-6, Upper Amazon Forastero, VTLC-57 and Kew collection and VTLC-86 were found to be relatively tolerant. Evaluation of cocoa genotypes for antixenosis mechanisms of resistance revealed that nymphal emergence and transformation to adults were the lowest in VTLC-65 and VTLC-77, indicating resistance to TMB. The Malaysian clone, VTLC-8 was the most susceptible showing higher emergence of nymphs and their survival to adults. Prolonged nymphal developmental period was observed in the resistant genotypes VTLC-65 and VTLC-28 followed by VTLC-49, VTLC-64, VTLC-115, VTLC-52, as compared to normal nymphal developmental period in the susceptible check VTLC-8.

The results of field efficacy of insecticides revealed significantly higher damage reduction at 30 days after first application with Lamdacyhalothrin.

Monitoring and surveillance for possible invasive and emerging pests

Brontispa longissima was not located in any of the surveyed areas in the mainland as well as in Lakshadweep, Bay Islands and North-East hilly regions. Coconut inflorescence moth, *Batrachedra arenosella* Walker was recorded from South and Northern region of Minicoy Island, Andamans, Bay Island and Kasaragod, Kerala. Larvae feed on the male flowers within the inflorescence (visible gummosis) as well as after blooming causing male flowers to remain intact even after drying. At times, larvae scraps female buttons too causing lesions on the developing buttons which falls immaturely.

Two species of mealy bugs infesting coconut viz., leaf mealy bug (*Pseudococcus cryptus*) and spadix mealy bug (*Dysmicoccus finitimus*) were reported from Kerala and Tamil Nadu. *P. cryptus* and *D. finitimus* were found to colonize at moderate level on the leaves and spadix of coconut, respectively. Incidence of mealy bug on coconut buttons and immature nuts under the perianth was reported for the first time in NEH, from CPCRI campus, Kahikuchi.

Presence of both eriophyid mite and mealy bug has also been observed beneath the coconut tepals in some instances. The colour of the mealy bug is white to creamish white and older nymphs are covered with fluffy white wax and they are seen to colonize under the perianth lobes of coconut. The nymphs are approx. 1mm long and 0.3mm wide while the adults are approx. 2mm long and 1mm wide and they are covered with white waxy coating. The presence of mealybug is more abundant in inner tepals (5-6) than outer tepals (1-4). Further studies on its taxonomic identity, morphology, biology, host plants and distribution are underway.

Surveillance at Colachel and Cumbum in Tamil Nadu revealed coconut scale insect (*Aspidiotus destructor*), star scale (*Vinsonia stellifera*), arecanut whitefly (*Aleurocanthus arecae*), soft scale (*Lecanium* sp.) and leaf mealybug (*Pseudococcus cryptus*) at low to moderate levels. These sucking insect pests were naturally suppressed by the lady beetles belonging to *Chilocorus nigritus* as well as cybocephalid beetles in the coconut system. Occurrence of coreid bug infestation was found to be higher in Southern Kerala.

Gradient outbreak of coconut scale insect (*Aspidiotus destructor*) was observed at Chingoli near Kayamkulam during August-September 2012. Though the pest attack was confined in a limited pocket on coconut leaflets along a homestead farm pond, rise in maximum temperature (0.8°C) and reduction in relative humidity (4.1%) and rainfall (91.8 mm) during June 2012 compared to that of June 2011 could be the major reasons for the immediate flare up of the pest which was otherwise not reported as a major pest of the region.

Surveillance in Morigaon, Kamrup and Goalpara districts of Assam, Ri-Bhoi and East Garo Hills of Meghalaya indicated minor incidence (<1%) of red palm weevil in coconut. Eriophyid mite incidence in coconut was found to be low to medium (20- 30%).

A database has been developed for recording the monthly disease and pest incidence in the institute farms by institute surveillance squad.

Source: Annual Report 2012-13, Central Plantation Crops Research Institute, Kasargod

Salient Achievements of Central Plantation Crops Research Institute, Kasargod

Coconut

Root (Wilt) Disease

- Etiology of root (wilt) disease of coconut was established as phytoplasma by electron microscopy, transmission through dodder, light microscopic staining techniques and molecular tools.
- Molecular detection of phytoplasma associated with the root (wilt) disease of coconut was achieved by PCR using custom designed primers.
- First reported the association of 16Sr XI group phytoplasma with coconut root (wilt) disease in India
- A rapid and sensitive sero-diagnostic test (ELISA) was standardized for the early detection of root (wilt) disease.

- *Proutista moesta* and *Stephanitis typica* were identified as the vectors and established their role in disease transmission.
- Integrated disease management strategies were developed for heavily diseased tracts and mildly disease affected areas. An increase in yield of 25-83% depending on severity of disease can be obtained by adopting IDM.

Leaf rot disease

- Elucidated the etiology and symptomatology of leaf rot disease. Leaf rot is caused by fungal pathogens, mainly *Colletotrichum gloeosporioides*, *Exerohilum rostratum* and *Fusarium solani*.
- Control measures were developed with fungicide (Mancozeb/ Hexaconazole) and biocontrol agents. Application of consortium of biocontrol agents (*Pseudomonas fluorescens* and *Bacillus subtilis*) along with phytosanitation was found to be effective in the management of leaf rot disease.

Bud rot

- Prophylactic treatment of pouring Mancozeb solution (5g/300ml of water) coupled with placing of 2 Mancozeb sachets (5 g/ sachet) in the inner most leaf axils was found to be effective in controlling the disease.
- A new eco-friendly method of prophylactic management of bud rot disease has been developed with phosphorus acid (Akomin) solution (0.5%) @ 300 ml/palm.
- Developed integrated bud rot management strategy involving removal of disease advanced and dead palms, field and plant hygiene, treatment with fungicide, INM, control of other pests and diseases especially rhinoceros beetle infestation etc.

Stem bleeding

- Root feeding with tridemorph and soil application of Neem cake were found effective in managing stem bleeding disease of coconut.
- *Trichoderma virens, T. harzianum, T. viride* and *T. hamatum* were found to be antagonistic to the stem bleeding pathogen *Thielaviopsis paradoxa*.

Fruit Rot and Immature Nut Fall

- *Lasiodiplodia theobromae* was identified as the major causal organism of rotting and nut fall of eriophyid mite infested coconut.
- Carbendazim or a combination product of Carbendazim and Mancozeb was found effective in controlling *L. theobromae* infection of coconuts.
- If coconut is cultivated under organic farming system spraying of garlic bulb extract (10%) can be recommended for the management of fruit rot and immature nut fall in coconut.

Arecanut

Yellow leaf disease

- Etiology of yellow leaf disease (YLD) of arecanut was established as phytoplasma by electron microscopy and transmission studies with vectors.
- Confirmed the phytoplasmal etiology of YLD through molecular characterization using custom designed primers.
- Confirmed plant hopper *Proutista moesta* as vector of YLD.
- Standardized sero-diagnostic test for detection of YLD.
- Management strategy is recommended in the initial stages of disease. Apply 220 g urea, 200g super phosphate and 230 g muriate of potash per palm / year in two split doses during pre and post monsoon period. In addition to the above, apply one kg of super phosphate in affected gardens. Apply organic manure @ 12 kg each of compost and green leaves per palm per year. Avoid water stagnation by providing drainage facility. Control other pests and diseases.

Phytophthora diseases

- *Phytophthora meadii* was found to be the causative agent of fruit rot and bud rot diseases of arecanut.
- Developed a linear model based on multiple regression analysis to predict the fruit rot incidence four days in advance.
- Covering bunches with poly bags prior to monsoon was incorporated as a component in IDM of mahali.
- Developed crown rot management strategy using 0.3% phosphorous acid.

Anabe or Foot Rot

- Established the association of *Ganoderma applanatum* with the disease incidence.
- Developed IDM practices for foot rot of Arecanut.

Cocoa

Phytophthora diseases

- *Phytophthora palmivora* (A_2 mating type) was found to be the predominant species causing black pod and stem canker diseases in India.
- Rare incidence of black pod disease caused by *P. capsici* and *P. citrophthora* has been reported from Kerala state.
- Cocoa accession C78 was found to be less susceptible to black pod disease caused by *P. palmivora.*
- Developed IDM for black pod and canker.
- A technique has been recently developed to control stem canker using the biocontrol agent, *Trichoderma harzianum.*

Cherelle rot

- Elucidated the symptomatology, etiology and seasonal cycle of cherelle rot of cocoa in India caused by *Colletotrichum gleosporioides.*
- Developed IDM for Colletotrichum pod rot.
- Developed low cost techniques for mass multiplication of bio-control agents utilizing plantation crop wastes like coconut water and cocoa sweating. Diluted cocoa sweating was found to be superior to conventional liquid media.
- Developed a simple and highly economical low cost technology for farm level mass production of biocontrol agents using diluted cocoa sweating, bean shell and pod husk.
- A very promising mycological culture medium was developed using cocoa bean shell, an industrial waste.
- Developed technologies for production of oyster mushroom utilizing arecanut wastes.
- Developed technologies for production of coir pith formulation of Trichoderma (Trichoderma coir pith cake).
- Pesticide slow release product was developed using coir pith

Pests

Coconut

Rhinoceros beetle (*Oryctes rhinoceros*)

- *Metarhizium anisopliae* var. *major* was identified as a potential pathogen on *Oryctes* grubs. The mass production of the fungus in solid substrates namely cassava chips and rice bran mixture supplemented with nitrogen source have been developed at CPCRI.
- The fungal spores are to be applied @ of 5 x 10^{11} spores / m^3 on the breeding pits in which the beetles breed.
- The mode of infection of *Oryctes* virus (OrV) on rhinoceros beetle grubs has been established. Mass production of the virus is done by rearing the grubs on virus inoculated food such as sterilized cow dung, saw dust, coir pith etc. or by mouth feeding the healthy grubs with viral suspension and rearing them on sterilized food materials. Release of 12-15 infected beetles per hectare was recommended to disseminate the virus in nature.
- Shade dried leaf powder of the common weed plant *Clerodendron infortunatum* proved its insecticidal property against *Oryctes rhinoceros* at 10% w/w. Incorporation of the whole plant in the breeding sites of rhinoceros beetle proved to be effective in checking the build up of the pest in the breeding sites.
- Application of naphthalene balls @ 10-12 g/ palm or leaf axil filling of powdered marotti cake (*Hydnocarpus*) @250g/palm+ fine sand 250 g during April-May, and September -October proved to be effective as a prophylactic control measure for rhinoceros beetle and red weevil.
- A modified PVC trap was developed to be used with pheromone to attract the rhinoceros beetle adults.

Red Palm weevil *(Rhynchophorus ferrugineus)*

- Developed IPM package for red palm weevil that includes cleaning of palm crown periodically to avoid decaying of organic debris, proper cutting, splitting and burning of red palm weevil infested palms, treating of any wounds on the palm with coal tar and stem injection with 1% Carbaryll.
- Prophylactic leaf axil filling with 250 g marotti oil cake (*Hydnocarpus* sp.) + 200 g of fine sand in leaf axils around spindle during May, September and December has been found to be effective.

- A red palm weevil lure ferrugineol was formulated at CPCRI. Unlike the sachets used in commercial lure the CPCRI lure was made of glass capillary of 6 cm in length and id. of 1.5 mm. Each glass capillary had 100 μl ie 78.5 mg of ferrugineol with the density of 0.76g/cc. CPCRI lure was cost effective as compared to the imported lures.

- The role of kairamones in attracting the red palm weevils was established. Pheromone lure when placed in isolation did not yield desirable results; hence they had to be associated with food baits. Among the various phagostimulants that were evaluated to be used in tandem with pheromone lure, it was observed that macerated plantain (200g) or sugarcane bits attracted more weevils (an average of 8 weevils/trap/week), whilst the pheromone trap without food bait had a lowest weevil catch of 1 weevil/ trap / week.

Leaf eating caterpillar *(Opisina arenosella)*

- Techniques were evolved at CPCRI for mass multiplication of the promising parasitoids namely *Goniozus nephantidis, Bracon brevicornis, Elasmus nephantidis* and *Brachymeria nosatoi* which are being currently recommended for biocontrol of *Opisina.*

- Release rate of the parasitoids have been fixed. *G. nephantidis*, *E. nephantidis* and *B. nosatoi* are to be released at fixed norms of 20.5, 49.4 and 31.9% of larval, prepupal and pupal stage of the pest respectively. In a multistage condition of the pest, a combined release of all the parasitoids @40% of each of the target pest stage is required. Mass production technique for *Apanteles taragamae* on early instar caterpillars of *O. arenosella* was evolved.

- Olfactory conditioning in bethylid, *Goniozus nephantidis* an ectoparasitoid of coconut black headed caterpillar *Opisina arenosella* resulted in increased host searching ability of the parasitoid released in the field.

Eriophyid mite *(Aceria guerreronis)*

- In the year 1998 CPCRI first identified and reported the occurrence of the coconut infesting eriophyid mite, *Aceria guerreronis*. This report was the first for the whole of Asia. Palms receiving root feeding of 10 ml Neemoil+ 10 ml water and palms base drenched with 10 ml Neemoilalong with soil application of Neemcake showed the highest percentage reduction in the eriophyid mite infestation.

- The acaropathogenic fungus *Hirsutella thompsonii* has been identified as potential bioagent against the eriophyid mite *Aceria guerreronis*. Talc based formulations of two isolates CPCRI -19 and CPCRI 51 (11) are virulent. Spraying two rounds of the fungal formulation caused a reduction to a tune of 44 – 73 %.
- Developed management strategies for eriophyid mite using botanochemicals. Spraying Neem formulation containing 1% azadirachtin @ 4 ml/ l of water or 2% Neem oil, garlic and soap mixture during April-May, Oct.-Nov. and Dec.-Jan., in such a way that all mite infested palms in an area should be covered at the shortest possible interval.

White grub *(Leucopholis coneophora)*

- Developed IPM technology that includes deep ploughing and digging of soil during pre- and post-monsoon period, collection and destruction of adult beetles during peak emergence period in May-June, setting up of the light traps to attract adult beetles and killing them during May-June and September-October.

White grub *(Leucopholis coneophora)*

- Developed IPM technology that includes deep ploughing and digging of soil during pre- and post-monsoon period, collection and destruction of adult beetles during peak emergence period in May-June, setting up of the light traps to attract adult beetles and killing them during May-June and September-October.

Coreid bug *(Paradasynus rostratus)*

- Applying three rounds of 0.1% Carbaryll was recommended for managing the coreid bug.

Rodents(arboreal black rat *Rattus rattus wroughtoni, burrowing rodents Bandicota bengalensis, B. indica, Gerbils Tatera indica)*

- Placing of single dose anticoagulant Bromodiolone (0.005%) 10 g blocks twice at an interval of 12 days on the palm crown of one tree out of every 5 trees was found effective against arboreal black rat.

Termites *(Odontotermes obesus)*

- Drenching the nursery with 0.05 % chlorpyriphos twice at 20-25 days internal or swabbing the affected portion of the trunk in adult palms with 0.05 % chlorpyriphos solution was found effective.

Arecanut

White grub *(Leucopholis lepidophera Blanch)*

- Collecting the adult beetles between 6.30-7.30 pm, during pre monsoon showers was identified as an effective measure.
- Placing light traps @ 2 to 3 / ha was an effective approach.
- Basin application of Neem cake @2kg/palm/year during June to July for re-generation of affected roots.
- Basin application of Chlorpyrifos @ 7 ml in one litre of water /palm during May - June and September – October was found to be effective in managing the grubs.

Spindle bug of arecanut *(Carvalhoia arecae)*

- Sucking of plant sap by the nymphs and adults of spindle bug result in typical linear, dark brown lesions which turn necrotic on the spindle. Damaged open leaves dry up and drop off. Severely affected spindle leaves fail to unfurl completely. Infestation in young exposed plantations results in severe manifestation of symptoms during summer months.
- Spray spindle leaf and inner most leaves of palms in infested plantations with Dimethoate (0.06%) 30 EC 2 ml/ lit of water.

Red and White Mites *(Raoiella indica and Oligonychus indicus)*

- Spraying Dimethoate (0.06%) 30EC @ 2 ml/litre of water to the lower surface of leaves was effective in reducing the pest incidence.

Immature fruit drop

- Feeding by *Halyomorpha marmorea* was identified as a cause for immature fruit drop with shrinking and browning of kernels.
- An egg parasitoid of the pentatomid bug, *Halyomorpha marmorea* which causes immature fruit drop in areca palm is identified as *Anastatus bangalorensis* (Hymenoptera: Eupelmidae).
- Spraying Dimethoate (0.06%) 30EC 2ml/litre of water to the tender bunches of the affected palms and surrounding few palms.

Scale Insect

- Spraying Dimethoate (0.06%) 30EC 2ml/litre of water to the tender bunches was found to be effective in containing scale insects.

- Cowpea, okra, bitter gourd, chillies etc, were identified as alternate host for this pest.
- Lady bird beetle, *Chilocorus nigrita* was identified as an effective predator against the scale insects.

Cocoa

Tea mosquito bug

- The damage symptoms like necrosis and dieback of shoots, bark roughening, destruction of flowers and cherelles were identified as symptoms due to damage.
- In cherelles, feeding punctures cause distortion during growth.
- Yield reduction was reported up to 30 to 40 %.

Rodents

- Placing 10g Bromadiolone (0.005%) wax cakes on the branches of cocoa trees twice at an interval of 10-12 days will check the rat damage was found as an effective technique to manage the rodents.

Squirrels

- Squirrels gnaw the pods in the center and damage only the mature ones. They feed on the mucilage covering of the beans. Trapping with wooden or wire mesh single catch 'live' trap with ripe coconut kernel as the bait is effective. Timely harvest of the pods will help in increasing the efficiency of poison baiting as well as trapping.

Source: Central Plantation Crops Research Institute, Kasargod

□□□□

16

Recent Initiatives in Post Harvest Technology

Technique to collect fresh and hygienic inflorescence sap (Neera) from coconut

In the traditional method of collecting sap, the inflorescence to be tapped is connected with an earthen pot/bamboo receptacle for the sap collection. Spadix to be tapped is tied with plastic rope and a small portion is sliced off from the top. The spadix is gently stroked with a mallet in the morning and evening. After a period of 8 to 10 days, sufficient sap starts oozing out from the cut end. A type of clay soil, a gummy material or leaf extract is pasted on the surface and sides of the cut end. Since the sap is highly nutritious, both enzymatic and microbial fermentation takes place. The total sugar reduces progressively and it becomes more acidic and develops a harsh odour and becomes unpalatable. In order to prevent fermentation, lime is coated inside the pot. The above method is not hygienic and attracts lot of insects, ants and contaminates from pollen and dust particles. In order to collect the sap fresh and hygienic, a simple device was developed.

Preservation studies on coconut inflorescence sap

Preservation studies in coconut inflorescence sap were performed using two commonly used preservatives, namely, citric acid and potassium metabisulphite with different concentrations. Addition of 1.0% citric acid showed better preservation compared to 0.5% citric acid and 1.0% potassium metabisulphite.

Preparation of natural coconut sugar from coconut inflorescence sap

Natural coconut sugar can be prepared from fresh sap without the addition of any chemicals. The sugar is brown in colour. Comparative studies on biochemical and mineral constituents revealed that there was a significant difference between sap sugar and cane sugar. The total sugar content was less in coconut sap sugar (91.57g $100g^{-1}$) compared to cane sugar (97.7g $100g^{-1}$). The protein content is nearly 9 times higher in coconut sap sugar (2.6 g) compared to cane sugar (0.31g). Coconut sap sugar is also rich in minerals like sodium, potassium and phosphorus compared to cane sugar.

Coconut haustorium – a rich source of biochemical and mineral nutrients

Haustorium is the spongy tissue developed in coconut during germination. The haustorium is of a pale yellow colour on the outside and inside is cream white. The yellow portion contains more oil while the inner white portion has carbohydrates and proteins. The total sugar, reducing sugar and protein content in the white portion ranged from 42 to 44 %, 26 to 27% and 6.8 to 7.9%, respectively. Phosphorous (P), potassium (K), sodium (Na), calcium (Ca), magnesium (Mg), iron (Fe) and zinc (Zn) content in white portions of coconut haustorium were found to be in the range of 0.32-0.35 g, 1.76-2.89 g, 1.79- 2.46 g, 91.4-132.6 mg, 14.6-34.8 mg, 2.6-2.9 mg, 1.1-1.2 mg, respectively. Contribution of coconut haustorium towards recommended dietary allowance (RDA) for mineral was calculated and found that 100g dry haustorium powder can supply 53.3 – 58.3 % P, 47.5 - 90.0 % K, 85.2 - 117.1 % Na, 15.2 – 22.1% Ca, 4.3 – 10.2% Mg, 15.2 – 17% Fe and 8.9 – 10.0% Zn requirements.

Physiological and biochemical basis to improve the shelf life of tender coconut

The quality of tender nut water gets deteriorated within 2 to 3 days of harvest of the tender nuts from the tree. The reasons behind the deterioration of tender coconut water are not well studied. In most of the fruits and vegetables, the senescence and deterioration is attributed to enhanced respiration (climacteric rise), ethylene synthesis and activity, enhanced loss of water through transpiration or combination of all these factors. In harvested tender nuts, it was observed that the total sugar and reducing sugar content of tender nut water increased within 3 days of its harvest and start declining after 3 days. On the other hand, the total polyphenols and ascorbic acid contents declined after harvest.

Chemical parameters and anti oxidant capacity of virgin coconut oil

Various chemical parameters such as free fatty acids content, saponification value, peroxide value and total polyphenols content were determined in hot processed virgin coconut oil (VCO) and traditional coconut oil (TCO). Free fatty acid content of VCO (0.23 ± 0.18 mg KOH g^{-1}) was found to be very low compared to TCO (1.066 ± 0.079 mg KOH g^{-1}). The saponification value of both VCO and TCO are almost same, with 271.80 for VCO and 253.57 for TCO. The peroxide value of 2 weeks old VCO and TCO was found to be 0.54 and 0.823 meq oxygen kg^{-1} respectively. The peroxide values were negligible in fresh sample. Total polyphenols content was high in VCO in terms of Catechol (18mg $100g^{-1}$oil) equivalent and Gallic acid equivalent (9.57 mg $100g^{-1}$oil) compared to TCO (below 2mg $100g^{-1}$).

The antioxidant activity of the methanol extracts of the VCO and TCO was investigated using the DPPH radical scavenging assay. The methanol extract of the VCO and TCO exhibited a significant dose-dependent inhibition of DPPH activity, with a 50% inhibition (IC 50) at a concentration of 44.15±1.02g g^{-1} BHT; 36.12±0.84g g^{-1} Vitamin C for VCO and 342.68±2.12g/g BHT; 355±3. 42g g^{-1} Vit. C for TCO, indicating that the antioxidant activity of VCO is almost nine times greater than that of the TCO.

Antimicrobial and anticariogenic activities of arecanut

The effect of ethanol extract of arecanut in different concentrations (0.025-100 mg ml^{-1}) was screened against *Staphylococcus aureus* and *Escherchia coli*. Ethanol extract of arecanut showed a zone of inhibition (15-20mm) against *S. aureus*. The antibacterial activity increased with increasing concentration of arecanut ethanol extract. In this study, minimum inhibitory concentration (MIC) and minimum bactericidal concentration for *S. aureus* were found to be 75 mg ml^{-1} and 100 mg ml^{-1}, respectively. Ethanol arecanut extract did not show strong antibacterial activity against *E.coli*.

Development of an electrical coconut flaking machine

Prototype of a coconut flaking machine has been designed and fabricated. The machine is provided with two sets of cutting blades to produce coconut flakes of 5 mm and 10 mm thickness. The machine has a capacity to make 3 kg coconut flakes in one hour, with an energy consumption of 0.03kwh (0.03units). The machine is women-friendly, can considerably reduce the drudgery and would be useful for entrepreneurs engaged in production of confectionaries and bakery products where coconut flakes is used as an ingredient.

Development of a double screw coconut milk expeller

Coconut milk expelling is an important process in the production of high value products like virgin coconut oil and coconut milk cream. Prototype of a double screw coconut milk expeller has been designed and fabricated to extract milk from coconut kernel. The expeller has a capacity to extract milk from 1000 coconuts in one hour. Maximum milk could be extracted in two runs itself. Milk extraction efficiency remains the same when the coconut kernel is fed with or without pulverization.

Source: Annual Report 2012-12, Central Plantation Crops Research Institute, Kasargod, Kerala

□□□□

17

Diseases of Major Plantation Crops

Common name, causal organism, symptoms and management of diseases of major plantation crops

Name of Disease and its causal organism	Pathogen	Symptoms	Management
Cashewnut			
Die – back or Pink disease *Corticium salmonicolor*	Fungus	• It is a very common disease of cashew, often assuming great importance during the south-west monsoon period • Whitish or pinkish growth of the fungus can be seen on the affected branches. • The fungus penetrates into the deeper tissues and causes the death of the shoots from the tip downwards and hence the name dieback. • After heavy rains a film of silky thread of the fungus is seen on the branches. • In advanced stages, the bark splits and peels off. Some times only one branch is affected, but often many branches turn yellow and shed giving a barren appearance to a portion of the tree.	• Prune the affected branches well below the site of infection and destroy them. • Protect the cut surfaces by applying Bordeaux paste. • If disease is severe, it is advisable to spray Bordeaux mixture (I %). • Spraying is to be done twice, once in. May-June and the second during October. • Collect all dried up and affected branches to reduce the source of inoculum.
Damping off *Phytophthora palmivora*	Fungus	• The Disease occurs in nurseries where drainage conditions are poor. • The organisms attack the roots or collar region of "seedlings or both the regions and cause their death. when seedlings are infected by *Phytophthora palmivora,* they become pale. • Water-soaked lesions can be observed at the collar region which turns dark and girdles the stem. • The seedlings droop and ultimately the plants die.	• Provide adequate drainage in the beds and polythene bags. • Drench the beds/bags with 1 % Bordeaux mixture.

Name of Disease and its causal organism	Pathogen	Symptoms	Management
		• On leaves, water-soaked lesions can be observed in severe cases. These lesions enlarge and coalesce, often covering the entire leaf lamina. • All the organisms in combination or alone may cause the disease.	
Anthracnose *Gloeosporium gloeosporioides*		• The disease has been reported in an epidemic form Tamil Nadu. It is known to cause severe loss in Brazil. • The fungus infests the tender leaves, twigs and forms reddish brown, water-soaked lesions. On the affected region exudation of resin can be seen. • The lesions enlarge and kill the shoots. The tender leaves are crinkled and fruits shrivelled. • The infected inflorescences turn black. Repeated Infection of the terminal shoots leads to the death of the tree in course of time.	• Destroy all affected branches • Spray the plants with 0.5% Bordeaux mixture or Mancozeb 0.25%.
Inflorescence blight *Gloeosporium mangiferae*/ *Phomopsis anacardii*/. *Fusarium* sp	Fungus	• This is also a common disease in Kerala especially during the monsoon period. • The characteristic symptom is the drying of floral branches. The symptoms appear as minute water soaked lesions on the main rachis and secondary rachis. • The lesions are pinkish brown, enlarge and soon turn scabby. Gummy exudates can be seen at the affected regions. • The lesions develop into bigger patches and result in drying up of the inflorescences. The incidence is very severe when cloudy weather prevails.	• A combination spray of a fungicide (Blitox 250 g in 100 litres of water) and an insecticide Chloropyriphos (200 ml in 100 litres) is recommended. • It is often claimed that the primary cause of the disease is the tea mosquito and fungi are only secondary organisms.
Shoot rot and leaf fall *P. nicotianae*	Fungus	• During the south west monsoon months of June - August extensive leaf fall and shoot rot symptoms	• Spraying with Bordeaux mixture (1 %) before the onset of monsoon will check the spread of the

Name of Disease and its causal organism	Pathogen	Symptoms	Management
var. *nicotianae*		are observed. • Black elongate lesions are firstdeveloped on the stem with exudation of gum. Later, infection spreads up and down, causing the tender stem to collapse and tender leaves to shrivel up. • The lower mature leaves are also infected with black elongated lesions on mid rib, which later spread to the main lateral veins and the leaf blade. • The infected leaves are soon shed.	disease.
Arecanut			
Mahali / kolerogo/ fruit rot *Phytophthora arecae*	Fungus	• Rotting and excessive shedding of immature nuts from the trees. • The water soaked spots initially develop at the base of the nut. Fruit stalks and rachis of inflorescence are also affected. • Nuts show large vacuoles and dark brown radial strands. • Very often the top of the affected trees dries up resulting in withering of leaves and bunches. • Affected nuts fall off and show the white mycelial growth of the fungus.	• Clean cultivation • Destruction of affected trees. • Collection and burning of fallen nuts. • Spray with Bordeaux mixture 1% before on set of the monsoon and second spray 40-45.
Bud rot *Phytophthora arecae/ P. meadii*	Fungus	• The first symptom is the change of spindle leaf colour from green to yellow and then brown. • The leaves rot and the growing bud rots causing death of the palm. • The affected young leaf whorl can be easily pulled off. • The outer leaves also become yellow and droop off one	• Destruction and removal of dead palms and bunches affected by mahali disease. • Drenching crowns with Bordeaux mixture 1% including surrounding healthy palms.

Name of Disease and its causal organism	Pathogen	Symptoms	Management
		by one leaving a bare stem.	
Foot rot or Anabe roga *Ganoderma lucidum*	Fungus	• The leaf lets in the outer wholes of leaves become yellow and spreads to the whole leaf and the leaves drooping down covering the stem. • Later, the inner whole leaves also become yellow. Subsequently all the leaves droop, dry up and fall off, leaving the stem alone. • Then the stem also become brittle and easily broken by heavy wind. • The base of the stem shows brown discoloration and oozing of dark fluid. • Bracket shaped fructifications of the fungus called 'anabe' appears at the base of the trunk. Roots become discoloured, brittle and dried. • When infected trunk is cut brown discoloration can be seen up to one metre from ground level.	• Clean cultivation • Destruction of infected trees and stumps • Maintaining optimal plant population without over crowding. • Providing good drainage facility and fertilizers and manures. • Digging trenches to avoid root contact from diseased to healthy. • Drench 1% Bordeaux mixture at frequent intervals.
Yellow leaf disease	Phytoplasma	• Yellowing of tips of leaflets in 2 or 3 leaves of outermost whorl. • Brown necrotic streaks run parallel to veins in unfolded leaves. • The yellowing extends to the middle of the lamina. Tips of the chlorotic leaves dry up. In advanced stage all the leaves become yellow. • Finally the crown leaves fall off leaving of a bare trunk. Root tips turn black and gradually rot.	• Maintain trees in seed vigour. • Growing resistant varieties like true Mangala and south Kanara • Application of potassium and magnesium more than the recommended level.

Name of Disease and its causal organism	Pathogen	Symptoms	Management
Inflorescence die back and button shedding *Colletotrichum gloeosporioides*	Fungus	• Disease appears on rachillae of the male flowers and then in the main rachis as brownish patches which soon spread from tip downwards covering the entire rachis causing wilting. • The female flowers of the infected rachis shed and the whole inflorescence show 'die back' symptom. • The fruiting bodies of the fungus, (conidia) appear as concentric rings in the discoloured areas. • The disease is severe mostly during dry condition (February – March).	• Spraying with COC 0.25% or Mancozeb 0.2%.
Coconut			
Root wilt	Unknown etiology	• Tapering of terminal portion of the trunk. • Reduction of leaf size. • Ribbing and flaccitation of leaf lets.	• Cut and remove disease advanced, uneconomical palms yielding less than 10 nuts per palm per year. • Grow green manure crops - cowpea, sunhemp (Crotalaria juncea), Mimosa invisa, Calapagonium mucanoides, Pueraria phaseoloides etc. may be sown in coconut basins during April-May and incorporated during September-October. • Irrigate coconut palms with at least 250 litre water in a week. • Adopt suitable inter/mixed cropping in coconut gardens. • Provide adequate drainage facilities. • Pseudomonas fluorescence 50g + Trichoderma viride 50g + Cow dung 5kg /tree.Mix well & apply around the tree. Root feeding with Calyxin 5ml

Name of Disease and its causal organism	Pathogen	Symptoms	Management
			+100ml water Intercrop with Sunhemp. • Calyxin Root Feeding (Calyxin 2ml+Water 100ml) Calyxin paste applied around the stem Bleeding place. Growing Sunhemp as intercrop around the tree.
Bud Rot *Phytophthora palmivora*	Fungus	• The earlier symptom is the yellowing of heart leaves. • In the later stages the heart withers and drops down. • Decay of basal region of the heart leaves • Infected heart leaves are easily come out with slight pulling.	• Remove all the affected tissue of the crown region and crown drenching with Copper oxychloride 0.25%. • Spray 0.25% Copper oxychloride on the crown of the neighbouring palms as a prophylactic measure before the onset of monsoon. • Spray with Copper oxychloride 0.25% after the onset of Monsoon.
Stem bleeding *Thielaviopsis (Ceratostomella) paradoxa*	Fungus		
Basal stem end rot *Ganoderma lucidum*	Fungus	• Exudation of reddish brown liquid through cracks at the base of the trunk and oozing spread upward. • Decaying of tissues at bleeding point • Drooping of leaves. • Bracket formation at the base of the trunk.	• Isolation trench around the tree, 4 ft away from the base of the trunk. • Apply Sulphur dust inside the trench. • Soil drenching with Bordeaux mixture 1% @ 40lit/ tree. • Trunk injection / root feeding with Calixin 3ml/ tree. • Apply FYM 50kg + Neem cake 5kg once in 6 month. • Raising of green leaf manure and in-situ ploughing

Name of Disease and its causal organism	Pathogen	Symptoms	Management
			at the time of flowering. • Root Feeding –Calyxin 5ml +100ml Water.
Leaf blight or Grey Leaf Spot *Pestalosia palmivora*	Fungus	• Minute yellow spots encircled by greyish bands appear on the surface of mature leaves of the outer whorl. • Later they become greyish white. These spots coalesce into irregular necrotic patches. • Complete drying and shrivelling of the leaf blade are common when the infection is severe.	• Removal of the older 2-3 disease affected leaves and spraying the foliage with 0.25% Copper oxychloride will check the spread of the disease. • Application of extra 1.5kg of Potash. Pseudomonas fluorescence 200g+Cowdung 5kg/ tree. Removed burning of affected leaves.
Oil palm			
Spear -rot	Phytoplasma		
Fruit rot	Fungus	*Marasmium palmivorus*	
Bud rot	Fungus	*Ermima* sp.	
Anthracnose *Botryodiplodia palmarum*	Fungus	• This disease occurs in the nursery. • It is recognized by regular or irregular brown to black leaf blotches surrounded by yellow haloes, which develop along the margin, centre or tip of the leaves. • It causes heavy seedling loss.	• The disease can be controlled by spraying Mancozeb or Captan at the rate of 200 g/100 litres of water. • Copper fungicides should not be used because of the extreme susceptibility of oil palm seedlings to Copper burn (scorching).
Rubber			
Abnormal leaf fall *Phytophthora palmivora and P. nicotianae var. parasitica*	Fungus	• It occurs during June – August, while general leaf fall occurs during December. • On leaves dull grey, circularspots appear which enlarge and become irregular. • The petiole exhibit sunken spot. Affected fruits rot. Leaves shed prematurely either green or after turning Coppery red.	• Prophylactic sprays prior to onset of south west monsoon with Bordeaux mixture 1%. Addition of ZnSo4 @ 0.2% improves the efficacy.

Name of Disease and its causal organism	Pathogen	Symptoms	Management
		· The affected leaf form a thick carpet of rotting foliage which emits bad smell.	
Secondary leaf fall *Glomerella cingulata*	Fungus	• Numerous minute circular brown spots appear on leaves which turn into grey spots with brown margin. • The spots coaleses and dry up. Leaves crinkle and shed prematurely. Green pods which are affected become rotted.	• Spraying with 1% Bordeaux mixture.
Birds eye spot *Drechslera heveae (Helminthosporium heveae)*	Fungus	• Circular spots with grey centre and with brown border occur in large number in nursery or on young plants in the field.	• Spray with 1% Bordeaux mixture.
Corynespora leafs spot *Corynespora cossicola*	Fungus	• Circular spots with brown or papery centre and with a dark brown ring appear on the young leaves. • An yellow hallo develop around the ring. Short holes occur. Affected leaves dries up.	• Spray with 1% Bordeaux mixture.
Pink disease *Corticium salminicolor*	Fungus	• Prevalent in wet areas. Young twigs and branches are mostly affected. • The fungal growth encircles the stem penetrate the bark and cortical tissues which eventually decay. • The park splits and peals off. If the infection in not recognized early the tree is killed after the rainy season.	• Affected parts should be pruned and burn and pasting with Bordeaux paste or painting tar. • Copper fungicides should not be used in rubber because they will be contaminated in latex.
Powdery mildew *Oidium heveae*	Fungus	• White powdery fungal growth appears on young leaves and also on matured leaves. • Infected leaves curl, crinkle, role inwards and fall off leaving the petiole attached to the tree giving a broom stick appearance. • The infected flowers and tender fruits shed.	• Dusting with sulphur 3-5 spray at 15 days interval. • Carbendazim 0.1% or Tridemorph 1.5% dusting alternated with sulphur dust 70%. • Carbendazim +wettable sulphur and microsul (52% EC) is also effective.
Brown root	Fungus	• The affected roots are encrusted with mass of soil and	• Removal and destruction of the infected trees.

Name of Disease and its causal organism	Pathogen	Symptoms	Management
diseases *Fomes noxius*		small stones held by net work of mycelium. • The affected tap root becomes rotted and the entire tree is killed.	• Apply lime @ 2.5t/ha • In the area of dead tree apply additional dose of lime @ 25 kg/ dead tree.
Coffee			
Rust *Hemileia vastatrix*	Fungus	• It affects mostly leaves and some time leaves and tender shoots. • Rust lesions develop as orange yellow spots on lower surface of leaves. • Later increase in size and cause entire leaf become necrotic. • Diseased leaves full of yield of tree are reduced.	• Collection and destruction of diseased leaves. • Spraying with Bordeaux mixture 0.5% or COC 0.25% Triadimefon 0.05%, Mancozeb 0.25%.
Brown leaf spot *Cercospora coffeicola*	Fungus	• Brown spots with grey centre and brown margin surrounded by yellow ralo is developed. • The lesion spreads to major leaf area. On berries dark blotches are developed shrink and drops.	• Spraying with Bordeaux mixture 0.5%.
Damping off / Collar rot *Rhizocotonia solani*	Fungus	• It caused pre emergence damping off and post emergence damping off. • In post emergence damping off collar region near soil level is infected leading the rotting of tissue and death of seedlings.	• Soil drenching with Copper oxychloride 0.25%.
Die back or Anthranose *Collectorichum coffeasum*	Fungus	• On leaves circular to greyish spots of 2-3 m in dia. • On berries small dark coloured sunken spots are farmed. Beans become brown. Die back also occurs.	• Spraying Mancozeb 0.25%.

Name of Disease and its causal organism	Pathogen	Symptoms	Management
Black rot or koleroga *Corticium salmonicolor*	Fungus	· The most important symptom is dark brown or black decaying leaves, twings and bessies from which black rot is derived. • The leaves hang sown by a strong fungal stranchs – myedial threads can be seen on twinges and petirles. • Sclerotia can be seen all over the affected areas. Defoliation and drop occurs.	· Removal of diseased portions. • Spraying with Bordeaux mixture 1%.
Cocoa			
Seedling blight *Phytophthora palmivora*		• The symptom develops on the leaves and stem of seedlings or budded plants. • On leaves, small water soaked lesions appear which later coalesce in the blighting of leaves. • On stem, water soaked lesions develop initially and later turn to black colour. • Stem infection develop any point on the stem causing the death of seedlings.	• Remove and destroy the affected seedlings. • Spray with 1 percent Bordeaux mixture or 0.2% Copper oxychloride just before the onset of monsoon and thereafter at frequent intervals.
Black Pod rot *Phytophthora palmivora*		• Infection appears as chocolate brown spot, which spreads rapidly and soon occupies the entire surface of the pod. • As the disease advances, a whitish growth of fungus consisting of fungal sporangia is produced over the affected pod surface. • Ultimately, the affected pods turn brown to black. • The internal tissues as well as the beans become discoloured as a result of infection. • The beans in the infected pods approaching ripeness	• Periodically remove the infected pods. • Spray 1 per cent Bordeaux mixture with on set of monsoon and also frequent intervals. • Provide frequent drainage and regulate shade to \ increases aeration. • *Pseudomonas fluorescens* (Pf1) liquid formulations @ 0.5% as soil and foliar spray (3 times per year- June, October & February) was found to be effective in reducing the Cocoa Pod rot and Stem canker.

Name of Disease and Pathogen its causal organism	Symptoms	Management
	may escape infection because they are separated from the husk on ripening.	
Cherelle wilt	• The shriveling and mummifying of some young fruits are a familiar sight in all cocoa gardens. • In the early stages the fruits lose their lustre and in four to seven days they shrivel. • The fruits may wilt but do not abscise. • Many other factors like insects, diseases, nutrient competition and over production may also be associated with this problem.	
Vascular Streak Dieback (VSD) *Oncobasidium theobromae*	• The first indication of the disease is a characteristic yellowing of one or two leaves on the second or third flush behind the growing tip. • Diseased leaves fall within a few days of turning yellow and the other leaves on the shoot show similar symptoms. • When the infected shoot is split lengthwise there is always a characteristic brown streaking.	• The disease can be controlled by disposing diseased branches and regular pruning of chupons on the trunk. • Cocoa nurseries should not be located near the diseased area. • Avoid getting seedlings from diseased tracts. • Kerala Agriculture University has developed some VSD resistant and high yielding varieties CCRP-1 to CCRP-7.
Tea		
Blister blight *Exobasidium vexans*	• Small pale or pinkish circular spots appear on leaves and attain a size of 2.5 cm dia. • The spots in the upper surface of the leaf becomes light brown in colour and depressed while in under surface of leaf it bulges farming a blister like swelling.	• Removal and destruction of the affected portion. • Spraying with Copper oxychloride 0.25 % in effective. • Spray with 210 g of COC + 210 g nickel chloride/ha at 5 days interval from June – September and

Name of Disease and its causal organism	Pathogen	Symptoms	Management
		• Lower budget portion is covered with a white powdery growth of fungus. When many spots coursers, curling of leaves will occur. • When it spreads to young succulent stems affected portion are withered. • The leaf yield is reduced vigour of the tea bush is affected.	11 days interval in October – November gives economic control. • Spraying with systemic insecticides like Atemi 50 SL at 400 ml/ha (or) Baycor (300 EC) at 340 ml/ha a weekly interval is also effective. • Chlorotalonil, Bayleton, tridemorph is also effective. Tridemorph at 340 and 60 ml/ha is satin factory under mild and moderate rainfall condition.
Red rust *Cephaleurus mycoidea*		• Orange yellow, circular patches appear on upper surface of leaves. • The spots become brown and dry up. When it affects the given stem it hardens prematurely.	• Removal of infected portion and spraying with Copper oxychloride 0.25 %.
Black root		• The fungus originates from the dead heaped leaves of 5–7.5 above the soil level. From there if spreads to roots region of tea bushes. • When bark is removed star like growth of mycelium can be seen. • At the surface of the soil the mycelium surrounds the stem and kills the bank for the length of 7.5 – 10.0 cm. • A swollen ring of tissue is formed round the stem above the dead patch.	• Removal and destruction of infected plant • Clean cultivation with out fallen leaves. • Dig a drench around the infected bush to provide sunlight in the drench which prevent the spread of mycelium.

□□□□

18

Insect Pests of Major Plantation Crops

Common name, symptoms of damage and management of pests of major plantation crops

Common and scientific name of pest	Symptoms of damage	Management	Remarks
Arecanut			
Spindle bug *Carvalhoia arecae*	• Sap sucking bug – damage the unopened spindle leaf. • Inhabit the inner most leaf axils, usually below the spindle. • Suck the sap from tender leaflets and spindle. • Severe infestation - blackish brown linear lesions on the spindle leaf. • Stunted growth and twisted. • Leaves become dried and shed.	• Spray application of Dimethoate 0.05%. • Filling the inner most leaf axils with Phorate 10% G (10g/palm)	
Root grub*Leucopholis burmeisteri*	• Grubs feed on growing roots. • Infested palms show a sickly appearance. • Yellowing of leaves • Tapering of stem and reduction in yield.	• Collection and destruction of adult beetles. • Digging and forking of the soil. • Addition of organic amendments and anti-feedants (Neem, pongamia and oilcake). • Application of Phorate (Thimet 10G) @ 15g per palm give effective control. • Soil application of Phorate around the plant twice a year before onset of southeast monsoon (May) and after the monsoon (Sep-October). • In severely infected gardens, the soil should be drenched with eco-friendly insecticides.	
Cholam or white mite *Oligonychus indicus*	• Adults and nymphs presence on the lower surface of leaves. • The colony is found under white webs. • Feed on lower surface of arecanut leaves.	• Removal of heavily infested and dried leaves and burning. • Spraying under surfaces of leaves and crown with Dicofol (2 ml/litre) or Rogor (1.5 ml/litre).	

Contd.

Common and scientific name of pest	Symptoms of damage	Management	Remarks
		• Repeat spraying at an interval of 15-20 days.	
Inflorescence caterpillar *Tirathaba mundella*	• Caterpillars feed on the inflorescences (tender female flowers) and rachillae. • Webbing and feeding the inflorescence. • Spathe opening is delayed. • Yellowing of spadices. • Presence of small holes with frass and drying on the spathe.	• Infected spadices may be forced open and sprayed with Malathion 0.05%.	
Coconut			
Rhinoceros beetle *Oryctes rhinoceros*	• Central spindle appears cut or toppled. • Fully opened fronds showing characteristic diamond shaped cuttings. • Holes with chewed fibre sticking out at the base of central spindle.	• Remove and burn all dead coconut trees in the garden (which are likely to serve as breeding ground) to maintain good sanitation. • Collect and destroy the various bio-stages of the beetle from the manure pits (breeding ground of the pest) whenever manure is lifted from the pits. • Incorporate the entomopathogen i.e., fungus (Metarrhizium anisopliae) in manure pits to check the perpetuation of the pest. • Soak castor cake at 1 kg in 5 l of water in small mud pots and keep them in the coconut gardens to attract and kill the adults. • Treat the longitudinally split tender coconut stem and green petiole of fronds with fresh toddy and keep them in the garden to attract and trap the beetles.	

Contd.

Common and scientific name of pest	Symptoms of damage	Management	Remarks
		• Examine the crowns of tree at every harvest and hook out and kill the adults. • For seedlings, apply 3 naphthalene balls/palm weighing 3.5 g each at the base of inter space in leaf sheath in the 3 inner most leaves of the crown once in 45 days. • Set up light traps following the first rains in summer and monsoon period to attract and kill the adult beetles. • Field release of Baculovirus inoculated adult rhinoceros beetle @ 15/ha reduces the leaf and crown damage caused by this beetle. • Apply mixture of either Neem seed powder + sand (1:2) @150 g per palm or Neem seed kernel powder + sand (1:2) @150 g per palm in the base of the 3 inner most leaves in the crown. • Place Phorate 10 G 5 g in perforated sachets in two inner most leaf axils for 2 times at 6 months intervals. • Set up rhinolure pheromone trap @ 1 per ha to trap and kill the beetles. • Maintain good sanitation soak castor cake at 1 kg in 5 litres of water in small mud pots and keep them in coconut gardens to attack and kill adults.	

Contd.

Common and scientific name of pest	Symptoms of damage	Management	Remarks
		• Apply 3 Naphthalene balls /palm weighing 3.5g each at the base of the inter space.	
Red palm weevil *Rhynchophorus ferrugineus*	• Holes on trunk with brownish ooze. • Yellowing of inner leaves. • Gradual wilting of central shoot in the crown.	• Remove and burn all wilting or damaged palms in coconut gardens to prevent further perpetuation of the pest. • Avoid injuries on stems of palms as the wounds may serve as oviposition sites for the weevil. Fill all holes in the stem with cement. • Avoid the cutting of green leaves. If needed, they should be cut about 120 cm away from the stem. • Fill the crown and the axils of top most three leaves with a mixture of fine sand and Neem seed powder or Neem seed kernel powder (2:1) once in three months to prevent the attack of rhinoceros beetle damage in which the red palm weevil lays eggs. • Plug all holes and inject pyrocone E or Carbaryll 1% or 10 ml of Monocrotophos into the stem by drilling a hole above the points of attack. • Setting up of attractant traps (mud pots) containing sugarcane molasses 2½ kg or toddy 2½ litres + acetic acid 5 ml + yeast 5 g + longitudinally split tender coconut stem/logs of green petiole of leaves of 30 numbers in one acre to trap adult red palm weevils in large numbers.	

Contd.

Common and scientific name of pest	Symptoms of damage	Management	Remarks
		• Install pheromone trap @ 12 per ha.	
Bark weevil *Dicalandra stigaticolis*	• Reddening of petioles and trunks especially around wounds. • Trees with stem bleeding disease.	• Stem injection through a stove wick soaked in 0.2% Dichlorvos and plugging the hole and repeating the treatment using the same wick and hole a month after.	
Shot hole or Bark borer *Xyleborus*	• Presence of lot on minute holes interconnected by fibrous galleries.	• Stem injection through a stove wick soaked in 0.2% Dichlorvos and plugging the hole and repeating the treatment using the same wick and hole a month after.	
Black headed caterpillar *Opisina arenosella*	• Dried up patches on leaflets of the lower leaves. • Galleries of silk and frass on under side of leaflets.	• Among the larval parasitoids, the bethylid *Goniozus nephantidis* is the most effective in controlling the pest. • The optimum level of release is 1:8 of host-parasitoid ratio. • The parasitoid should be released @3000/ha under the coconut trees when the pest is in the 2nd or 3rd instar larval stage. Parasitoid release trap may be used to release the parasitoid at the site of feeding. • Parasitoids should not be released in the crown region since they will be killed by predators like spiders and reduviid bugs. • Remove and burn all affected leaves/leaflets. • Spray Malathion 50 EC 0.05% (1mi/lit) to cover the under surface of the leaves thoroughly in case of severe epidemic outbreak of the pest in young palms.	

Contd.

Common and scientific name of pest	Symptoms of damage	Management	Remarks
Slug caterpillar *Parasa lepida* and *Contheyla rotunda*	• Defoliation.	• Dichorvos 76 WSC 2 ml/lit. • *Bacillus thuringiensis* 2 g/lit. • Triazophos 40 EC 5 ml. • Root feeding with Monocrotophos 15 ml + 15 ml of water.	
Coconut skipper *Gangara thyrsis*; *Suastus gremius*	• Tip of the leaflets rolled.	• Collect and destroy the immature stages of the insects by conducting study (or Neem compaign) wherever possible and spray Carbaryll 50 WP 2 g/lit. • Root feeding with Monocrotophos 36 WSC @ 10 ml + 10 ml water at 45 days interval for 3 times for control of leaf caterpillar. • Set up light traps to trap and collect adult moths. • Spray Dichlorvos 76 WSC 2 ml / lit.	
Leaf caterpillars *Turnaca acuta*	• Feeds on leaflets leaving stick.	• Collect and destroy the immature stages of the insects by conducting study (or Neem compaign) wherever possible and spray Carbaryll 50 WP 2 g/lit. • Root feeding with Monocrotophos 36 WSC @ 10 ml + 10 ml water at 45 days interval for 3 times for control of leaf caterpillar. • Set up light traps to trap and collect adult moths. • Spray Dichlorvos 76 WSC 2 ml / lit.	
Bag worm	• Small and irregular hole on the leaves.	• Collect and destroy the immature stages	

Contd.

Common and scientific name of pest	Symptoms of damage	Management	Remarks
Manatha albipes		of the wherever possible and spray Carbaryl 50 WP 2 g/lit. • Root feeding with Monocrotophos 36 WSC @ 10 ml + 10 ml water at 45 days interval for 3 times for control of leaf caterpillar. • Set up light traps to trap and collect adult moths. • Spray Dichlorvos 76 WSC 2 ml / lit.	
White grub *Leucopholis coneophora*	• Leaves turn yellow. • Premature nut shedding. • Flowering delayed. • White grubs are exposed when base of tree dug.	• Collect and destroy the adult beetles attracted to trees like Neem, *Ailanthus* and *Accasia* on the receipt of monsoon showers. • Plant Neem twigs with leaves in coconut gardens after rain to attract and kill adult beetles. • Set up light trap @ 1 / ha or bonfire. • Application of Malathion 5 D 25 kg/ ha at the time of planting.	
Termite *Odontotermes obesus*	• Wilting of seedlings. • base of trunks plastered with runways made of soil and fibres.	• Lime paste applied around the trunk of the coconut tree. • Locate termite mounds in or near the coconut nursery or garden and destroy. • Swabbing with Neem oil 5% once on the base and upto 2 m height of the trunk for effective control. • Spray Copper sulphate 1% or cashew nut shell oil 80% or spray Chlorpyriphos @ 3ml/lit of water, Neem oil 5% or NSKE 20% to preserve plaited coconut leaves from the termite attack.	

Contd.

Common and scientific name of pest	Symptoms of damage	Management	Remarks
Lacewing bug *Stephanitis typicus*	• White spots on the upper surface of the leaves.	• Remove leaflets harbouring these insects and destroy them.	
Scale insect *Aspidiotus destructor*	• Yellowing of leaves in patches, later coalescing together.	• Pluck mature nuts and spray Monocrotophos 36 WSC @ 2 ml/ha. • Do not harvest nuts for 45 days after spraying.	
Mealy bug *Pseudococcus longispinus*	• Central leaves stunted , deformed and suppressed. • Shedding of buttons.	• Remove leaflets harbouring these insects and destroy them. **Spray any one of the following** • Malathion 50 EC 2 ml/lit • Dimethoate 30 EC 1 ml/lit • Monocrotophos 36 WSC 1 ml/lit • Neem oil 3%	
Coconut eriophyid mite *Aceria guerreronis*	• Triangular pale or yellow patches close to perianth. • Necrotic tissues. • Brown colour patches, longitudinal fissures and splits on the husk. • Oozing of the gummy exudation from the affected surface. • Reduced size and copra content. • Malformed nuts with cracks and hardened husk.	• Apply urea 1.3 kg, super phosphate 2.0 and muriate of potash 3.5 kg/palm/year. • Neem cake @ 5 kg and organic manure 50 kg/palm/ year. • Borax 50 g + gypsum 1.0kg + Manganese sulphate 0.5 kg/palm/ year. • Grow intercrop (sun hemp, four crops/year) and shelter belt with *casuarina* all round the coconut garden to check further entry. • Triazophos 40 EC 5 ml/lit or Monocrotophos 36 WSC @ 2 ml / lit in alternation with Neem oil1% 5ml/lit as spot application.	

Contd.

Common and scientific name of pest	Symptoms of damage	Management	Remarks
Palmyrah			
Rhinoceros beetle *Oryctes rhinoceros*	• Central spindle appears cut or toppled. • Fully opened fronds showing characteristic diamond shaped cuttings. • Holes with chewed fibre sticking out at the base of central spindle.	• Remove and burn all dead palmyrah trees in the garden. • Collect and destroy the various bio-stages of the beetle from the manure pits. • Incorporate the entomopathogen fungus *Metarrhizium anisopliae* in manure pits. • Soak castor cake at 1 kg in 5 l of water in small mud pots and keep them gardens to attract and kill the adults. • Examine the crowns of tree at every harvest and hook out and kill the adults. • For seedlings, apply 3 naphthalene balls/palm weighing 3.5 g each at the base leaf sheath. • Set up light traps to attract and kill the adult beetles. • Field release of Baculovirus inoculated adult rhinoceros beetle @ 15/ha. • Apply mixture of either Neem seed powder + sand (1:2) @150 g per palm or Neem seed kernel powder + sand (1:2) @150 g per palm. • Set up rhinolure pheromone trap @ 2 /ha to trap and kill the beetles.	
Black headed caterpillar *Opisina arenosella*	• Dried up patches on leaflets of the lower leaves. • Galleries of silk and frass on underside of leaflets.	• Among the larval parasitoids, the bethylid *Goniozus nephantidis* is the most effective. • The optimum level of release is 1:8 of host-parasitoid ratio.	

Contd.

Common and scientific name of pest	Symptoms of damage	Management	Remarks
		• The parasitoid should be released @3000/ha (2nd or 3rd instar larval stage). • Remove and burn all affected leaves/leaflets. • Spray Malathion 50 EC 0.05% (1mi/lit) to cover the under surface of the leaves.	
Coffee			
White stem borer *Xylotrechus quadripes*	• Larvae enter into the hardwood and make the tunnels may extend even into the roots. • Tunnels - tightly filed with the excreta of the grubs. • Infested plants show visible ridges around the stem. • Yellowing and wilting of leaves. • Young plants (7 to 8 years old) attacked by the borer may die in a year.	• Maintain optimum shade on the estates. • Trace the infested plants - flight periods (i.e, during March and September). • Every year - looking for ridges on the main stem and thick primaries. • Collars prune the infested plants, uproot and burn the affected parts. • Remove the loose scaly bark of the main stem and thick primaries using coir glove or coconut husk – kill the eggs. • Scrubbing during flight periods - kills the eggs and grubs present in the bark region. • Deep scrubbing should be avoided (sharp implement may injure the green wood and eventually kill the plant). • Spraying the main stem and thick primaries with Neem kernel extract may afford good control of the pest. • Field release of predators *Apenesia* sp. for effective control of white stem borer. • Field release of white muscardine fungus	

Contd.

Common and scientific name of pest	Symptoms of damage	Management	Remarks
		Beaveria bassiana (2.5 % field infection reported).	
Coffee berry borer *Hypothenemus hampei*	• Pin hole at the tip of the berries (novel region). • Severe infestation - two or more holes may be seen. • Female beetle bores into the berries through the navel region make tunnelling and feed inside content. • Powdery substance pushed out through the holes.	• Proper adoption of cultural practices and phytosanitary measures important for management of coffee berry borer. • Transportation of infested coffee to uninfected areas is the main reason for spread. • Crop bags should be fumigated before delivery to estates to avoid cross infestation. • Timely harvest. • Gleaning - spreading gunny bags or polythene sheets on the ground after picking the berries - minimize gleaning. • Maintain optimum shade and good drainage. • Dipping infested berries in boiling water for 2-3 minutes kills all the stages inside. • Drying of coffee beans – prevents breeding of beetles in stored coffee based on moisture content level. • Arabica (10% moisture content) - 16kg/lit • Robusta (11% moisture content) - 18kg/lit • Install attractant trap to collect and kill the beetles. • Install multiple funnel trap and pitfall trap to collect and kill the adults.	

Contd.

Common and scientific name of pest	Symptoms of damage	Management	Remarks
		• Chlorpyriphos is also effective in controlling the beetle. • Field release of *Cephalonomia stephendarix* and *phymastichus coffea* for effective control these pests. • Applications of white muscardine fungus, *Beauveria bassiana* effectively reduce the borer attack. **Field application of *Beauveria bassiana*** • Mix 500 ml of groundnut oil with 500 ml of any agricultural wetting agent thoroughly. • Add this suspension to 20 l of water and three bags of *Beauveria* culture the above solution mix well and strain through a muslin cloth. • Application of spore suspension on the infested coffee bushes using a bakpak sprayer, targeting the berries. • After 5 days application spores germinate and kill the insect.	
Shot hole borer *Xylosandrus compactus*	• Infests green succulent branches. • Young plants - main stem may be attacked. • Presence of shot hole. • Initial symptom - drooping of leaves. • Withered (faster in young branches and delayed in older twigs) or dried branches.	• Insecticide applications do not provide for shot hole borer control. • Grub entered into twig and make tunnelling – Ambrosia fungus developed from the tunnelling. Female lays eggs on the tunnelling.	

Contd.

Common and scientific name of pest	Symptoms of damage	Management	Remarks
	• Attacked Leaves fall and prematurely. • Terminal leaves wilt, droop and dry up. • Severe infestation - loss of considerable number of productive branches.	• After egg hatching, the white milky larvae feed on the ambrosia fungus. • Prune the affected twigs 2 to 3 inches beyond the shot-hole and burn (September). • Remove and destroy all the unwanted/in fested suckers during summer (avoid breeding). • Maintain thin shade and good drainage.	
Mealy bugs *Planococcus* spp.	• Young plants – susceptible for heavy infestation. • Infest tender branches, nodes, leaves, spikes, berries and roots. • Both nymphs and adults suck the sap from the leaves. • Severe infestation - Chlorotic leaves, aborted flower buds and small berries. • Honey dew excrete – development of sooty mould fungus (affects photosynthesis).	• Maintain adequate shade. • Destroy nets of red ant and cocktailed ant. • Control ants by dusting Quinalphos 1.5% or Malathion 5% around the base of coffee and shade trees. • Spray the affected patches with Quinolphos 25 EC 300 ml using a gator sprayer. • Infested roots - drench the soil near the root zone with any one of the above insecticides at the same dosage. (Add 300 ml of Plantvax 20 EC or 160 g of Bayleton 25 WP in 200 l water). • In the case of young plants (2-4 year old) drenching with Dimethoate 30 EC @ 3 ml per litre of water effective in root and shoot mealy bugs. • Field release of two introduced natural enemies viz., the predatory lady bird beetle *Cryptolaemus montrouzieri* and the parasitoid *Leptomastix dactylopii.*	• Mealy bugs are the most important sucking pests of coffee. • ***P.citri*** (Risso) and ***P. lilacinus*** (Cockerell) are the most common species infesting coffee.

Contd.

Common and scientific name of pest	Symptoms of damage	Management	Remarks
Green scale *Coccus viridis*	• Nymphs and adults sucks sap from under surface of the leaves, green shoots, spikes and berries. • Upward curling and tender twigs drop. • Honeydew excrete – development of sooty mould fungus. • Heavy loss of the sap causes debility or even death of the plant.	• Control ants as in the case of mealy bugs. • Remove and burn weeds which harbour the scale. • Spray the affected patches Quinolphos 25 EC @ 120 ml using a gator sprayer in 200 l of water and add 200 ml of any agricultural wetting agent in 200 l of the spray solution.	
Brown scale *Saissetia coffeae*	• Nymphs and adults suck sap from under surface of the leaves. • Honeydew excrete – development of sooty mould fungus.	• Control ants as in the case of mealy bugs. • Remove and burn weeds which harbour the scale. • Spray the affected patches Quinolphos 25 EC @ 120 ml using a gator sprayer in 200 l of water and add 200 ml of any agricultural wetting agent in 200 l of the spray solution.	
Cockchafers or white grubs *Holotrichia* spp.	• Grown up coffee plants normally attacked. • Young plants (1 to 5 years old) attacked by white grubs. • Yellowing of leaves. • Stunted growth. • Wilting of plants and die (summer period). • Attacked plants can be easily pulled out.	• In white grub-infested areas – at the time of planting apply Phorate 5 G 10 g or Carbofuran 3G 10 g into the soil in the pit. • Soil application of Phorate 10 G @ 20 g per plant. • Field sanitation. • Collect and kill the grubs. • Install light traps and kill the trapped adults (March – June).	
Hairy caterpillars *Eupterote* spp.	• *Eupterote canaraica* - important pests of Arabica coffee. • Feed on leaves and cause severe damage by denuding the plants.	• Collect and kill the caterpillars. • Collect and burn pupae from January to May. • Install light traps (June/July) - collecting	

Contd.

Common and scientific name of pest	Symptoms of damage	Management	Remarks
	• Badly affected plants take 2 to 3 years to recover.	and killing months. • Spray any one of the insecticides such as Quinalphos 25 EC @ 320 ml, along with 100 ml of an agricultural wetting agent in 200 I of water.	
Coffee bean beetle *Araecerus fasciculatus*	• Severe infestation occurs only in storage. • Absorption of moisture content. • Infested coffee beans shows circular holes (bigger than made by the coffee berry borer). • Attacked fruits in the field shrink and become black. • Infestation results in loss of weight increase in triage and reduction of market value.	• Maintain optimum temperature, relative humidity and moisture content (less than 8%). • Fumigate the stored coffee with a 1:1 mixture of ethylene dibromide and Methyl bromide. • Impregnate the gunny bags with a mixture of Malathion 50 EC at 10ml + pyrethrum colloid at 2 g per litre of water.	
Snail *Ariophanta solata*	• The snail feeds on the leaves of Arabica coffee and bark of tender braches. • Skin of fruits.	• Keep the estates clean. • Hand picks the snails and dip in hot water or salt solution. • Use heaps of leaves to trap the snails. • Broadcast the poison bait 'snail kill' (metaldehyde) of 25-35kg per hectare. • Spreading of lime, soot and wood ash - repels snails.	
Red borer Zeuzera coffeae	• Caterpillar bores into the stem or branches to feed on the wood. • In early stages of attack, young plants or braches show wilting. • Pellet-like excrement of the larva hangs out and accumulates at the base of the plant. • In advanced cases, the branch dries up.	• Affected plants or twigs - cut and burnt. • Field release of white muscardine fungus *B. bassiana* and braconid parasite *Amyosoma zeuzerae* for control of the red borer.	

Contd.

Common and scientific name of pest	Symptoms of damage	Management	Remarks
Tea			
Tea mosquito bug *Helopeltis theivora*	• Adults and nymphs suck the sap from buds, young leaves and tender stems. • Puncturing the plant tissues with their needle like rostrum and inject toxic saliva. • Punctures appear as reddish brown spots. • Leaves curl up, badly deformed and shoots dry up.	• Spray Phosalone (or) Monocrotophos 2ml / lit of water.	• It is considered to be a serious pest of a tea. • The damage more in most shaded areas.
Looper cater pillar *Biston supressaria*	• Young caterpillars feed on the tender leaves - making punctures. • Mature larvae prefer older leaves. • Grown up larvae feed entire leaf. • Severe infestation - tea bushes are completely denuded.	• Spray NSKE 5% (or) Neem based oil formulations 3%. • Spray any one of the following insecticides Phosalone (or) Chlorpyriphos (or) Malathion 2ml/lit.	
Lobster Caterpillar *Neostauropus alternus*	• Two or three caterpillars can completely devour all the leaves from a small plant in two or three days.	• Hand collection and destruction. • Spray Chlorpyriphos 2 ml/lit.	
Flush worm *Cydia leuocostoma*	• Caterpillar ties up the margin of tender leaves and forms a case enclosing the bud. • Feed on the upper epidermis of leaves. • Affected leaves - rough, crinkled and leathery. • Bud - Shoot growth is arrested.	• Hand collection and destruction. • Spray Chlorpyriphos 2 ml/lit.	
Tea tortix *Homona coffearia*	• Caterpillars make leaf nests by webbing the leaves using silken threads. • Feed from inside the leaf nest. • Young larvae prefer tender leaves. • Older larvae are seen in mature leaves.	• Hand collection and destruction. • Spray Chlorpyriphos 2 ml/lit.	

Contd.

Common and scientific name of pest	Symptoms of damage	Management	Remarks
Shot hole borer *Envallacea fornicatus*	• Presence of round shot holes in primary branches. • The attack results in mortality of buds and dieback symptoms in branches. • Presence of circular or longitudinal tunnels inside the stem.	• Remove and destroy affected twigs.	
Red spider mite *Oligonychus coffeae*	• Infest the upper surface of mature leaves. • Infestation starts along midrib and veins and spreads to the entire upper surface of leaves. • Affected leaves - bronzed, dry and crumpled.	• Spray Dicofol 18EC 2 ml/lit or Ethion 50 EC 2 ml/lit or Monocrotophos 1ml/lit • Application of wettable sulphur 80 WP 2g/lit using hand operated sprayer.	
Scarlet mite *Brevipalpus californicus*	• Mites congregate on the under surface of mature tea leaves. • Feeding by scarlet mites leads to brown discolouration of leaves.	Same as above	
Purple mite*Calacarus carinatus*	• Mites feed on the under surface of mature leaves. • Assam type of tea - more susceptible to purple mite. • Damaged leaves - Coppery brown discolouration.	Same as above	
Pink mite (or) Orange mite*Acaphylla theae*	• Mites are found on the under surface (abaxial) of young leaves. • Affected leaves turn pale and upward curling. • In severe infestation, leaves become leathery and brown. • Damage - restricted to top 10–15 cm of	Same as above	

Contd.

Common and scientific name of pest	Symptoms of damage	Management	Remarks
	tender leaves. • Assam type of tea is susceptible.		
Yellow mite *Polyphagotarsonemus latus*	• Mite is seen on young leaves especially the top two to three leaves and the bud. • Affected leaves become rough and brittle and corky lines. • Downward curling • Internodes get shortened. • Shoots - stunted and deformed.	Same as above	
Thrips *Scirtothrips bispinosus*	• Thrips prefer young leaves and buds. • Leaf surface becomes uneven, curled and matty. • Feeding marks on the unopened buds - parallel brown lines on the leaves.	• Spray Dimethoate 30 EC (or) Chlorpyriphos 20 EC 2 ml/lit	
Aphid*i* *Toxoptera auranti*	• Pruning - colonies of aphids are seen on tender shoots of tea immediately after pruning. • Leaves curl up and shoot growth is stunted. • Aphids are attended by ants. • Honey dew secreted development of sooty moulds.	• Collect and destroy the infested plant parts • Spray Dimethoate 30 EC (or) Chlorpyriphos 20 EC 2 ml/lit. • Spray Phosalone 2 ml/lit.	
Cocoa			
Mealy bugs *Planococcus lilacinus, Planococcus citri, Paracoccus marginatus* and *Rastrococcus*	• It colonizes on the tender parts of the plant such as growing tips of the shoots, the terminal buds, the flower cushions, the young cherelles and mature pods. • Feeding of mealy bugs induces cherelle wilt.	• When the infestation is lesser: Spraying of Neem Oil 3% or fish oil rosin soap 25g/litre. • In case of severe incidence, spraying of any one of the following chemicals is recommended : Dimethoate (2 ml/litre),	

Contd.

Common and scientific name of pest	Symptoms of damage	Management	Remarks
iceryoides		Profenophos (2 ml/litre), Chlorpyriphos (5 ml/litre), Buprofezin (2 ml/litre), Imidacloprid (0.6 ml/lit), Thiamethoxam (0.6g/litre). • In the area where *P. marginatus* alone occurs, field release of *Acerophagus papayae*, the encyrtid parasitoid @ 100 per hamlet is recommended as the best management strategy.	
Tea mosquitoe bugs *Helopeltis antonii*	• Infested pods develop circular water soaked spots around the feeding punctures. • These punctures subsequently turn pitch black in colour. • Deformation of pods occurs because of multiple feeding injuries.	• When the infestation is lesser: Spraying of Neem Oil 3% is recommended. • In case of severe incidence, spraying of any one of the following chemicals is recommended: Imidacloprid (0.6 ml/lit) , Thiamethoxam (0.6g/litre), Profenophos (2 ml/litre), Carbaryll (2g/lit)	
Aphids*Toxoptera aurantii* and *Aphis gossypii*	• They colonize on the underside of tender leaves, succulent stem, flower buds and small cherelles. • Heavy infestation may occur during hot summer and after rainy season which brings about premature shedding of flowers and curling of leaves.	• Spray Dimethoate @ 2 ml per lit.	
Hairy caterpillars*L ymantriya* sp.,	• They cause serious leaf damage on seedlings and young trees.	• Foliar spray of Acephate @ 2g/litre of water.	
Stem Girdler	• Damage was done by female beetle which	• Swab Coal tar + Kerosene @ 1:2 or Carbaryll	

Contd.

Common and scientific name of pest	Symptoms of damage	Management	Remarks
Sthenias grisator	girdles the branches and inserts whitish spindle shaped eggs singly into the tissue in a slanting manner. • Due to mechanical injury caused by girdling and oviposition, the branches above the girdle wither and dry.	50 WP 20 g / litre (basal portion of the trunk - 3 feet height) after scraping the loose bark to prevent oviposition by adults. • Hook out the grub from the bore hole and apply Monocrotophos 36 WSC 5 to 10 ml/ bore (or) apply Carbofuran 3G 5 g/bore hole and plug with mud. • Injection of Dichlorvos (DDVP) + Monocrotophos solution into bore holes after removing the webs and subsequently sealing of the holes with clay gives satisfactory control of the pest.	
Oil Palm Spindle bug *Carvalhoia arecae*	• Spindle bug - generally noticed in nursery seedlings and field planted young seedlings. • Adults and nymphs of spindle bug live in the innermost two to three leaf axils. • Suck sap from the spindle of leaves. • Necrotic lesions which later on turn into dry brown patches. • In severe infestation the spindle fails, to open.	• Ploythene sachets - filled with Phorate 10 G (Thimet 10G), @ 2g per sachet, placed in the innermost two leaf axils (one sachet each in a leaf axil). • The same sachets can be used for 6 – 8 months.	
Rhinoceros beetle *Oryctes rhinoceros*	• Adult beetles bore into the palms and chew the tender tissues. • Spindle to break and droop. • Entry holes - presence of chewed-up tissues. • Infested leaves - shortened, broken and distorted. • Fruits - the upper portion of the bunch get underdeveloped and dried.	• All possible breeding sites are to be eliminated from the plantation. • The adult beetles which burrow deep into the crowns of young palm. • Extracted by means of a hooked out (metal rod). • After extraction of the beetle, the leaf axils around the injured spindle are to be filled with a mixture of Mancozeb (fungicide) and fine sand at the ratio	

Contd.

Common and scientific name of pest	Symptoms of damage	Management	Remarks
	• Damage to the heart of the palm - way for the entry of fungi and bacteria. • Induce rotting in the bud.	of 3g: 1kg. • Release of Baculovirus infected adult beetles (@ 10 – 15 beetles/ha. It is the most economical, effective and easy method for dissemination of the inoculum. • Application of *M. anisopliae* in the breeding sites is very effective for the control of grubs. • Rhinolure vane trap hanged at a height of 10ft. It is found to be very effective in trapping adult beetles. • Castor seed powder fermented in rice water and kept in mud pots at ground level. It is an efficient method for trapping adult beetles.	
Red palm weevil *Rhynchophorus ferru gineus*	• Rhinoceros beetle attacked palms are more prone to infestation by red palm weevil. • Grubs bore into stem and feed on softer tissues of stem, meristem and mesocarp of fruits. • Infested palms - gradual wilting and drying of outer whorl of fronds. • In severe - rotting of spear. • Red palm weevil lays eggs on injuries made by rhinoceros beetle on bunches. • Larvae feed on the mesocarp of fruits and penetrate into peduncle for pupation. • Infestation causes qualitative and quantitative loss of the fruits.	• Ripe bunches are to be harvested in time. Avoid making wounds on the palm and bunches. • Prophylactic leaf axil filling is to be done for young palms as in the case of rhinoceros beetle. • Trunk injection with Carbaryll 50% WP at 1% is an effective curative treatment. • Ferrolure placed in specially designed traps and hanged on the palms. • Weevils are trapped and destroyed using split logs of coconut/oil palm or petioles (50 cm long) treated with attractants such as coconut toddy, macerated grapes, cashew apple, pineapple and molasses.	

Contd.

Common and scientific name of pest	Symptoms of damage	Management	Remarks
Mealy bugs *Dymicoccus brevipes*	• Mealy bugs infest the spear leaves of oil palm seedlings in the nursery and main field. • Yellowing of young leaves and stunted growth of palms. • It is also infests the unripe and ripe oil palm fruits. • Pre anthesising of male and female inflorescences.	• Collection and destruction of infested plant parts. • Collect planting material from unaffected plantation. • Spraying with Dimethoate at 0.03% or Methyl demeton at 0.025%. • Release coccinellid beetle, *Cryptolaemus montrouzieri* @ 10 / tree.	
Nettle caterpillar *Thosea andamanica*	• Nettle caterpillars - stinging spines which can cause nettle rash on contact with the skin. • Caterpillars feed on the leaves. • Infestation – mostly occur in outer whorl of fronds and middle whorl of fronds.	• Cutting and burning the badly affected and dried leaves. • Spraying with Carbaryll 50% W.P. 0.1% is recommended (younger larvae are more susceptible).	
Aphids *Mysteroneura setariae, Schizaphis rottundiventris*	• Colonies of aphids occur on the under surface of leaves and suck sap. • In nursery and young palms - cause quite marked twisting and distortions of the spears.	• Spraying with Dimethoate 0.03% or Malathion 0.05% on the under surface of leaves.	
Mites *Tetranychus piercei*	• Mites suck the sap from the leaves. • Attacked leaves - discoloured and look speckeld. • Nursery and young palms are severely disrupted.	• Application of Dicofol 2ml/lit or wettable sulphur 2g/lit.	

□□□□

3. Medicinal and Aromatic Plants

19

Importance of Medicinal and Aromatic Plants

Introduction

- Medicinal Aromatic Plants (MAPs) play a valuable and important role in economic, social, cultural and ecological aspects of local communities the world over.
- According to World Health Organization (WHO) "a medicinal plant is plant which, in one or more of its organs, contains substance that can be used for therapeutic purpose, or which are precursors for chemo-pharmaceutical semi-synthesis".
- However, a simplest definition of the medicinal plant would be "Medicinal plants are those plants which are used in official and various traditional systems of medicines throughout the world".
- Other definition could be "Medicinal plants are plants that provide people with medicines - to prevent disease, maintain health or cure *ailments"*.
- In Indian traditions, all the plants in this earth are considered as medicinal [*Jivak* in *Astanga Hriday* (Sutra: 9-10)].
- Aromatic plants are a special class of plants used for their aroma and flavour. Many of them are exclusively used also for medicinal purposes in aromatherapy as well as in various systems of medicine.

- Similarly a number of medicinal plants also produce essential oils as well as being used for perfumery e.g. *Petroselinum sativum, Daucus carota, Anethum graveolens* and *Pimpinella anisum,* etc.
- According to a report from an international organization i.e. WHO (World Health Organization), about 85% of the population in developing countries depend on herbal plants for curatives, medicinal and other medico related applications.
- India being one of the twelve mega biodiversity centers of the world is fully fledged with diverse array of herbal and medicinal plants which makes it "Botanical Garden of World".
- India has 15 Agroclimatic zones and 17000-18000 species of flowering plants of which 6000-7000 are estimated to have medicinal usage in folk and documented systems of medicine, like Ayurveda, Siddha, Unani and Homoeopathy.
- About 960 species of medicinal plants are estimated to be in trade of which 178 species have annual consumption levels in excess of 100 metric tones.
- Medicinal plants are not only a major resource base for the traditional medicine & herbal industry but also provide livelihood and health security to a large segent of Indian population.
- There is global resurgence in traditional and alternative health care systems resulting in world herbal trade which stands at US$ 120 billion and is expected to reach US$ 7 trillion by 2050. Indian share in the world trade, at present, however, is quite low.
- An estimated 400 000 tonnes of MAPs are traded annually and more than 70 percent of the plant species used in herbal medicines, cosmetics, and other plant-based products are harvested from the wild, and the demand for them is globally increasing (Leaman, 2008).
- India has been considered as treasure house of a large number of valuable medicinal and aromatic plant species.

Distribution of Medicinal and Aromatic Plants (MAP)

- An analysis of distribution of MAP in natural habitat showed that about 70% of India's MAPs are found in tropical forests of Western and Eastern ghats, the Vindhyas, Chotta Nagpur plateau, Aravalis and the Himalayas.

- Studies also showed that a large percentage of known MAPs occur in the dry and moist deciduous vegetation area compared to evergreen and temperate regions.
- Habit-wise classification showed that about 33% are trees, 32% herbs, 20% shrubs, 12 % creepers and 3% others.
- An analysis of 386 families and 2200 genera in which MAPs are recorded showed that Asteraceae family posses highest number of species.

Cultivation

- Cultivation of MAPs is different than that of other agricultural crops.
- These plants are grown unlike the other crops for secondary metabolites and do not desire to have even stress quantity of pesticides or heavy metals.
- Therefore, WHO has insisted to develop Good Agricultural Practices (GAP) for MAPs to guarantee quality of raw drug and to facilitate the standardization of quality of ISM drugs for acceptance to science led world.
- A large number of State Agricultural Universities, ICAR and CSIR institutes are engaged in developing GAP for MAPs. However, the cultivation is not gaining due momentum in the country due to prevailing unorganised marketing arrangements.
- Agro technologies for cultivation of several such crops have been developed by CSIR and ICAR institutions.

Conservation

- Conservation of MAPs is necessary for the posterity of the future generation. Several issues have consistently been raised in the various sets of recommendations compiled to promote the conservation of MAPs. These include:
- The need for a coordinated conservation action based on both "*in situ*" and "*ex situ*" strategies.
- The inclusion of community and gender perspectives in the development of policies and programmes.
- The development of sustainable harvesting practices.

- The need for more information on the medicinal plant trade - establishing systems for inventorying and monitoring the status of stocks of medicinal plants.
- Encouraging micro-enterprise development based on sustainable resource use by indigenous and rural communities.
- Protection of traditional resources and the intellectual property rights.

GACPs (good agricultural and collection practices) for medicinal plants

Certain basic guidelines are to be followed under GACPs for cultivation and harvesting of the crop, some of which are described below:

- Select proper site for cultivation of a particular medicinal plant.
- Select correct time for cultivation.
- Select proper variety.
- Adopt organic farming.
- Prune and collect only desired mature part(s), without harming the mother plant.
- Do not collect the whole population; leave at least 30%–40% for regeneration.
- Do not cut twigs/branches for collection of plant parts.
- Use proper equipment for cutting, shearing, peeling, and so on.
- Start drying process immediately after collection; ensure complete drying before packing and storage.
- Dry aromatic herbs and delicate fruits in shade, and do not dry two or more herbs in close vicinity.
- Sift the herbs with appropriate sifter to remove dust and other undesirable matter.
- Pack the herbs in suitable packaging material to avoid losses due to external factors.
- Store the herbs in proper storage conditions to minimize loss on storage.

For collection of underground part(s), bark, and whole plant, following guidelines should be adhered to:

- Collect after the seeds are shed to facilitate regeneration.
- Do least digging for collection of underground parts and leave some underground part to facilitate regeneration.
- Collect underground parts when the mother plant is fully matured.
- Dry fleshy parts before packing and storing; cut large parts into smaller pieces.
- Do not harvest bark from immature plant; instead, collect from the branches of main trunk.
- Strip the bark longitudinally and not from all over the circumference of trunk/branches.
- Cut into small pieces to facilitate complete drying.
- Harvest only mature branches for stem.
- Dry the herbs properly before packing or storing.

For collection of leaves, flowers, fruits, seeds, floral parts, and so on, follow the following recommendation:

- Harvest only mature parts from healthy plants.
- Do not collect all material of the plant at a time.
- Do not cut branches for collecting leaves, fruits, flowers, and so on.
- Leave some floral parts on the plants to facilitate natural regeneration.
- Fleshy flowers may be dried in the sun, but should preferably be dried in shade.
- Parts like stiga, anthers, buds, and so on should be collected at appropriate time.
- Harvest the seeds once the fruits are completely mature.

For collection of gums, oils, resins, galls, and so on

- Make incisions only vertically on some portions of the tree and not horizontally.

- Treat the incisions after collection of the desired material.
- Do not collect the gum or resin from a tree continuously and collect them in precisely right season.
- Do not leave gum/resin exposed in the field. Pack them in appropriate containers or drums with polyethylene lining.
- Collect the galls only from prescribed species (for example, Karkatshringi from *Pistacia integerrima*).
- No live insect should be present inside the galls.

National Medicinal Plants Board (NMPB)

- The National Medicinal Plants Board (NMPB) set-up in November 2000 by the Government of India has the primary mandate of coordinating all matters relating to medicinal plants and support policies and programmes for growth of trade, export, conservation and cultivation.
- The Board is located in the Department of Ayurveda, Yoga & Naturopathy, Unani, Siddha & Homeopathy (AYUSH) of the Ministry of Health & Family Welfare.

Central Institute of Medicinal and Aromatic Plants (CIMAP)

- The Central Institute of Medicinal and Aromatic Plants, popularly known as CIMAP, is a frontier plant research laboratory of Council of Scientific and Industrial Research (CSIR).
- Established originally as Central Indian Medicinal Plants Organisation (CIMPO) in 1959, CIMAP is steering multidisciplinary high quality research in biological and chemical sciences and extending technologies and services to the farmers and entrepreneurs of medicinal and aromatic plants (MAPs) with its research headquarter at Lucknow and Research Centres at Bangalore, Hyderabad, Pantnagar and Purara.
- The Organization started functioning with effect from 26 March 1959 with the appointment of late Shri P.M. Nabar its first Officer Incharge.

ICAR-Directorate of Medicinal Aromatic Plants Research (ICAR-DMAPR)

- The ICAR-Directorate of Medicinal Aromatic Plants Research (ICAR-DMAPR) is located at Anand, Gujarat.

- The Indian Council of Agricultural Research established Directorate of Medicinal & Aromatic Plants Research (formerly it was National Research Centre for Medicinal & Aromatic Plants) on November 24, 1992 in a 20.2 hectare irrigated land at Boriavi in Anand district of Gujarat.

□□□□

20

Technical Information on Medicinal Plants

Common name, botanical name, family, economic part, major phytochemical constituents and uses of medicinal plants

S. No	Common name	Botanical name	Family	Economic part	Major Phytochemical constituents	Uses
1.	Kalmegh	*Andrographis paniculata* (Burm. f.)	Acanthaceae	Whole herb	Andrographolide	Treating fever
2.	Adusa	*Adhatoda vasica* Nees *A. zeylanica* Medic.	Acanthaceae	Leaves	Vascine and vasicinone	Treatment of bronchitis, asthma, cold, cough and whooping cough
3.	Mandookaparni	*Centella asiatica* L.	Apiaceae	Whole plant	Asiatic acid, Asiaticoside, Madecassic acid, Brahmic acid, Thankuniside, Centellose	Treatment of leprosy, skin diseases and to improve memory
4.	Periwinkle (Sadabahar)	*Catharanthus roseus*	Apocynaceae	Leaves and roots vincustine (VCR)	Vinblastine (VLB) and present in its leaves, and ajmalicine, present in its roots are medicinally important	Used to treat a various types of cancers
5.	Sarpagandha	*Rauvolfia serpentina* (L.) Benth.ex. Kurz.	Apocynaceae	Roots	Reserpine, desrpidine and reseinamine	Used to control high blood pressure
6.	Vach	*Acorus calamus*	Araceae	Rhizomes	Calamus' α and β asarone, acoric acid and choline	Treatment of diarrhoea, dysentery, abdominal obstructions and colic
7.	Esarmool	*Aristolochia indica*	Aristolochiaceae	Roots	Aristolochic acid	Antitode for snake bite
8.	Madhunasini	*Gymnema sylvestis*	Asclepiadaceae	Leaves	Gymnemic	Treatment of diabetics
9.	Bankakri	*Podophyllum hexandrm*	Berberidaceae	Rhizomes and roots	Podophyllin or podophyllin	Curing skin diseases and tumour growth.

Contd.

S. No	Common name	Botanical name	Family	Economic part	Major Phytochemical constituents	Uses
10.	Guggal	*Commiphora wightii* (Arn.) Bhandari	Burseraceae	Oleo gum resin	Guggulosterol and guggulosterone	Treatment of arthritis, rheumatism, yperlipidemia, thrombosis, and hypercholesterolemia
11.	Kalihari	*Gloriosa superba* L.	Colchicaceae	Whole plant	Colchicine	Antitode for snake bite
12.	Behda	*Terminalia bellerica*	Combretaceae	Fruits	Beta sitosterol, gallic acid and bellericanin	Used against heart diseases, anaemia and rheumatism.
13.	Harde	*Terminalia chebula*	Combretaceae	Fruits	Chebulin, chebulinic acid, tannic acid and behinic acid	Extensively in a number of ailments
14.	Stevia	*Stevia rebaudiana*	Compositae	Leaves	Stevioside (3-10 %), Rebaudiside-A (13 %), Rebaudiside-B and C	Calorie free sweetener, sugar substitute for diabetic patients
15.	Shankhapushpi	*Evolvulus alsinoides* Linn. Syn. *Convolvulus alsinoides* Linn.	Convolvulaceae	Whole plant	Evolvine; β-sitosterol, stearic, oleic and linoleic acid, betaine and an uncharacterized base.	Used in form of decoction in nervous debility and loss of memory
16.	Bhui Aonla	*Phyllanthus amarus* Schum. & Thonn.	Euphorbiaceae	Whole plant	Phyllanthine and hypophyllanthine	Treatment of jaundice
17.	Aonla	*Emblica-officinale* Geaertn. *Phyllanthus emblica* L.	Euphorbiaceae	Fruits	Emblicannin A and Emblicannin B	Treatment of vomiting, biliousness, urinary discharges, constipation, leprosy, piles and diseases related to eyes
18.	Senna	*Cassia angustifolia* Vahl	Fabaceae	Leaves, flowers and pods	Sennosides	Constipation

Contd.

S. No	Common name	Botanical name	Family	Economic part	Major Phytochemical constituents	Uses
19.	Kauch	*Mucuna pruriens*	Fabaceae	Seeds	L-dopa	Treating of Parkinson disease, sexual disorders, cholera, urinary troubles and liver and gall bladder diseases
20.	Asoka	*Saraca asoca*	Fabaceae	Flowers	Steroidal component and calcium salt	Treatment of bleeding piles and skin diseases
21.	Liquorice (Mulathi)	*Glycyrrhiza glabra* Linn.	Fabaceae	Roots	Glycyrrhizin	Used to cure cough and peptic ulcers
22.	Chiraita	*Swertia chirayita* (Roxb.ex Flem.) Karst.	Gentianaceae	Whole plant	Ophelic acid, glucosides	Treatment of bronchial asthma, liver disorders, and anaemia
23.	Indian Gentian	*Gentiana kurroo*	Gentianaceae	Roots	Gentiopicrin and tannins	Treatment of fever, abdominal pains, and for the purification of blood
24.	Pathar Choor	*Coleus forskohlii* Briq	Lamiaceae	Roots	Forskholin	It is used for treating analgesic, ophthalmic, and febrifuge
25.	Safed musli	*Chlorophytum . borivilianum* Santapu & Fern	Liliaceae	Roots	Saponins	It is used as a general tonic and is a well-known aphrodisiac
26.	Aloe	*Aloe barbadensis* Mill	Liliaceae	Leaves	Aloins	Skin tonic and herbal cosmetics
27.	Satawar	*Asparagus racemosus* Willd.	Liliaceae	Roots	Saponins	Sexual weaknesses, leucorrhoea and increase lactation in feeding mothers

Contd.

S. No	Common name	Botanical name	Family	Economic part	Major Phytochemical constituents	Uses
28.	Giloi	*Tinospora cordifolia* (Willd.) Miers ex HK. f. & Th.	Menispermaceae	Stem	Tinosporon, tinosporic acid, tinosporoside, tinosporin and cordifolide	Treatment of general debility, dyspepsia, fevers and urinary diseases
29.	Opium poppy	*Papaver somniferum* L.	Papavaraceae	Capsule	Morphine, codeine, thebaine, narcotine and papaverine	Used in medicine as a pain reliever, analgesic, hypnotic, sedative, antispasmodic, antitussive & sudorific purpose in modern medicine
30.	Long pepper	*Piper longum* Linn.	Piperaceae	Fruits and roots	Piperine and piplartine	Treatment of diseases related respiratory tracts viz., bronchitis, asthma, cough, etc.
31.	Isabgol	*Plantago ovata* Forsk.	Plantagonaceae	Husk	Mucilage	Laxative, used against irritation in gastrointestinal tract, constipation, dysentery
32.	Atis	*Aconitum heterophyllum* Wall.ex Royle	Ranunculaceae	Roots	Heterophyllisine, atidine and atisine	It is used as a bitter tonic to combat debility after malaria and other fevers
33.	Cinchona	*Cinchona officinalis*	Rubiaceae	Bark	Quinine	Treatment of malaria
34.	Brahmi	*Bacopa monnieri* (L.) Pennel	Scrophlariaceae	Whole plant	Bacoposide	Improveing mental alertness, learning performance, and memory as well as in the treatment of insomnia

Contd.

S. No	Common name	Botanical name	Family	Economic part	Major Phytochemical constituents	Uses
35.	Purple foxglove	*Digitalis purpurea*	Scrophulariaceae	Leaves and seeds	Digitoxin	Treatment of various heart troubles
36.	Belladona	*Atropa belladona*	Solanaceae	Leaves	Hyoscyamine and Atropine	Treatment of eye diseases
37.	Ashwagandha	*Withania somnifera* (Linn.) Dunal	Solanaceae	Roots	Withanoloids and withanins	Stress-induced disorders, To improve male potency
38.	Henbane	*Hyocyanus niger*	Solanaceae	Leaves	Hyocyamine	Used to cure ashthama
39.	Indian valerian	*Valerianawallichii* DC. *V. jatamansi* Jones	Valerianaceae	Roots	Valepotriates	Treatment of eye, blood, liver and spleen related diseases

□□□□

21

Production Techniques of Medicinal Plants

Kalmegh

Botanical name	*Andrographis paniculata* (Burm. f.) Wall. ex Nees Syn. *Justicia paniculata* Burm. f.
Family	Acanthaceae
Centre of origin	South-East Asia
Economic part	Whole herb
Main Phytochemical constituent	Andrographolide
Uses	It is useful in burning sensation, chronic fever, malaria and intermittent fever, inflammation, cough, bronchitis, skin diseases, intestinal worm, dyspepsia, flatulence, colic, diarrhoea, dysentery, haemorrhoids and vitiated condition of *pitta*

- The species is a native of tropical South-East Asia and occurs throughout hotter parts of India.
- The species is distributed in India, Sri Lanka, Bangladesh and Malaysia. In India it is found in the plains of Himachal Pradesh to Assam and Mizoram and also in Peninsular India.
- It is also cultivated as kharif season crop in Gujarat, Uttar Pradesh, West Bengal, Madhya Pradesh, Odisha, Andhra Pradesh and Tamil Nadu. Harvesting is done in the month of October-November.

- It is an erect, annual herb and 30-90 cm tall with upper part of stem quadrangular while the lower part nearly rounded stem.
- Leaves are opposite sessile or subsessile, linear-lanceolate or lanceolate, 3-8 cm long, acute, glabrous or minutely puberulous beneath and base cuneate, margin slightly undulate.
- Flowers are pedicelled, biliped, white purple or spotted purple and solitary. Pedicel is 2.5–10 mm in size, slender and glandular pubescent. Bracts are acicular and 2.5 mm long.
- Calyx lobes are subacute, 2.5-3.7 mm long and glandular. Corolla is 7.5-12.5 mm in size, tube about half as long as the corolla. Filaments are hairy and anthers are purple beared at base.
- Fruit is a capsule, oblong, 18-20X4.5–5.0 mm, young ones sparsely glandular and hairy; when mature it is glabrous.
- Seeds are sub quadrate, yellow to brownish in colour and rugose. Flowering and fruiting occurs in October – December (North India).
- The whole herb is medicinally useful.
- **Andrographolide** is the active principle having the therapeutic action.

Production technology

Climate	The plant comes up well in tropical and subtropical regions all over India.
Soil	It is a hardy species, therefore, can be grown in medium fertile sandy loam to clay-loam soils, possibly with irrigation.
Propagation	Propagation through seeds
Planting	Transplanting of seedling is done in second fortnight of June at a row and plant spacing of 60 to 45 cm and 45 to 30 cm respectively. Beds should be irrigated immediately after planting.
Manuring	FYM @ 20 t/ha is given as basal application. It is given NPK (75:75:50 kg/ha) in two split doses *i.e.* first at planting stage and second 40 days after plantation. Use of 5 kg *Azospirillium* + 5 kg Phosphobacteria per hectare has also given good results.
Irrigation	4-6 light irrigations are required till harvesting the crop.
Weeding	Since it is an herbaceous plant, the field should be free from weeds. Two to three weedings are essential during the crop season *viz.* at 20 days and 60 days after transplantation.
Crop maturity	The crop matures after 120 days of sowing.
Yield	A well-maintained crop grown during monsoon season yields 2.5 to 3.0 tons of dried herb per hectare.

Mandookaparni

Botanical name	*Centella asiatica* L.
Family	Apiaceae
Economic part	The whole plant has therapeutic values
Main Phytochemical constituents	Asiatic acid, Asiaticoside, Madecassic acid, Brahmic acid, Thankuniside, Centellose
Uses	It is used as nervine tonic, for improving memory and mental disorders. It is anti-leprosy, diuretic, stomachic and used in insomnia, asthma, abdominal disorders and fever.

- The plant is a small trailing herb. It is the only species of *Centella* found in India.
- Stem is glabrous, pink striated and rooting at nodes. Leaves are fleshy, orbicular to reniform and dentate. Petiole is long, smooth on upper surface and hairy below.
- Flowers are pink and white in fascicled umbels. The fruits are oblong, dull brown, laterally compressed, pericarp hard, thickened and woody white.
- The plant occurs in marshy places throughout the country in tropical and subtropical regions.
- The species is common in moist shady places all over India, particularly during rainy season.
- It is propagated both by **rooted suckers** as well as by **seeds.**
- It is used for the treatment of leprosy, skin diseases and to improve memory.
- It is also used against cholera, ulcers, bronchitis, leucorrhoea and kidney troubles.
- Asiatic acid, Asiaticoside, Madecassic acid, Brahmic acid, Thankuniside, Centellose.
- Total triterpinoids are in leaves and they are approximately 1.0% of leaves.

Production technology

Climate	Thrives best in monsoon periods in well drained beds.
Soil	Plant naturally grows over moist, fertile, loose, sandy loam and clayey soil.
Propagation	Rooted suckers and seeds.
Seed rate	300 kg rooted suckers are needed to plant one hectare land.
Time of sowing	February-March

Planting distance	45X45 cm
Manuring	Manure (FYM) at the rate of 20 t/ha should be mixed thoroughly with the soil at the time of field preparation. NPK fertilizers @ 100:50:50 kg/ha in 4 split doses are given.
Irrigation	During dry months fortnightly irrigation is needed and needs drainage during rainy season
Maturity	The crop matures in 90 days period after planting.
Harvesting	It is harvested through hand-cutting at fully grown leaf stage. The leaves are harvested in sunny weather to facilitate drying.
Yield	10-12 t/ha/years yield is obtained by 3 harvests in a year.

Periwinkle

Botanical name	*Catharanthus roseus* (Linn.) G. Don Syn. *Vinca rosea* Linn.
Family	Apocynaceae
Centre of origin	West Indies
Economic part	Leaves and roots
Main Phytochemical constituents	At present, more than 100 alkaloids have been isolated from the various parts of the plant of these, vinblastine (VLB) and vincustine (VCR) present in its leaves, and ajmalicine, present in its roots are medicinally important. The foliage contains 0.003-0.004%VLB alkaloid.
Uses	Plant is used in cancer and diabetes; root paste is used in septic wounds; root decoction is used in fever; leaves are used in menorrhagia; leaf juice is used in blood dysentery. The decoction of leaf is used for babies in gripping pain while the latex is useful in scabies. Plant contains hypotensive, sedative and antiviral activities.

- ***Catharanthus roseus*** is a perennial small herb or sub-shrub, up to 90 cm in height.
- Stem is erect, lax branching with flexible long branches, purple or light green.
- Leaves are simple, cauline, opposite, ex-stipulate, petiolate, elliptic ovate to oblong, 4-10 by 2-4 cm glabrous to pubescent, base acute or cuneate, apex obtusely apiculate and lateral nerves 10-12 pairs. Petiole is 1.0- 1.5 cm long.
- Inflorescence is recemose or axillary or terminal cyme or solitary/paired and shortly pedicillate.
- Flower colour is pink/white and tubular, swollen in the region of anthers, throat of corolla-tube hairy.
- Androecium contains 5 stamens included in the corolla tube, filaments are very short, epipetalous, anthers forming a cone-like structure above the stiga.

- Gynoecium contains two carpels, which are free below, but united in the stylar region. Ovaries are two, free and single style with dumbbell stiga.
- After fertilization, the carpels separate and form two fruits which form a pair of elongated follicles. Seeds are small in size and black in colour. 1000 seeds may weight 1.2 g.
- It is a native of West Indies. Plant is distributed throughout tropical and sub-tropical parts of India.
- Tamil Nadu, Karnataka, Andhra Pradesh, Gujarat, Madhya Pradesh and Assam are ideally suited for the cultivation of the plant.

Production technology

Climate	However, its growth is better in tropical and sub-tropical areas. The plant also grows in sub-tropical areas of northern India but the growth is slow due to extremely low temperature during the winter. A well distributed rainfall of 100 cm or more is ideal if the plant is to be grown as a rainfed crop.
Soil	It can be grown in any type of soil except those which are highly saline, alkaline or water-logged. Light soils, which are rich in humus, are preferable for large scale cultivation, since harvesting of the roots become easy.
Varieties	White flowered, Purple flowered and Nirmal (white flowered variety: from CIMAP) are the three types of periwinkle. 'Nirmal' and 'Dhawal' are the released cultivars of white flowered type.
Propagation	Propagated through **seeds** either by direct sowing or through transplanting.
Seed rate	About 2.5 kg of seeds are required per hectare for direct sowing or 0.5 kg/ ha through transplanting.
Time of sowing	During the monsoon months.
Planting distance	45X45 cm
Manuring	Apply FYM at 20 t/ha and N, P and K at 25:40:40 kg/ha as basal dose. On 60 days after transplanting apply 50 kg N as top dressing.
Irrigation	4-5 irrigations once in fifteen days during February, March and April months are needed to get optimum yield.
Harvesting	The crop becomes ready for harvest of roots after one year. Two leaf strippings can be taken, the first one after 6 months and the second after 9 months of sowing. Aerial parts are cut and the soil is ploughed for harvest of roots. Fruits are collected without damage.
Yield	Under irrigated conditions, about 1.5 tonnes of leaves, and 0.5 tonnes of roots on air-dry basis are obtained per hectare.

Sarpagandha

Botanical name	*Rauvolfia serpentina* (L.) Benth.ex. Kurz.
Family	Apocynaceae
Origin	India
Economic part	Roots
Main Phytochemical constituents	Reserpine, desrpidine and reseinamine
Uses	It is used to control high blood pressure

- Sarpagandha a member of family **Apocynaceae** is a perennial under shrub with irregular tubular roots distributed through out India.
- The plat grows wild upto 1200 m in deciduous forests of the sub-Himalayan tracts up to Nepal, the Gangetic plains, and Eastern and Western Ghats.
- The species attain a height of about 75 cm –1 m with inflorescence arranged in cymes with deep red flowers.
- Roots contain alkaloids (reserpine, desrpidine and reseinamine) which a sedative and used to control high blood pressure.
- It is also used for the treatment of insomnia, asthma and acute stomach-ache.
- Ruthless collection of the species from its wild habitats and the Government of India has prohibited its collection from the wild.
- The crop is under cultivation and propagated by seeds.

Production technology

Climate	The temperature varying from 10°-38° C is ideal.
Soil	The crop grows luxuriantly in deep, loamy to clay-loam soils rich in organic matter and of 6-8.5pH.
Propagation	Through seeds
Seed rate	6 kg/ha
Sowing	Seeds should be sown in furrows 2-3 cm deep, 20cm apart in April-May.
Manuring	NPK 40:40:40 kg/ha
Harvesting	The roots are dug out in (November to February) during second year and both tap root and its lateral branches besides attached fibrous roots are collected.
Yield	An average yield of 16-20 q/ha of dry root is obtained.

Guggal

Botanical name	*Commiphora wightii* (Arn.) Bhandari
Family	Burseraceae
Centre of origin	Africa and Asia
Economic part	Oleo-gum resin
Main Phytochemical constituents	Guggulosterol and guggulosterone are the important active ingredient of the gum resin
Uses	The gum of guggal is anti-inflammatory and efficacious in the treatment of arthritis, rheumatism, hyperlipidemia, thrombosis, and hypercholesterolemia.

- Guggal is one of the important medicinal plants belongs to family Burseraceae.
- Africa and Asia are the centers of origin of *Commiphora* spp.
- Guggal is a xerophyte and grows naturally in arid and rocky zones of India, that is, Gujarat, Karnataka, Madhya Pradesh, and Rajasthan, and also in Pakistan.
- It is a perennial shrub or small size tree of about 5 m height.
- It is a slow growing plant with crooked and knotty branches ending in sharp spines.
- The species contains male, female and hermaphrodite plants in the population.
- Flowers are sessile and appear singly or in groups of two to three. Fruits are ovoid, reddish brown to purple in colour.
- Two types of seeds, that is, black and yellowish-white, are produced.
- The flowering and fruiting take place throughout the year. However, maximum fruiting is observed from January to April.
- Propagation is done by seed as well as stem cuttings.

Production technology

Climate	It prefers arid and semi-arid climates.
Soil	The plant grows well in arid, sandy, and rocky tracts of tropical India. Sandy or sandy loam soils are best for its cultivation.
Varieties	CIMAP (Central Institute of Medicinal and Aromatic Plants), Lucknow, has developed a good oleo-gum-resin-yielding cultivar called 'marusudha'.
Propagation	It can be successfully propagated vegetatively by stem cuttings.

Planting distance	The plants are kept in nursery for 6 months and during the monsoon rooted cutting are transplanted in the field at a spacing of 2 x 2 meters.
Tapping and Collection	The plant should be allowed to grow for at least five to six years before commencing incision of thick branches for extracting oleo-gum resin. The oleo-gum resin is tapped during winter, from November–February, by making a 7–10-cm long incision in the main stem near the base. The cut part is completely covered with resin in about a month's time. The exuded gum secreted is collected every week up to one month after which further exudation of gum stops.
Yield	Starting from sixth year the guggul gum yield increases from 200 g to 500g per plant/season.

Flame lily (Kalihari)

Botanical name	*Gloriosa superba* L.
Family	Colchicaceae
Origin	Tropical Africa and Asia
Economic part	Whole plant
Main Phytochemical constituents	Colchicine, The colchicine content varies from 0.15 to 0.3% in the tubers and 0.7 % to 0.9% in the seeds
Uses	Treatment of acute gout, intestinal worms, infertility, wounds and other skin problems. The roots and leaves used as an antidote for snake bite, as a laxative, and to induce abortion

- *Gloriosa superba* an annual climbing perennial herb with tuberous roots.
- It grows between 3.5 to 6 m in length. They have showy flowers, many with distinctive and pronouncedly reflexed petals, like a Turk's cap lily, ranging in colour from a greenish-yellow through yellow, orange, red and sometimes even a deep pinkish-red.
- The plant grows from an underground tuber and is easy to cultivate in a warm conservatory in summer.
- All parts of the plant are poisonous and the tubers particularly so, since they contain the toxic alkaloid **colchicine.**
- Flame lily has a wide variety of uses, especially within traditional medicine as practised in tropical Africa and Asia (including Ayurvedic medicine in India).
- It contains the alkaloid **colchicine**, which has been used effectively to treat acute gout, intestinal worms, infertility, wounds and other skin problems.

- The roots and leaves used as an antidote for snake bite, as a laxative, and to induce abortion.
- It has proven useful in the treatment of chronic ulcers, arthritis, cholera, colic, kidney problems and typhus.
- Colchicine, an alkaloid extracted from the tubers and seeds gives high price in the market and used in scientific research.
- **Colchicine** is widely used as an experimental tool in the study of cell division, as it can inhibit mitosis (a type of cell division), induce polyPloidy level (cells containing more than two sets of chromosomes).
- It is also the state flower of Tamil Nadu.

Production technology

Climate	This can be cultivated up to 600 m from mean sea level with an annual rainfall of 70 cm.
Soil	A well drained red loamy soil is good for cultivation. Water logged soils are not suitable. The ideal pH should be 6.0 – 7.0.
Propagation	Propagation generally occurs from **seeds,** although mature plants can be divided and grown from tubers. The hard seeds can remain dormant for 6-9 months.
Planting	Planting can be done at 60 × 45 cm spacing during the rainy seasons and required the seed rate of nearly 1800 - 2000kg of tubers per ha.
Manuring	NPK (Nitrogen, Phosphorus and Potash) at 120:50:75 kg/ha is applied in two split doses. Half of N and the entire dose of P and K are applied as basal dressing. Remaining quantity of N is applied at one and two months after planting.
Harvesting	Flowering during September -October and pods can be harvested 170 to 180 days after planting. Seeds can be dried in shade for 10 to 15 days. The tubers are finally harvested after 5 to 6 years.
Yield	In initial years, the yield is low and gradually increases year by year. First two years about 100 kg per acre but after 3 years 200-300 kg of dried seeds per acre can be achieved in well maintained fields.

Stevia

Botanical name	*Stevia rebaudiana* (Bert.)
Family	Compositae
Economic part	Leaves
Phytochemical constituents	Stevioside (3-10 %), Rebaudiside-A (1-3 %), Rebaudiside-B and C
Uses	Calorie free sweetener, sugar substitute for diabetic patients

- ***Stevia rebaudiana* (Bert.)** is one of the 154 members of the genus *Stevia.*
- This is widely grown for its sweet leaves.
- The sweet herb Stevia is becoming a major source of natural sweetener as an alternate of sugar.
- It is rapidly replacing the chemical sweetener like Splenda, Saccharine and Aspartame.
- The plant is native to **tropical and subtropical regions of North America and South America.**
- Stevia leaves are 30 times sweeter than normal sugar.
- There are two compounds in stevia leaves 1.) **Stevioside(3%-10%)** and 2.) **Rebaudioside-A(1-3%).**
- The Stevioside is stable at 100 °C this is main advantage of Steviocide over other sweetener.
- Today stevia is ready to become King in the market of sweetener. As demand of low carbohydrate sweetener is rising day by day.
- A good market is being developed for stevia domestically as well as internationally. This is a ray of hope for the **diabetic patient.**
- As it is natural source of sweetener. It is also used for the treatment of obesity and high blood sugar patient.
- There is not increase in sugar level after consumption of stevia as sweetener as substitute of normal sugar.

Production technology

Climate	Annual average temperature of 31o C with a rainfall of 140 cm per year has been found optimum for its good growth. The day temperatures should not exceed 48°C and the night temperatures should not fall below 4°C.
Soil	Stevia grows well in sandy loam soils with an ample supply of water. Stevia prefers acidic to neutral soil with a pH range of 6.5-7.5 for its best growth.
Propagation	Stem cuttings of 15 cm length taken from leaf axils of the current year's growth have been given better results. The best months for propagation are February-March. The cuttings will be ready for transplanting after 25-30 days of rooting.
Planting distance	Rooted cuttings are planted at a spacing of 45X45 cm, with a total plant density in the order of 30,000 plants per ha.
Manuring	27.5:112.5:112.5 kg NPK/ha is recommended for higher yields.
Harvesting	It can be harvested when plants are 40-60 cm in height. Shorter days induce flowering. Optimum yield (biomass) and stevioside quality and quantity is best just prior to flowering.

Yield	An estimated 3-3.5t/ha dry green leaves are obtained from three-four cutting of every year. It is cultivated up to 5 years after a one time plantation.
Drying and Threshing	Depending on weather conditions and density of loading, it generally takes 24 to 48 hours to dry Stevia at 40°C to 50°C. Immediately following drying, a specially designed thresher/separator is necessary to separate dry stevia leaves from its stem.

Shankhapushpi

Botanical name	*Evolvulus alsinoides* Linn. Syn. *Convolvulus alsinoides* Linn.
Family	Convolvulaceae
Centre of origin	India
Economic part	Whole plant
Main Phytochemical constituents	Plant contains an alkaloid – evolvine; â-sitosterol, stearic, oleic and linoleic acid, betaine and an uncharacterized base. Also obtained were pentatriacontane and triacotane
Uses	The whole plant is used in form of decoction in nervous debility and loss of memory. The plant is also useful as blood purifier and in bleeding piles.

- It is an annual/perennial herb with prostrate spreading branches in all directions.
- The root system is deep. Leaflets are 2.5-5 cm long, elliptical, oblong and rounded at base.
- Leaves are densely clothed with white appressed and long spreading hairs.
- In nature, seedlings appear after 2-3 showers during July-August. Generally, the seedlings are abundant under the canopy of trees or shrubs.
- The young seedling grows straight and produces lateral branches during first week of August. The lateral branches spread in all directions.
- The flowering starts during first week of August. Fruit formation takes place in the last week of August, while flowering still continues.
- The dispersal of the seeds takes place from September to December. Plant dries during November – December.
- Flowers are pretty blue in colour and axillary in position. The flower colour varies from dark purple to whitish purple.
- They are either solitary or in pairs on the long stalk.

- Styles are two and each is bifid. Fruit is globose with four-valved capsule containing dark brown to black smooth seeds.
- Plant is commonly found growing in open and grassy places, almost throughout India, ascending upto 1500 meter in the Himalayas.
- It is cultivated also in the sub-tropical parts of India, ascending upto 1800 meter elevation.

Production technology

Climate	The plant prefers shady and humid climatic conditions.
Soil	It is easily grown in organically rich, consistently moist, well-drained soils in full sun to part shade. Performs well in sandy soils with good drainage.
Propagation	Seeds; collected during October-November from natural habitats.
Seed rate	190 g seeds are required for planting one hectare area.
Time of sowing	The seedlings are transplanted during late July-August.
Planting distance	25X25 cm
Manuring	Farm yard manure at the rate of 10 tonnes per hectare should be applied.
Irrigation	Five days irrigation schedule was found to be suitable for obtaining maximum growth and biomass yield under dry field conditions.
Harvesting	It is a rainy season crop and takes about four months to complete its life cycle. The seedlings transplanted in the month of July attain its full growth during September-October.
Yield	As a pure crop, 18.6 quintal (fresh herbage) per hectare is obtained which is reduced to 1/3rd after drying.

Bhui Aonla

Botanical name	*Phyllanthus amarus* Schum. & Thonn.
Family	Euphorbiaceae
Economic part	Whole plant
Phytochemical constituents	The herb contains three crystalline lignans including phyllanthine and hypophyllanthine (non-bitter part). In addition, five flavonoids have been identified viz. quercetine, astralgin, quercitrin, isoquecitrin and rutin. Four alkaloids have also been separated. The total phyllanthin lignans range between 1 to 1.2% in the dry herb.
Uses	The plant is widely used to tone-up sluggish liver and also given in chronic liver condition and jaundice. In Unani medicine, the plant is used in jaundice as deobstruent, diuretic, cooling and astringent.

- ***Phyllanthus amarus*** is an annual herb 60 to 75 cm tall, quite glabrous. Root is stout and woody.

- Stems are often branched at base and angular.
- Leaves are numerous, sub-sessile, disticthious, stipulate and paripinnate with small leaflets.
- The leaflets are oblong, having nerve obscure and base rounded. Root is stout tortuous and woody.
- Flowers are very minute, shortly pedicelled numerous and axillary and yellowish in colour.
- Sepals are 5-6, ovate-oblong outer acute, coriaceous with pale margins; disk in both sexes of glands; male flowers 1-3 pedicelled; female flowers are solitary, larger and erect. Stamens are 3, sessile on a short column didynamous, styles minute, reflexed very short.
- The fruit is capsule, minute, globose and dehiscent. Seeds are with strong parallel and transverse ribs.

Production technology

Climate	The plant grows in tropical and subtropical climate.
Soil	Well drained sandy -loam soil for its luxurious growth.
Varieties	Navyakrit (CIMAP) - High herbage and active constituents.
Propagation	**Seeds** (They are viable upto six months from the time of harvest).
Seed rate	About 1 kg of seeds is required for raising seedlings to plant in one ha.
Time of sowing	After the first monsoon shower.
Planting distance	25X25 cm
Manuring	To encourage good vegetative growth, apply 10-20 tonnes of FYM, 50 kg N, 25 kg each of P2O5 and K2O are applied per hectare.
Irrigation	Irrigation is required during dry season if monsoon rains is scanty. The frequency of irrigation depends on the moisture content of soil.
Harvesting	The crop matures in 80-90 days when it should be harvested; it has maximum active chemical ingredients at fruiting. However, seeds collection is done after 110-120 days old crop.
Yield	A yield of 2 to 3 t/ha of dry herb is obtained.

Aonla or Indian gooseberry

Botanical name	*Emblica-officinale* Geaertn. *Phyllanthus emblica* L.
Family	Euphorbiaceae
Origin	Indo-China
Chromosome number	X=7, 2n= 28
Ploidy level	Auto-tetraploid
Type of self incompatibility	Sporophytic

Fruit Bud Differentiation	March-April
Bearing habit	Flowers are born as axillary cyme in the axil of leaves in determinate shoots
Pollination	Cross pollination
Pollinators	Wind, honeybees
Type of fruit	Berry
Edible portion	Mesocarp and Endocarp
Nutritional value	Vitamin C- 600-700mg/100g
Value added products	*Trifla* and *Chavanprash*
Economic part	Fruit
Main Phytochemical constituents	Emblicannin A and Emblicannin B
Uses	The fruits are used for the treatment of vomiting, biliousness, urinary discharges, constipation, leprosy, piles and diseases related to eyes

- The plant is a small to medium sized deciduous tree belongs to family **Euphorbiaceae.**
- The species is a native of **Indio-China.**
- In India the plant is found wild as well as in organized cultivation.
- Flowers are **unisexual.** Flowering mostly occurs during **March to June** and fruits ripen in the coming winter.
- The fruit is used as a major constituent in several Ayurvedic preparations such as Chyavanprash and Rasayana which promotes health and longevity.
- Aonla is one of the richest sources of **Vitamin C** known. In addition to this, potent antioxidant, several active tannoid principles (Emblicannin A, Emblicannin B, Punigluconin and Pedunculagin) have been identified which appear to account for its health benefits.
- The fruits of ***Emblica officinalis*** are rich in tannins. The fruits have 28% of the total tannins distributed in the whole plant.
- The fruit contains two hydrolysable tannins **Emblicanin A and B**, which have antioxidant properties, one on hydrolysis gives gallic acid, ellagic acid and glucose wherein the other gives ellagic acid and glucose. The fruit also contains **Phyllemblin.**
- The fruits, leaves and bark are rich in tannins. The root contains ellagic acid and lupeol and bark contains leucodelphinidin.
- The seeds yield a fixed oil (16%) which is brownish-yellow in colour. It has the following fatty acids: linolenic (8.8%), linoleic (44.0%), oleic (28.4%), stearic (2.15%), palmitic (3.0%) and myristic (1.0%).

- The fruits are used for the treatment of vomiting, biliousness, urinary discharges, constipation, leprosy, piles and diseases related to eyes.
- Triphala, an ayurvedic formulation includes Aonla as one of the three constituents.

Production technology

Climate	Arid and semi arid conditions, sub-tropical climate. A mature aonla tree can tolerate freezing as well as high temperature of 46°C but not heavy frost in winter. Warm temperature is conducive for the initiation of floral buds and ample humidity is essential for initiation of fruit growth of dormant fruit lets during July-August.
Soil	Aonla is a hardy plant, which is successfully grown in variable soil conditions. Slightly acidic to saline, pH 6.5-9.5. Heavy soils or high water table areas are not suited for its cultivation.
Propagation	Wedge grafting: January to April and October to DecemberPatch budding: Mid may to September
Root stock	*Phylanthus indofischeri*- Resistant to fungal spot
Planting	8X8m
Manuring	10kg FYM, 100g N, 50g P, 100g K for one year plant. This dose should be increased up to 10 years.
Time of application	January-February and August
Training	Modified central leader system
Maturity indices	Change in seed colour from creamy-white to brown.
Yield	1-3q/tree, giving 15-200q/ha for 60-75 years.

Varieties

Particular	Varieties
Improved varieties	Banarsi, BSR 1, Chakaiya, Francis, Goma Aishwariya, Kanchan (NA-4), Krishna (NA-5), NA-6, NA-7, NA-9, NA-10, NA-20 Kanchan (NA-4): Seedling selection from Chakaiya. Suitable for processing Krishna (NA-5): Seedling selection from Banarsi. Suitable for processing Amrit (NA-6): Seedling selection from Chakaiya. Suitable for processing Neelum (NA-7): Seedling selection from Francis. Suitable for processing Balwant (NA-10): Seedling selection from Banarsi. Suitable for processing NA-20: Seedling selection from Banarsi. Suitable for processing Goma Aishwariya- Early, drought tolerant

Senna

Botanical name	*Cassia angustifolia* Vahl
Family	Fabaceae (Leguminosae)
Centre of origin	Yemen and parts of Pakisthan, Gujarat and Saudi Arabia
Economic part	Leaves, flowers and pods
Main Phytochemical constituent	Sennosides
Uses	The drug is used as a cathartic in giving relief to constipation

- The herb is a member of family, **Fabaceae** and native to Yemen and parts of Pakisthan, Gujarat and Saudi Arabia.
- In India the species is cultivated both as irrigated and rain fed crop in Tamil Nadu, Rajasthan and Gujarat.
- Dried young leaves, flowers and 3-5 days old pods are used as source of raw drug.
- It contains **sennosides**, which is responsible for the purgative action.
- The drug is used as a cathartic in giving relief to constipation.
- In European Countries, it is used as herbal tea.
- In world market, there are two sources of senna raw drug, one is Tinnevely Senna, which has gone from India and the other is Alexandrian senna that has gone from Arabian Countries.
- The former species is *Cassia angustifolia* and the latter is *Cassia acutifolia*.

Production technology

Climate	It is a hardy warm weather crop grown under rainfed and irrigated conditions.
Soil	It is grown in sandy or sandy loam or laterite soils.
Varieties	KKM - 1, Anand Late Selection, ALFT-2 and Sona
Propagation	Seeds
Seed rate	6kg/ha
Time of sowing	June - July
Planting distance	45 x 30 cm
Manuring	**Basal:** Apply FYM 10 - 15 t/ha and N, P and K at 25, 25 and 40 kg/ ha.**Top dressing:** Apply 25 kg N in two splits at 40 and 80 days after sowing.

Harvesting	The first harvest of leaves and pods are done 2 months after sowing and subsequent harvests at 30 days interval. Leaves and pods are dried for 7 - 10 days.
Yield	**Dried leaves:** 2 t/ha., **Dried pod :** 150 - 200 kg/ha.

Liquorice (Mulathi)

Botanical name	*Glycyrrhiza glabra* Linn. Syn. *Loquiritae officinalis* Moench
Family	Fabaceae
Centre of origin	Iraq
Economic part	Roots
Main Phytochemical constituents	Glycyrrhizin
Uses	The plant root is a demulcent, mild expectorant and anti-inflammatory agent. An extract of the root provide relief in treating peptic ulcers. It has glycyrrhizic acid as main constituent and this has showed anti-viral and anti-inflammatory actions. The plant extract is used as a sweetener in tonic, laxative and given in sore throat and in cough remedies.

- It is a perennial under shrub, reaching up to 120 cm height under cultivation.
- The stolon crown gives rise to a number of long semi-woody stems which bear compound pinnate leaves.
- Stolon is nearly cylindrical, upto 2 cm in diameter. Outer surface is yellowish brown or longitudinally wrinkled with patches of cork.
- Its odour is characteristics and taste is sweet.
- Flowers are pale blue in colour and flowering occurs from 2-3 years of planting onwards. Pod is 2.0 - 2.5 cm long with 2 to 5 seeds.
- This species is widely distributed in the world from 5°W to 100°E longitude and 20° to 50° N latitude.
- The plant has been found to grow well in Patiala, Hissar, Jhansi and Lucknow. It is also cultivated in Punjab and Sub-Himalayan tracts in India.

Production technology

Climate	In nature, it has wide distribution from dry cold temperate parts of Asia to Mediterranean climates, where annual temperature varies from 25°C in summer and 5°C in winter season. The plant thrives in cultivation, where the locality receives 50- 100 cm rainfall annually and cultivation is supported with irrigation.
Soil	Mulethi is a hardy plant and grows over rich forest soils, ranging from pH 5.5 to 8.2. Sandy-loam fertile soils with pH 6.0 to 7.5 have been found to promote good root development in India
Varieties	Haryana Mulhatti-1,
Propagation	Propagation is usually carried through stolons cuttings of about 10-15 cm.
Seed rate	It requires 300 kg of planting material for one hectare land.
Time of sowing	During spring seasons
Planting distance	60X45 cm or 90X45 cm
Manuring	Farm Yard Manure (FYM) has been found useful for good development and growth of underground roots and should be applied at the rate of 10 t/ha at the time of field preparation.
Irrigation	The crop requires irrigation at an interval of 30-45 days in dry summer season.
Harvesting	Manual digging is carried for harvesting of roots. The crop is harvested in winter season i.e. November or December to obtain roots containing high glycyrrhizic acid.
Yield	The yield of dry root at Hissar (Haryana) is recorded around 7 t/ha.

Chiraita

Botanical name	*Swertia chirayita* (Roxb.ex Flem.) Karst.
Family	Gentianaceae
Economic part	Whole plant
Main Phytochemical constituents	Ophelic acid, glucosides
Uses	Treatment of bronchial asthma, liver disorders, and anaemia

- It is an erect annual herb belongs to family **Gentianaceae**.
- Chirayata is an erect, annual, branched herb, up to 1.5 m high. stem is robust and cylindrical below, but four-angled upwards. Leaves are broadly lanceolate, opposite, sessile, about 10 cm long, acute at tip, and five-nerved.
- **Inflorescence is a large leafy panicle**. Flowers are numerous, greenishyellow, and tinged with purple. Sepals and petals are four in number.

- Each petal lobe has a pair of green, honey-secreting glands. Capsules are minute, ovate, about 6 mm in diameter, and sharp pointed. Seeds are smooth and many angled.
- The species is distributed throughout the temperate Himalayas, between 1200 m and 3000 m altitudes, extending from Kashmir to Bhutan. The species may be cultivated at lower elevations in north-eastern Himalayas as compared to western Himalayas.
- The plant is propagated by seeds. It grows well in moist, temperate forests of Himachal Pradesh.
- Dried herbage portion is used as raw drug. Flowering occurs in July to October and the raw drug is collected when the capsules are fully formed.
- The drug is extremely bitter in taste.
- Chiraita is also known as brown or white chiraita to distinguish it from 'green chaiaita', which is the dried herbage of *Andrographis paniculata.*
- The bitter tonic made from the raw drug improves bile secretion and used for the treatment of bronchial asthma, liver disorders, and anaemia.
- The juice of fresh plant or infusion of dry plants is prescribed as a " blood purifier" in skin diseases.
- The active ingredient of the raw drug includes **ophelic acid, glucosides**, etc.
- Cultivation also has been initiated in the species and **seeds** are used for propagation.

Production technology

Climate	The crop can be grown in areas having mild rainfall (100 cm) in rainy season and in areas with long cold winter, receiving snowfall frequently.
Soil	The plant can be grown in variety of soils with sandy loam rich in carbon and humus.
Propagation	Through seeds
Sowing	Nursery beds are prepared in selected areas of suitable climatic conditions. Seeds are sown during May-June.
Planting	After 3-4 months seedlings are transplanted in the field in rows at a distance of 45-60cm. The distance between two rows is also kept about 60 cm.
Manuring	The medicinal plants have to be grown without chemical fertilizers and use of pesticides. Organic manures like, Farm Yard Manure (FYM), Vermi-Compost, Green Manure etc. may be used as per requirement of the species.

Weeding	Periodical weeding and hoeing is required in nursery and field.
Irrigation	The nursery beds and field after plantation should be irrigated periodically as and when required weekly or fortnightly. The plants are irrigated till it flowers.
Harvesting	The Crop is usually ready for harvesting after 6-8 months. The plants are collected and dried in shade.
Yield	About 3.75 tonnes/hectare of dried herbage yield is estimated in cultivation in two years.

Patther Chur or Pashan Bhedi (Coleus/Hriver)

Botanical name	*Coleus forskohlii* Briq
Family	Lamiaceae
Economic part	Roots
Main Phytochemical constituent	Forskholin
Uses	It is used for treating analgesic, ophthalmic, and febrifuge

- The root of *Coleus* species is considered analgesic, ophthalmic, and febrifuge. It is very useful in epilepsy.
- Fresh aromatic root is offered in temples for its characteristic odour and is also used as a flavouring agent.
- Patharchur is a large, succulent, densely hairy, herb with aromatic roots.
- It grows up to a height of 70 cm. The roots are perennial, while the shoots are annual. Stems are much branched, scandant, hairy, fragile, and knotted at nodes. Leaves are thick, fleshy, ovate, hairy, with serrate margins.
- Pink-white flowers appear in winter season, that is, from October to February. This is also the best growth period for shoots.
- Patharchur is found wild in tropical moist forests of Tamil Nadu and Karnataka. It is quite rare in nature but very commonly cultivated in moist tropical regions and southern states.

Production technology

Climate	Plants grow well in hot, humid climate and tropical and sub tropical situation under irrigation.
Soil	It grows well in red sandy loam soil. A soft soil having pH 5.5-7 with low moisture contains is reported to be ideal for rich growth of plants. Water stagnation should be avoided.
Varieties	Maimul, Manganiperu, Garmai are some of the popular varieties.

Propagation	The crop is mainly propagated through stem cuttings of length 15 to 20 cm, containing four to five nodes.
Season	Planting is done during June – July.
Planting	Planting is done at 60 x 45 cm spacing (37,030 plants/ha). In low fertile soils, planting is done at 60 x 30 cm which requires 55, 500 plants/ha.
Manuring	Incorporate 15 t/ha of FYM during last ploughing. NPK at 30:60:50 kg/ha is applied in two split doses at 30 and 45 days after planting. Apply 10 kg ZnSo4/ha to avoid micronutrient deficiency.
Irrigation	Irrigation is given immediately after planting and subsequently at weekly intervals of time.
Weeding	Weeding and hoeing is to be done at regular interval.
Harvesting	Harvest can be done at 5 – 6 months after planting. Harvesting must be done without any damage to the tubers.
Yield	Yield of roots is about 250 kg per hectare.
Post harvest handling	Harvested tubers are cut into small pieces and dried under shade until it reaches 8 % moisture level.

Safed musli

Botanical name	*Chlorophytum borivilianum* Santapu& Fern.
Family	Liliaceae
Centre of origin	Africa
Chromosome number	x=7, 2n=4x=28
Ploidy level	Tetraploid and octoploid
Economic part	Roots
Main Phytochemical constituent	Saponins
Uses	It is used as a general tonic and is a well-known aphrodisiac

- There are a number of *Chlorophytum* species, which are known under the trade 'safed musli' of which ***C. borivilianum*** is the commercially utilized species.
- The plant, a member of family **Liliaceae** is a perennial herb with condensed stem disc and a whirl of sessile leaves.
- Cytological studies of *Chlorophytum borivilianum* revealed a new chromosome number, 2n=4x=28 and also the presence of polyploids (tetraploid and octoploid levels) in the population with a basic number of x=7.
- The leaves are 15-35 cm long and oblanceolate.
- The plant is considered endangered species in the country.

- Inflorescence is dense; flowers are arranged in raceme and shortly branched.
- Flowers white, anthers as long as or longer than the filaments and yellow in colour.
- Bracts are usually long and over topping the shortly pedicelled buds. Cells of the orbicular capsule are 3-4 seeded and black coloured.
- Fruit: **Capsule**, 4-seeded with a slender beak and spongy septa, seeds oblong, black and shiny with crustaceans testa. Root stock stout, short or elongate, more or less cylindrical, 4 to 15 cm. long, average being 7.4 cm., 0.5- 1. cm. wide.
- The upper portion of the root stock remains clothed with the withered leaf bases and with copious lateral roots which are long, almost whitish or yellowish gray, shrieked. Externally the root stock is yellowish brown, thin. Fracture is starchy, odour indistinct; taste slightly bitter and mucilaginous.
- Fasciculated roots contain **saponins** and are medicinally important.
- It is used as a general tonic and is a well-known aphrodisiac.
- The species is naturally distributed in the forest areas of Maharashtra, MP, Rajasthan and Gujarat.
- The plant is propagated by the **stem disc** with the attached fleshy roots as well as by seeds.

Production technology

Climate	A tropical and subtropical climate with humid atmosphere is suitable for its growth. Temperature 20°C-25°C, rainfall 150-200 cm and relative humidity around 70% is suitable for its better performances.
Soil	Sandy-loam and organic matter rich in clay loam soil is suitable for the plant.
Varieties	JS405, RVSM 414
Propagation	**Stem disc** with the attached fleshy roots as well as by seeds.
Seed rate	6-8q discs/ha
Planting	June-July at the distance of 30X20 cm
Manuring	FYM @ 20 t/ha may be applied during land preparation. NPK @ 90:30:30 kg/ha is recommended; of this, N is split in 3 doses.
Weeding	Generally two weeding of the crop are required to free it from the weeds. The first within 20-25 days of sowing and the other 20-25 days later. The fields must be kept free from weeds to obtain the proper yield for the crop.

Irrigation	Largely rain fed irrigated in dry season only or as required.
Harvesting	After 3-3.5 months of sowing (generally in the month of Oct/Nov the leaves of Musli start yellowing. Subsequently they become dry and fall off and get detached from the tuber/disc.
Yield	4-5q dry roots/ha.

Aloe vera

Botanical name	*Aloe barbadensis* Mill
Family	Liliaceae
Economic part	Leaves
Main Phytochemical constituent	Aloins
Uses	It is used in cosmetic industries

- Aloe – a member of the family **Liliaceae** is indigenous to African countries and naturalized in India.
- The plant is perennial herb with fleshy leaves and condensed stem.
- Flowering occurs in winter and the inflorescence stack is about 90-150 cm long with orange coloured flowers.
- Leaves contain gel (polysaccharides) and leaf exudates contain **aloins,** which are commercially useful.
- Gel has a cooling and moisturizing action and hence used in cosmetic industries and the leaf exudates contains **aloins** and aloe **emodine,** which are used as pain killer and purgative.
- The crop is under cultivation in Gujarat, Rajasthan, Madhya Pradesh and Uttar Pradesh. Raw material is collected both from wild and cultivation for the industry.

Production technology

Climate	Commercial cultivation can be done in regions having 25 – 40°C.
Soil	Well drained laterite to loamy soils is suited for aloe cultivation. The soil pH must be ranged from 7.0 to 8.5.
Propagation	It is propagated through suckers.
Seed rate	For planting one hectare of land, 10,000 suckers are required.
Time of sowing	The planting can be done during two seasons namely June – July and September – October.
Planting distance	Field is thoroughly ploughed and beds of suitable sizes are formed. Suckers are planted at 90 x 90 cm.

Manuring	Incorporate 10 t/ha of FYM during last ploughing. For fertile soils manuring is not required at all. But for low fertile soils, apply 30 kg/ha of Nitrogen + 120 kg/ha of Gypsum during field preparation. In addition, 30 kg N can be applied after 20 days of planting.
Irrigation	It is mainly grown as rainfed crop. During summer months 4-5 irrigations are required
Harvesting	Harvesting can be done 6 – 7 months after planting. Plants are uprooted and processed (gel extraction) within 6 hours after harvest.
Yield	An average leaf yield of 15 t/ha can be obtained.

Satawar

Botanical name	*Asparagus racemosus* Willd.
Family	Liliaceae
Economic part	Tuberous root
Main Phytochemical constituent	Saponins
Uses	Sexual weaknesses, leucorrhoea and increase lactation in feeding mothers

- Satavary is a perennial spiny woody climber belongs to the family Liliaceae.
- The plant is a perennial, prickly climber excessively branched prickly under shrub found throughout the tropical & subtropical parts of India upto 1200 m.
- Flowers are white, fragrant, and minute, about 3 mm long and occur in solitary or fascicled, 2.5–5 cm long, racemes. Fruit is a three-lobed, red coloured berry, up to 6 mm in diameter, with mottled seeds and oily endosperm.
- Flowering and fruiting occur in December–January.
- The roots are fascicled, fleshy, spindle-shaped, light ash-coloured externally and white internally, more or less smooth when fresh, but on drying, develop longitudinal wrinkles and lack any well-marked odour. Branches are modified into cladodes with long basal decurved spines.
- The plant is common at low altitudes in shade and in tropical climate through out India.
- Fasciculated roots are medicinally important. It contains **saponins** and is used for the treatment of dysentery, tumours, rheumatism and kidney and liver disorders.
- Powdered roots are a common ingredient of a number of vital tonics, which are believed to cure sexual weaknesses, leucorrhoea and increase lactation in feeding mothers.

Production technology

Climate	It can be easily grown in sub-tropical & Sub-temperate agro-climatic regions up to 1400 m
Soil	Plant usually grows in a variety of soils including medium black having pH 7-8, electrical conductivity 0.15, organic carbon 0.79%.
Propagation	Through seeds and vegetative propagation is by division of rhizomatous disc present at the base of the aerial stem.
Seed sowing	Seeds are sown in April in raised beds at 5 cm apart to facilitate decay of its hard seed coat by the time monsoon commenced. Germination starts in 8 to 10 days after the first shower of monsoon in June.
Transplanting	The seedlings were transplanted on ridges at 60 x 60 cm apart and provided bamboo stakes when the plants attained a height of 45cm.
Manuring	The medicinal plants have to be grown without chemical fertilizers and use of pesticides. Organic manures like, Farm Yard Manure (FYM), Vermi-Compost, Green Manure etc. may be used as per requirement of the species.
Weeding	Two weedings are carried out during the rainy months, thereafter one in next 2-3 months
Irrigation	Irrigation is given after the rainy season is over, at the rate of two irrigations in winter season and one per month in summer season.
Harvesting	The crop matures in 12 months after planting; however, for seed harvesting, it is recommended to be harvested only after 20 months. Rabi season, that is, November–December, is the best time for harvesting tuberous roots when the above-ground parts start turning pale yellow.
Processing	It is observed that in case the roots are not peeled off within a few days, it is a bit difficult to remove the skin as such. In such a condition the roots are kept in boiling water for about 10 minutes, followed by cold-water treatment to facilitate peeling. After removing the skin, it is cut transversely into small pieces and dried in shade.
Yield	The crop, when harvested in 12 months, yields about 4–5 tonnes/hectare, while harvesting after 20 months yields about 6 tonnes/hectare of tubers along with 35 kg/hectare of seeds, which are not obtained in the 12-month-old crop.

Giloi

Botanical name	*Tinospora cordifolia* (Willd.) Miers ex HK. f. & Th. Syn. *Meninspermum cordifolium* Willd.
Family	Menispermaceae
Type of fruit	Drupe
Economic part	Stem
Main Phytochemical constituents	Tinosporon, tinosporic acid, tinosporoside, tinosporin and cordifolide
Uses	Treatment of general debility, dyspepsia, fevers and urinary diseases

- It is a deciduous perennial climber belongs to the family **Menispermaceae.**
- It is distributed throughout tropical India.
- A large extensively spreading, perennial woody climber with succulent stems.
- Leaves simple, alternate, cordat ovate; flowers unisexual dioecious, yellow.
- Flowers are yellow, unisexual, minute, and less than 2 mm in size. Male flowers are grouped in axillary racemes, while female flowers are solitary.
- Fruit is an ovoid and succulent **drupe**, lustrous, red in colour, and of the size of a large pea, having a single seed.
- Seed is fleshy and curved.
- Flowering occurs in **May–June**, while fruiting is witnessed in **September–October.**
- The stem and leaves are medicinally used as raw drug.
- ***Tinospora*** stem is a common constituent of a number of ayurvedic vital tonics for the treatment of general debility, dyspepsia, fevers and urinary diseases.
- Starch called as **"Sat Giloe"** present in the stem along with alkaloids is the active principle of the species.
- Major constituents **are tinosporon, tinosporic acid, tinosporoside, tinosporin and cordifolide**.
- Giloe is a tonic and has alterative, diuretic, and aphrodisiac properties.
- It is a **febrifuge used in malarial and chronic fever**. It is also a liver tonic. The plant is used in general debility, loss of appetite, fevers, urinary disorders, diabetes, rheumatism, and dyspepsia. Fresh plant is more efficacious than dried plant.
- Leaf also contains a number of alkaloids. Leaf is used for the treatment of gout, jaundice and rheumatism.

Production technology

Climate and	The plant grows in subtropical and tropical climate. It does not tolerate high rainfall or waterlogged conditions.
Soil	Light medium sandy loam soil rich in organic matter, and with adequate drainage, is suitable for its cultivation.
Propagation	The plant is propagated by stem cuttings. The cuttings can be obtained from mother plants in June–July. The plant can also be raised using seeds.
Planting	The stem cuttings with nodes are sown directly in the field. An optimum spacing of 3 m × 3 m is recommended for better yield. The plant requires support to grow, which can be provided by raising wooden stakes or trellis. Already growing shrubs or trees can also support the plant.
Manuring	The medicinal plants have to be grown without chemical fertilizers and use of pesticides. Organic manures like, Farm Yard Manure (FYM), Vermi-Compost, Green Manure etc. may be used as per requirement of the species.
Irrigation	The field after plantation should be irrigated periodically as and when required weekly or fortnightly.
Weeding	Periodical hoeing is done, both in the nursery and field as per requirement.
Harvesting	The stem is harvested during autumn when it develops to a diameter of more than 2.5 cm. Basal part is left for further growth.
Yield	The plant yields about 1500 kg of fresh woody stem, reduced to 300 kg of dry weight per hectare in about two years.

Opium poppy

Botanical name	*Papaver somniferum* L.
Family	Papaveracae
Chromosome number	Basic number x=11, and 2n=22.
Economic part	Whole plant
Main Phytochemical constituents	Morphine, codeine, thebaine, narcotine and papaverine
Uses	Used in medicine as a pain reliever, analgesic, hypnotic, sedative, antispasmodic, antitussive & sudorific purpose in modern medicine

- Opium poppy (***Papaver somniferum*** **L.**) belongs to the family **Papaveracae** has its importance as a plant based natural **pain reliever** from the time dating back to early civilization till today.
- It is **indigenous to Asia Minor**, and is cultivated largely in European and Asiatic Turkey, Persia, India and China for the production of Opium.

- The word opium is derived from the **Greek opos (juice).**
- It is a sparsely branched, annual herb growing up to 150cm height.
- The leaves are ovate, oblong or linear oblong; flowers are large usually bluish with a purplish base or white, purple or variegated.
- It produces large, spherical to globular, greyish-green capsules 90-110 days after sowing, studded with numerous light flat seeds.
- The capsules vary much in shape and size. They are usually hemispherical, but depressed at the top, where the many-rayed stiga occupies the centre; they have a swollen ring below where the capsule joins the stalk.
- It produces capsular type of fruits from which **the latex known as opium** is obtained on lancing.
- The fruits are about 2.5 cm in diameter, globose in shape.
- Seeds are reniform with white or black in colour.
- Though nearly all parts of the poppy plant contain a white milky latex, the unripe capsules contain large amount.
- The small kidney-shaped seeds, minute and very numerous, are attached to lateral projections from the inner walls of the capsule and vary in colour from whitish to slate.
- The seeds of opium poppy are highly nutritious as it contains **protein upto 24%** and other vital nutrients beneficial for human health.
- The seed oil of poppy is also important for health point of view due to having high percentage of **linoleic acid (68%)** which helps in lowering blood cholesterol level in human body and is also used in the treatment of cardiovascular diseases in human system.
- The heads are of a pale glaucous green when young. As they mature and ripen they change to a yellowish brown, and are then cut from the stem if the *dried* poppy heads are required.
- *Opium* is extracted from the **poppy heads** before they have ripened, and from Poppies grown in the East, those grown in Europe yielding but little of the drug.
- The **whole plant parts** of opium poppy are valuable in terms of food, medicine, vegetable and as brew ages.
- It is an important crop of pharmacopeial uses.

- Opium is one of the oldest known **pain killer** and is source of several alkaloids used for analgesic, hypnotic, sedative, antispasmodic, antitussive & sudorific purpose in modern medicine.
- **Morphine** is the major alkaloid present in opium ranges from **7-14%** mainly used to relieve from almost all type of severe pains.
- Other important alkaloids are **codeine (1.25-3.75%)** used as mild analgesic, respiratory sedative; **thebaine (0.2- 0.5%)** is used as analgesic & sedative; **papaverine (0.4- 1.0%)** is used as antispasmodic and also have strong effect on blood vessel & abdominal viscera and **narcotine (5.0-7.0%)** is beneficial in allaying cough & headache.
- Today our pharmaceutical industries solely depend on opium poppy for their crude resources for manufacturing of pain killing drugs.
- The medical practitioners around the world routinely prescribe important life saving drugs, are the secondary metabolites produced as a result of complex plant metabolism.
- The important life saving drugs are mostly derived from five major alkaloids viz., **morphine, codeine, thebaine, narcotine and papaverine** which are present in **opium latex** in ample amount.
- The dry latex obtained by scratching the capsule is called **opium.**
- These valuable alkaloids are mainly extracted in India from green unripe capsules by making incision upto 1-2 mm in the epidermal wall of the capsule, but globally it is extracted from the dried capsule which is called CPS (Concentrated poppy straw) system.
- In CPS system, the dried capsules along with eight inches of peduncle are harvested and seeds are threshed. The remaining husk is used to extract various alkaloids.

Economic importance of opium poppy and its derivatives

- Opium poppy belongs to the family **Papaveracae** and has been attracting the interest of researchers because of its pharmaceutical, decorative and alimentary attributes.
- At present our scientists have been able to discover more than **40 alkaloids** of immense medicinal importance.
- Despite of their best possible efforts to identify more and more alkaloids, the demand for five major alkaloids i.e. **morphine, codeine, thebaine,**

narcotine and papaverine have elevated due to major application in medical field.

- However, several of these medicinal compounds can be made synthetically but alkaloids belonging to various groups viz., **Phenanthredene (Morphine, Codeine, Thebaine)**, the true **Benzylisoquinilone (Papaverine)** and **Phthalideisoquinilone (Narcotine)** are only obtained from opium which place opium poppy at the highest place among the diverse array of medicinal plants.
- The most important and potent alkaloid is **morphine** which can be used for both short term as well as long term pain control, is widely used in many prescriptions of pain medications.
- The drug occurs as a white crystalline powder or colourless crystals and is available for legal medical use.
- Another important alkaloid is **codeine** which is considered as a prodrug because it is converted into morphine and codeine-6-glucuronide (C6G) in *in vivo*.

Production technology

Climate	The crop prefers moderately cool weather and light showers to foggy nights at rosette stage which support good vegetative growth; but open sunny weather is needed at capsule formation stage till their maturity. Dry, warm, weather (32-38°C) during February-March favours good flow of latex during lancing and consequently produces high latex yield.
Soil	Opium poppy prefers a well drained, highly fertile light black or loam soil with an optimum pH around 6- 7.5.
Propagation	Through seeds
Sowing	The seed is either sown broadcast or in lines. Before sowing, the seeds may be treated with fungicides like Mancozeb@ 3 g per Kg of seeds.
Season	The best time for sowing is late October or early November.
Seed rate	Seed rate is 7-8 Kg/ha for broadcast method and 4-6 Kg/ha for line sowing.
Spacing	A spacing of 30 cm between lines and 30 cm between plants is normally adopted.
Manuring	Farm yard manure @ 20-30 t/ha is generally applied by broadcasting while the field is prepared for sowing. Besides, 50- 80 Kg of N and 40-50 Kg of P_2O_5 per hectare is recommended.
Irrigation	A light irrigation is given immediately after sowing followed by another light irrigation after 7 days when the seeds start germinating. Three irrigations at an interval of 12-15 days are given till pre flowering stage and then irrigation frequency is reduced at 8-10 days during flowering and capsule formation stage. Normally, 12-15 irrigations are given during the entire crop period.

Lancing and latex collection	Opium starts flowering in 95-115 days after sowing. The petals start shedding after 3-4 days of flowering. The capsule is ready for lancing after 20-25 days of the fall of petals. Lancing of the capsules exudes maximum latex at this stage. At this stage, the capsules is green, immature but is of fully grown size. It contains 70% of the total morphine synthesized by the plant. Lancing may be done with a knife having three or four equi spaced pointed ends which does not penetrate more than 1-2 mm in the capsule. Too deep or too shallow incision is not advisable. Lancing may be done early in the morning before 8.00 a.m. at two days interval in each capsule. The length of the incision should be 1/3 or less than the full length of capsule.
Harvesting	The crop is left for drying for about 20-25 days when the last lancing on the capsules stops exudation of latex. The capsules are then picked up and the plant is removed with sickles. Harvested capsules are dried in open yard and seeds are collected by beating with a wooden rod.
Yield	The yield of raw opium varies from 50 to 60 kg/ha.

Improved varieties

Variety	Salient features
Varieties developed by National Botanical Research Institute, Lucknow	
BROP-1	In this variety the plants are medium sized having 3-4 capsules/plant and capsules are of three types viz. oily, parrot coloured and black peduncle. Flowers are white. It is a synthetic variety stabilized after hybridization/ intermating between three high yielding cultivars viz. kali dandi (black peduncle), suga pankhi (parrot colour) and sufaid dandi (white peduncle) followed by selection. The average opium yield, seed yield and morphine content are up to 54kg/ha, 1000-1200kg/ha and 13% respectively. Geographically it can be cultivated mainly in Northern Indian plains.
NBRI-1	This variety is developed through selection. The plants are medium tall having large fringed leaves and white flowers. The average opium yield, seed yield and morphine content are up to 52kg/ha, 1000kg/ha and 12-13% respectively. Geographically it can be cultivated mainly in Northern Central India.
NBRI-2	This variety has intermediate tall plants, thick stem, broad leaves, long peduncle with big capsules and flowers are white. This variety is also developed through selection amongst local collection having above characters. Average opium yield, seed yield and morphine content are up to 52 kg/ha, 1200kg/ha and 15%respectively. Northern Central part of India is recommended for its cultivation.
NBRI-6	In this variety, plants are medium tall with narrow leaves and white flowers. It is developed by hybridization between two germplasm lines BR007 and BR008 (BR007 x BR008) followed by rigorous selection generation after generation up to eight generations. Average opium yield, seed yield and morphine content are up to 55kg/ha, 1200kg/ha and 13-14% respectively. Geographically Northern Indian plains are mainly recommended for its cultivation.

Variety	Salient features
NBRI-9	In this variety, plants are intermediate sized with white flowers and large capsules. The variety is high yielding (seed yield) and is developed by hybridization between germplasm lines S-10 x S-18 followed by rigorous selection until the variety is stabilized (up to eight generations). It's average opium yield, seed yield and morphine content is upto 52kg/ha, 1400kg/ha and 12% respectively. For the cultivation of the variety, Northern Indian Plains are recommended as most suitable.
NBRI-10	The plants in this variety are medium tall, having dark green leaves and white flowers. The development of variety was done through hybridization germplasm lines (IC-30 x S-10) followed by rigorous selection up to eight generations. Average opium yield, seed yield and morphine content are up to 50kg/ha, 1200kg/ha and 12% respectively. Geographically Northern Central plains are recommended for its cultivation.
Madakini	It is a high yielding variety for opium poppy, have multiple disease resistance and is granted US patent no.7,442,854B2 in 2009. The variety is developed by hybridization germplasm lines (BR007 x BR008) followed by rigorous selection. Plants of the variety are vigorous having dark green leaves, white flowers with blackish flowering stalk at the bottom of capsule at maturity. Average opium yield, seed yield and morphine content are up to 64kg/ha, 1200kg/ha and 15% respectively. Northern Central India is recommended for its cultivation.
High thebaine lines	As we know that thebaine is a non-narcotic alkaloid and can be used in making pain killing drugs. Thus for fulfilling the increasing worldwide demand of thebaine, with the help of interspecific hybridization (*P. somniferum* x *P. setigerum*) and mutation breeding experiments NBRI has succeeded in the development of few stable high thebaine lines. Thebaine content in these lines ranges 8-10% which is much higher than pre-existing varieties and germplasm
Varieties developed by Central Institute of Medicinal and Aromatic Plants, Lucknow	
Rakshit	It is a disease resistant and morphine rich variety in CPS (concentrated poppy straw). The plants are 106-112 cm tall with 20-26 cm long green peduncle and oblong capsules with waxy surface. The variety is developed by hybridization and selection generation after generation up to eight generations. Average seed yield and straw yield of the variety are up to 1200-1400kg/ha and 900-1100kg/ha respectively.
Sanchita	In this high yielding variety, plants are 107 cm tall and have 2-3 capsules/plant. The average seed yield and straw yield are 840kg/ha and 640kg/ha respectively. Morphine content of this variety is very low in CPS (approx. 0.74%).
Vivek	The plants of this variety are 112 cm tall with 2-3 capsules/plant. It is also a high yielding variety having seed yield and straw yield up to

Variety	Salient features
	840kg/ha and 760kg/ha respectively. Morphine content of this variety is also very low in husk (approx. 0.73%).
Sweta	It is high yielding variety with 66.5kg/ha opium yield and about 18% morphine content in latex.
Subhra	In this variety plants are medium sized having 3-4 capsules/plant. The average seed yield and husk yield of the variety are approx. 910kg/ha and 790kg/ha respectively. The morphine percentage in husk is approx. 0.77%.
Shyama	In this variety plants are 105 cm tall with black peduncle and also has erect incised leaves. It is a high yielding variety having seed yield, husk yield and morphine content up to 720kg/ha, 650kg/ha and 0.75% respectively.
Sujata	In this variety plants are 80-100 cm tall having 3-4 flat glabrous capsules with 18-20 cm long erratic black peduncle. It is an opium less, alkaloid less and non narcotic variety.
Variety developed by Narendra Dev University of Agriculture and Technology, Faizabad	
Kirtiman (NOP-4)	The plants of this variety are quite tall having white flowers and 1-2 oval capsules. Opium yield, seed yield and morphine content of the variety ranges 35-46kg/ha, 900-1100kg/ha and up to 12% respectively. Eastern U.P. region is best suited for its cultivation. It is downy mildew resistant variety developed by selection.
Variety developed by National Bureau of Plant Genetic Resources, New Delhi	
Trishna (IC42)	It is for resistance to frost, root rot and downy mildew through inbreeding and selection. The plants of the variety are tall with 5-7 capsules/plant and pink flowers. Opium yield, seed yield and morphine content ranges up to 49-53kg/ha, 1000kg/ha and 12-14.78% respectively.
Variety developed by Rajasthan Agricultural University, Udaipur	
Chetak (UO 285)	The plants of this variety are average tall with big capsules. Flowers are white with smooth petals. Opium yield, seed yield and morphine content ranges up to 54kg/ha, 1000-1200kg/ha and 12% respectively. Geographically Rajasthan is most suitable for its cultivation.
Variety developed by Jawaharlal Nehru Krishi Vishwavidalaya, Jabalpur (M.P.)	
Jawahar Aphim 16 (JA-16)	Plants of this variety are tall having white flowers and 1-3 big capsules plant. The variety has opium yield, seed yield and morphine content up to 45-54kg/ha, 900-1000kg/ha and 12% respectively. Madhya Pradesh is geographically recommended for its cultivation. It is a downy mildew resistant variety developed through pure line selection.

Long Pepper (Pippali)

Botanical name	*Piper longum* Linn.
Family	Piperaceae
Economic part	Fruits and roots
Main Phytochemical constituents	Fruits contain volatile oil, resin, alkaloids (4-5% piperine) and a terpenoid substance. Root contains piperlongumine as major alkaloid in addition to piperine.
Uses	Plant root is used in Ayurveda as a carminative, tonic to the liver, stomachic, emmenagogue, abortifacient and aphrodisiac. Fruits contain haematinic, diuretic, digestive and general tonic properties, besides being useful in inflammation of the lever, pains in the joints, snakebite, scorpion sting and night blindness. The plant is also used in dyspepsia, abdominal pain and diuretic splenopathy, anorexia, asthma, fever and act as anti-haemorrhoidal and appetiser.

- Long pepper is the fruit of ***Piper longum*** which is a slender, much branched, ascending herb and needs support for its proper growth.
- The leaves are 5-9 cm long and 5 cm wide; lower leaves are broadly ovate, deeply cordate with big lobes at the base, sub acute, entire and glabrous; upper leaves are dark green and cordate with short petiole or nearly sessile.
- Flowers are unisexual arranged in erect spikes. Female spikes are 1.25-2.00 cm long arising singly from leaf axil, are cylindrical, short and stout.
- It gives rise to multiple fruit, which is shining dark green when immature and blackish green when fully mature.
- Male spikes are longer, slender and are 2.5-7.5 cm long. The male spikes are dehiscent and non-productive.
- Long pepper is a native of the Indo-Malaya region. It is found growing wild in the tropical rain forests of India.
- Indian long pepper is mostly derived from the wild plants, but is also grown in small area in the Khasi hills, the lower hills of West Bengal, Eastern Uttar Pradesh, Madhya Pradesh, Maharashtra, Kerala, Karnataka and Tamil Nadu.
- It occurs wild in the forests of Andhra Pradesh and Andaman and Nicobar Islands as well.

Production technology

Climate	The plant requires hot, humid climate and an elevation between 100 to 1000 MSL. Higher elevations are not conducive to high yields. It needs partial shade for its ideal growth. Partial shade of about 20-25 % intensity is found to be optimum.
Soil	The crop thrives well in a variety of soils. It is cultivated successfully in laterite soils with high organic matter content, water holding capacity and well drained fertile black cotton soil. However, light, porous and well-drained soil rich in organic content is most suitable for its cultivation.
Varieties	The important varieties for cultivation are Vellanikara -1 (Visvum Tippili) and Yercaud (Pl)-9.
Propagation	Long pepper is propagated through **stem/vine cuttings** at the beginning of rainy season.
Time of sowing	The nursery can be raised during March and April. The cuttings planted in March-April will be ready for planting in the main field by the end of May.
Planting distance	60X60 cm
Manuring	About 20 t/ha FYM or any other organic manuring is applied at the time of area preparation.
Irrigation	Irrigation is utmost essential during summer months. One or two irrigations in a week depending upon the water holding capacity of the soil, is needed.
Harvesting	Vines start flowering six months after planting. Fruits take about two months to mature from its formation. Full-grown mature fruits are harvested before ripening, when it is firm and blackish green.
Yield	Yield of dry fruits in first year is about 100-150 kg/ha and it attains up to 0.75 -1.0 t/ha in third to fourth year. Thereafter, yield starts declining and gradually becomes uneconomic after fifth year. Therefore, it is usually cultivated as a 4 to 5-year crop.

Isabgol

Botanical name	*Plantago ovata* Forsk.
Family	Plantaginacea
Centre of origin	Persia
Economic part	Husk, seed
Main Phytochemical constituent	Mucilage
Uses	Laxative, used against irritation in gastrointestinal tract, constipation, dysentery

- Isabgol belongs to family **Plantaginacea** is an annual herb grown during the rabi season.

- ***Plantago ovata*** Forsk, the common isabgul has gained importance throughout the world for its wide variety of medicinal properties.
- It is a native of the **Persia.**
- In India it is cultivated in Gujarat and Rajasthan.
- Plantago ovata has recently claimed economic importance for its immense use in agricultural, pharmaceutical, cosmetic and food industries worldwide.
- The **seed husk** of *Plantago ovata* are used as laxatives. The **mucilage** contained in the seed coat is used as a stabilizer in ice cream; chocolate etc.
- Seed coat, which is known as **isabgol husk** under trade, is medicinally important.
- **Isabgol husk** consists of the episperm and collapsed adjacent layers removed from the seeds of *Plantago ovata* Forssk.
- **Isabgol husk** traditionally used against constipation, diarrhoea, intestinal irritation, etc., is a very good source of dietary fibre and has hypo cholaesterolemic activity.
- **Isabgol husk** consists of 85% water-soluble fibre. The active polysaccharidic fraction comprises 65% D-xylose, 20% L-arabinose, 6% rhamnose and 9% D-galacturonic acid (3).
- The polysaccharide is shown to be a highly branched acidic arabinoxylan, the xylan backbone having both $1 \rightarrow 4$ and $1 \rightarrow 3$ sugar linkages.
- The swelling property of the husk after absorption of water is the cause of its use as a famous medicine against **constipation and gastrointestinal** irritations.
- The seeds contain **hydrocolloidal polysaccharide (mucilage)** in the **outer seed coat (20-30%),** fixed oils, tannin, aucubin glycoside (iridoid), sugars, sterols and protein.
- Seeds are demulcent, cooling, diuretic and used in inflammatory conditions of mucus membrane of gastrointestinal and gestourinary tracts.
- Average seed yield of the crop is about 10-15 q per ha.
- **India** is the leader in Isabgol production and largest exporter of husk.

Production technology

Climate	Isabgol is predominantly grown as cold season crop. It requires cool climate and dry sunny weather.
Soil	Well-drained, light sandy to sandy-loam soils having a pH of 7-8.
Varieties	Gujarat Isabgol- 1, Gujarat Isabgol- 2, Haryana Isabgol-5, Jawahar Isabgol-4.
Propagation	Through seeds.
Seed rate	4 kg/ha.
Time of sowing	November to mid December
Planting distance	Seed should be sown in lines Line 30 cm apart.
Manuring	25-30 kg NP/ha.
Harvesting	The crop mature during March-April 110-130 days after sowing.
Yield	8-10q/ha.

Atis

Botanical name	*Aconitum heterophyllum* Wall.ex Royle
Family	Ranunculaceae
Economic part	Tuberous root
Main Phytochemical constituents	Heterophyllisine, atidine and atisine
Uses	It is used as a bitter tonic to combat debility after malaria and other fevers

- It is a biennial herb with fleshy roots belongs to family **Ranunculaceae.**
- Atees is an herbaceous, erect, Biennial; leaves more or less hetronorphom; flower blue or violet, fruits follicles.
- The plant was common at 2000-3000 m all over the western Himalaya but now because sparsely distributed.
- **Root tuberous** in pairs whitish or grey, breaks very easily and taste very bitter.
- Roots are used medicinal purposes. It is used in diarrhoea, dysentery and gastric pain. It is used as a bitter tonic to combat debility after malaria and other fevers. It is also used against hysteria, dyspepsia, vomiting and cough.
- Wild habitats are the sole source of the raw drug since the plant is not in cultivation.
- Alkaloids (**heterophyllisine, atidine** and **atisine**) present in the rhizome are the active ingredient.

Production technology

Climate	In general, cultivation up to 2200m elevation.
Soil	Sandy loam and acidic soil is best for seed germination, survival, better growth and yield.
Propagation	Through seeds and vegetative.
Seed rate	About 44,000 plants planted 1 acre of land.
Time of sowing	Seedlings raised at lower altitude during winter months are transplanted in nursery beds at higher altitude during April-May, which reduce their vegetative growth period.
Manuring	Organic manures like, Farm Yard Manure (FYM), Vermi-Compost, Green Manure etc. may be used as per requirement of the species.
Irrigation	Irrigation requirement also depends on the texture of soil. Frequent watering is required once at 24hrs interval for 6 months old seedlings at lower altitudes (1800- 2200m) in dry season.
Weeding	Weeding during rainy season is required at weekly interval. During winter months irrigation is needed once in a week to retain moisture and weeding at 15-20 days interval is required when plant is cultivated at lower altitude.
Harvesting	Harvesting of tubers is recommended after the completion of reproductive phase and maturation of seeds during October-November.
Yield	Per acre production from mature strands in natural pockets is estimated as 440 kg.

Brahmi

Botanical name	*Bacopa monnieri* (L.) Pennel
Family	Scrophulariaceae
Economic part	Whole plant
Main Phytochemical constituent	Bacoposide
Uses	It is a potent tonic used to improve mental alertness, learning performance, and memory as well as in the treatment of insomnia

- Brahmi, a member of **Scrophulariaceae** family is a creeping, branched succulent perennial herb distributed in wet and marshy lands through out India.
- It is a reputed intellect promoter, hypotensive, and neuropathic sedative drug. It is a potent tonic used to improve mental alertness, learning performance, and memory as well as in the treatment of insomnia. It is the principal ingredient of classical Ayurvedic preparations like brahmighritam and brahmirasayanam.
- The succulent leaves are sessile, opposite decussate, obovate-oblanceolate in shape, 1.0- 2.5 cm x 0.4-1.0 cm in size.

- The whole herb is the source of the ayurvedic drug 'Brahmi'.
- **Bacoposide** is considered as the major active ingredient in this plant. Raw drug is mainly collected from the wild.

Production technology

Climate	The plants grow faster at high temperatures (33-40°C) and humidity (65-80%) and should be cultivated as summer-rainy season crop.
Soil	The plant is known to grow under varying soil and climatic conditions. It grows exceptionally well in poorly drained soils and waterlogged areas under sub-tropical conditions.
Varieties	CIMAP (Central Institute of Medicinal and Aromatic Plants), Lucknow, has also developed three varieties of this crop, namely, subodhak, pragyashakti, and CIM-jagriti, which can be grown as perennials with at least two harvests per year.
Propagation	Freshly collected shoot cuttings of 5–10 cm length with internodes and rootlets are the best planting material for cultivation.
Planting	Propagules, that is, stem cuttings with rootlets, are transplanted in field at a spacing of 20 cm × 20 cm. Best time for direct planting or transplanting in the field is beginning of rainy season commencing in last week of June to first week of August.
Manuring	Organic manures like, Farm Yard Manure (FYM), Vermi-Compost, Green Manure etc. may be used as per requirement of the species.
Irrigation	Irrigation immediately after transplanting is essential for the successful survival of the plants. Subsequently, the fields are irrigated by flooding as per requirement usually every 7-8 days. There is no need for irrigation during the monsoon.
Weeding	Initially hand weeding is required every 5-20 days. Later as the plant proliferate and forms a dense mat of vegetation, weeding may be required sporadically.
Harvesting	The crop can be harvested 75–90 days after planting. September–October is the best time for harvesting.
Yield	As a pure crop, fresh yield is 22.5 tonnes/ hectare, reduced to approximately 5.5 tonnes/hectare on drying.

Ashvagandha

Botanical name	*Withania somnifera* (Linn.) Dunal
Family	Solanaceae
Centre of origin	Africa
Economic part	Roots
Main Phytochemical constituent	The main constituents of ashwagandha are alkaloids and withanolides (steroidal lactones), the major groups of secondary metabolites of medicinal interest. Among the various alkaloids, withanine is the main constituent.

Uses	The plant has anti-tumor, anti-inflammatory, anti-bacterial, fungicidal, anthelmintic, anti-convulsant, anti-stress, immunomodulatory and anti-pyretic properties. It is also used in insomnia, weakness, ulcers and painful swellings as aphrodisiac and in leukoderma.

- It is a dense, hairy, erect, grayish-tomentose herb or under-shrub, grows up to a height of 1.5 meter.
- Its all parts are covered with whitish, stellate trichomes.
- Branching is extensive; leaves are simple, alternate or sub-opposite, ovate, entire, basis cunate, 10 cm long. The roots are stout, long tuberous, fleshy, whitish brown.
- The flowers are greenish-yellow and found in few flowered clusters in axils; pedicels 0-4 mm long.
- Calyx is 5 mm long and stellately tomentose; teeth 2.5 mm long, linear, acute and form a deltoid base.
- Corolla is 8mm long, divided rather more than ½ - way down; lobes lanceolate, acute and pubescent outside.
- Filaments are 3mm long, slender, glabrous and anthers are broadly elliptic (almost orbicular), 1.25 mm long. Ovary and style are glabrous.
- The fruit is red yellow berry, smooth, 6mm in diameter, enclosed in the inflated calyx which reaches more than 25 mm diameter and is globose, slightly 5-angled, pointed with the connivent calyx-teeth and scurfy-pubescent outside.
- Seeds are 2.5 mm in diameter, yellow and somewhat scurfy.
- It is found throughout the drier parts in subtropical regions and upper Gangetic Plains.

Production technology

Climate	It can be cultivated between upto 1000 meter altitudes. Ashwagandha prefers a sub-tropical climate. The semitropical areas receiving 500-750 mm rainfall are suitable for cultivation of this crop. The crop requires dry season during the growing period. Temperature between 20°C to 35°C is most suitable for its cultivation. Late winter rains are conducive for the proper development of the plant roots.
Soil	Ashwagandha is grown on sub-marginal waste lands and low fertility areas. Plant grows well in red, sandy, black and loamy soil with pH 6.5- 8.0 with good water drainage.

Varieties	Poshita and Rakshita are high yielding varieties released by CSIR-CIMAP, Lucknow. Jawahar Asgand-20 and Jawahar Asgand-134 is cultivated in Madhya Pradesh. WSR is another variety released by CSIR-Regional Research Laboratory, Jammu.
Propagation	Through seeds
Seed rate	A seed rate of 10-12 kg/ha.
Time of sowing	July to August
Planting distance	25X25 cm
Manuring	200-300 kg FYM/ha
Irrigation	Light shower after transplantation ensures establishment of seedlings. There is no need of irrigation if rainfall is at regular intervals.
Harvesting	The crop is ready for harvest in January – March *i.e.* 150 to 180 days after sowing. The maturity of crop is judged by drying out of leaves and yellow red berries. The entire plant is uprooted for roots, which are separated from aerial parts by cutting the stem 1 – 2 cm, above the crown. The roots are cut transversely into small pieces (7 to 10 cm). Occasionally, the roots are dried as a whole. The berries are plucked from the dried plants and are threshed to obtain the seeds.
Yield	On an average yield from one hectare area under commercial cultivation is an approx 5- 7 q of dried roots and 30- 40 kg seeds.

□□□□

22

Technical Information on Aromatic Plants

Common name, botanical name, family, origin, chromosome number, economic part, main phytochemical constituents and uses of aromatic plants

S. No	Common name	Botanical name	Family	Origin	Chromosome number 2n=	Economic part	Main Phytochemical constituents	Uses
1.	Geranium	*Pelargonium graveolens* (L) Hervitt.	Geraniaceae	South Africa	88	Leaves and tender shoots	Geraniol	Manufacturing of perfumes, creams, talcum powder and body lotion
2.	Lavender	*Lavandula angustifolia*	Lamiaceae	Mediterranean region	50	Flowers and leaves	Linalool and linayl acetate	True lavender is one of the most important essential oils used in the perfumery industry
3.	Mint	*Mentha arvensis* L.	Lamiaceae	Mediterranean regions	96	Leaves and tender shoots	Menthol, carvone, linalyl acetate and linalool	It is used in cold and cough drops, mouth washes, cosmetics, tobacco flavouring
4.	Basil (Tulsi)	*Ocimum* spp.	Lamiaceae	Tropical and subtropical regions of Asia, Africa, and Central and South America	X = 8, 10, 12, or 16	Leaves and tender parts of the shoots	Cineol, linalool, citral, methyl chavicol (estragole), eugenol and methyl cinnamate	treatment of heart diseases, leucoderma, asthma, bronchitis and fever
5.	Patchouli	*Pogostemon patchouli* Pellet.	Lamiaceae	Philippines	32	Leaves	Pogostol	The oil is used in scenting of soaps, cosmetics, after-shave lotions, detergents and many fancy products. The oil is also used in major food products as a flavouring agent
6.	Jasmine	*Jasminum* spp.	Oleaceae	Tropical and sub-tropical regions	26	Leaves and flowers	Fats, glycosides, carbohydrates, flavonoids, steroids, saponins, proteins and amino acids, tannins and phenolic compounds	The flowers are used in perfumery and leaves are used for treatment of fever, ulcer, diarrhea, diabetes and skin diseases like itching, leprosy etc.
7.	Lemon grass	*Cymbopogon flexuosus* Nees	Poaceae (Graminae)	India	40, 60	Leaves	Citral	The oil is used in flavouring and medicinal industry and also for

S. No	Common name	Botanical name	Family	Origin	Chromosome number 2n=	Economic part	Main Phytochemical constituents	Uses
								production of Vitamin A and synthetic violets
8.	Palmarosa	*Cymbopogon martini* Stap f. var. *Motia*	Poaceae (Graminae)	India	20		Geraniol	The oil has high demand in perfumery, soap, cosmetics and blending tobacco products industries
9.	Citronella	*Cymbopogon winterianus* Jowitt	Poaceae (Graminae)	Sri Lanka	20	Leaves	Citronellol, citronellol, gereniol	Its oil from leaves has a strong characteristic lemonic odour extensively used in scenting of soaps, cosmetics, producing deodorants and mosquito repellent products
10.	Vetiver	*Vetiveria zizanioides* (L.) Nash.	Poaceae (Graminae)	Indian subcontinent	20	Roots	Vetiverol and vetiverone	The oil is used as a valuable fixative in blending of perfumes, cosmetics and scenting of soaps
19	Scented rose	*Rosa damascena* Mill.	Rosaceae	Central Asia	14	Petals, fruits (hips) and buds	Rose oil , containing 34% 1- citronellol,, 35% geraniol and 12-13% nerol besides a large number of minor aroma compounds	Different perfumery products obtained from rose are rose attar, Gulkhand, Gul roghan, Punkhuri and Otto rose

23

Production Techniques of Aromatic Plants

Geranium

Botanical name	*Pelargonium graveolens* (L) Hervitt.
Family	Geraniaceae
Centre of origin	South Africa
Chromosome number	2n=88
Economic part	Leaves and tender shoots
Main Phytochemical constituent	Geraniol
Uses	It is used in manufacturing of perfumes, creams, talcum powder and body lotion

- The geranium of trade is the common name given to a number of species belonging to the genus *Pelargonium* belonging to the family Geraniaceae.
- The major species that is cultivated for the commercial geranium oil is *P. graveolens* L' Hervitt.
- It is a bushy aromatic herb. Stem is cylindrical, woody at base. Pubescent, green when young and turning brown with age leaves are alternate, stipulate, simple with about 5 palmatisect primary lobes and pinnatisect secondary lobes, pubescent on both lobes.
- Flowers are pentamerous, bisexual with pink corolla.

- The crop is grown as rain fed crop in hilly areas and as irrigated crop in plains.
- There are two types of *P. graveolens* i.e. Algerian or Tunisian and Bourbon or Reunion.
- Egyptian type is another strain identified for cultivation in plains, which is rich in **geraniol** and resistant to wilt compared to Bourbon.
- The former is with dark pink flowers, not suitable for wet conditions and cultivated in higher elevations and the latter is with light pink flowers, suitable for wet conditions and cultivated in plains.
- The plant is bushy in nature growing to a height of 1 m. Stem is cylindrical, green when young and turned to brown when matured.
- Flowering is more profuse in higher altitudes. Rose scented essential oil, extracted from its leaves and tender shoots are commercially important.

Production technology

Climate	It is grown cooler and subtropical climate. An elevation ranging from 1000 - 2400 m with an annual rainfall of 100 to 150 cm evenly distributed throughout the year is ideal.
Soil	A deep light porous well drained soil, rich in organic matter and acidic in reaction with a pH 5.5 to 6.0 is suitable.
Varieties	Algerian type: PG 7, Reunion, IIHR-8, Kodaikanal 1, Hemanti and Egyptian.
Propagation	The crop is propagated by stem cuttings.
Time of sowing	April - May.
Planting distance	60 x 45 cm or 60X60 cm.
Manuring	FYM 25 t/ha and NPK 25:150:100 kg/ha are applied as basal. Remaining 125 kg N/ha in 5 equal splits is applied at every 2 months interval. During June-July of every year, ZnSo4 20 kg/ha and Boron 10 kg/ha are applied.
Irrigation	Geranium is usually grown as a rainfed crop. Irrigation during dry periods increases the yield.
Pruning	Pruning of the bushes is necessary when the bush shows signs of decline. The branches are cut back leaving 15 - 20 cm once in 4 - 5 years.
Crop maturity and harvesting	In the first year of planting, only one harvest can be done at 7 – 8 months and thereafter 3 or 4 harvests can be done in a year. The tender tip portion with 6 – 12 nodes constitutes the materials for harvest. The harvested material is withered in shade for two to three hours and distilled.
Yield	**Herbage:** 20 - 25 t/ha., **Oil yield:** 15 - 20 kg/ha

Improved varieties

Variety	Salient features
Algerian or Tunisian	It is a slender plant with dark pink coloured flowers. It is cultivated in Nilgiris. This variety yields 50-60% more oil than the reunion type.
Reunion or Bourbon	It is grown in Nilgiri and Annamalai hills. The plant is sturdy with light pink flowers and suited for wet conditions. The oil content is higher during summer season and the terminal 6-10 leaves contain more oil than the middle and basal portion.
Sel-8 (Reunion Type)	It is a variety released by IIHR, Bangalore.
Hemanthi	This variety is released by CIMAP, Lucknow for cultivation in the plains of north India. It is same as Algerian type rich in citronellol.
Bipuli	This variety is released by CIMAP, Lucknow. It is the same as Bourbon type equally rich in both geraniol and citronellol.

Lavender

Botanical name	*Lavandula angustifolia*
Family	Lamiaceae
Centre of origin	Mediterranean region
Chromosome number	2n = 50
Type of Pollination	Cross pollination
Economic part	Flowers and leaves
Main Phytochemical constituents	Linalool, camphor and linayl acetate
Uses	True lavender is one of the most important essential oils used in the perfumery, flavour and cosmetic industries as well as medicinal purpose industry.

- Lavenders belong to the Lamiaceae, a family which contains square stemmed plants such as thyme (*Thymus* spp.) and mint (*Mentha* spp.).
- Lavenders are contained within the genus *Lavandula* which contains 39 species, mostly of Mediterranean origin.
- It is a perennial aromatic herb or sub shrub 50 to 80 cm in height. The leaves are opposite, oblong linear or lanceolate with small flowers borne in terminal spikes.
- In India, Kashmir and low rainfall areas and slopes of hills of Himachal Pradesh and Uttar Pradesh are the suitable locations for cultivation, as the climate prevailing there is favourable to lavender.
- There are two main species and hybrids thereof which are grown commercially for their oil. These are *L. angustifolia* Miller (common

name: Lavender), *Lavandula latifolia* Medicus (common name: Spike lavender) and hybrids of the two termed *Lavandula* x *intermedia* (common name: Lavandin) (Tucker, 1981).

- The oils are used mainly in the fragrance and perfume industry. Commercially, the most valuable of lavender oils is extracted from L. angustifolia. A lower grade oil (lavandin oil) is obtained from the hybrids and a third oil is produced is from *L. latifolia,* 'spike oil'.
- Major components of the essential oil are linalool (24.63%), camphor (13.58%) and linayl acetate (8.89%).
- The lavender species can be exploited in a number of ways. As a medicinal plant, its dried inflorescence (Lavandulae flos) has traditionally been considered to be an antispasmodic, a carminative, a diuretic, a stimulant, and a tonic.
- The essential oil of lavender (*Aetheroleum lavandulae*) has antiseptic, carminative, and spasmolytic activity.
- The flowers are considered to be an insect repellent.
- Lavender oil is frequently used in cologne toilet waters, lotions and a wide range of high-quality perfumes, often after extensive bending or 'stretching'. Spike lavender oil scents cheap soaps, polishes, detergents and liquid cleaners, although the more delicate.
- French oil is also used in room's sprays, deodorants, disinfectants and insecticides. The major consumer of lavandin oil is the soap industry but the oil also has much application in men's colognes and numerous inexpensive perfumes, detergents, cleaning and washing-up liquids, polishes, talcum powders and hair preparations.

Production technology

Climate	It is a temperate plant and does well only in those areas, which have cold winter and cool summer. It is resistant to frost and drought. . High altitude areas are favourable for the crop. Lavender needs a great deal of sunshine. If the duration of the sunshine is not sufficient, then the formation of flowers, the essential oil content of the flowers, and the ester content of the essential oil, are significantly reduced.
Soil	Light well-aerated dry and calcareous soils rich in nutrients are the best for cultivation of lavender. Poor carbonate and sandy soils are also suitable but such soils need mineral fertilizers annually. Waterlogged soils are not suitable for its cultivation. It grows very well on sloppy lands, thus checking soil erosion to a great extent. The best crop and oil yields are obtained in neutral and alkaline soils (pH 7.0 to 8.4).

Propagation	Lavender is propagated through **seeds** as well as through **rooted cuttings.**
Transplanting	Transplantation can be done either in autumn or in spring. Lavender is planted at a distance of 1.20 or 1.40 m in between rows and 30-40 cm inside the rows. About 20,000 plants per ha give the highest yield. Earthing up of soil around the seedlings is considered good for the rooting.
Earthing	Earthing up of soil around the seedlings is considered good for the rooting.
Manuring	Generally 100kg N, 40kg each of P O and K O per hectare is recommended of which 20 kg N and 2 5 2 the entire P O and K O are applied as basal dose. The balance 80 kg of N is applied in four split 2 5 2 doses during each year. CIMAP centre at Srinagar has obtained highest spike yield by application of 150kg N/ha.
Weeding	The field of lavender is kept weed-free by regular weeding and hoeing. Regular hoeing keeps the soil loose which is responsible for proper plant development.
Crop maturity	Plants start flowering earlier in warmer and low altitude areas and later at higher slopes. Flowers are cut off with stem lengths not greater than 12 cm because of absence of oil in the stems and leaves. In the Kashmir Valley, the crop is ready for harvesting in August and September.
Harvesting	Harvesting should begin at 50 % blossoming and continued till the blooming is finished. Harvesting should be done on dry, warm and sunny days. Very cold weather prevents development of the esters and rain is fatal for harvesting. It is important to cut the flower axis completely at a uniform height because the flowers remaining on the shrub would affect flowering in the next year. The flower is cut with well-sharpened sickles.
Yield	Studies carried out by CIMAP indicated that oil content range was from 1.2% to 1.5% while the yield was 15 kg per ha is obtained.

Important cultivars and species

Cultivar/species	Salient features
True Lavender (*Lavandula Angustifolia* Aub Sp. *Angustifolia*)	It is the longest cultivated lavender species and is native to the Mediterranean region. It can be found as high as 1700 m above the sea level. The flowering shoots of true lavender as 0.20-0.40 cm long and are single stalked. The greyish green felty pubescent leaves are 30-50 mm long. The essential oil contents is 0.5-0.1%.
Spike Lavender (*L. latifolia*)	It is native to the Mediterranean region and grows at 200-700m above the sea level.
Lavandin (*L. Intermedia* (Syn. *L. Hybrida)*	It is a spontaneous hybrid of true lavender and spike lavender. It can be found frequently at heights of 700-1000m above the sea level. It is a shrub, which reaches a height of 0.8-0.1m and in natural condition it has a hemi-spherical shape. The spikes of lavendin may grow to 0.6-0.9m and generally it is branched. The leaves are linear or thin, lanceolate, opposite and bright green in colour without pubescent. The essential oil contents is 0.9-3.0%.
Sher-e-kashmir (*Lavandula Officinalis*)	A superior clone with about 100% higher oil yields than the Karlovo strain, which was released by CIMAP, Lucknow.

Mint

Botanical name	*Mentha arvensis* L.
Family	Lamiaceae
Centre of origin	Mediterranean regions
Chromosome number	2n=96
Type of Pollination	Cross pollination
Economic part	Leaves and tender shoots
Main Phytochemical constituents	Menthol, carvone, linalyl acetate and linalool
Uses	It is used in cold and cough drops, mouth washes, cosmetics, tobacco flavouring

- Mentha or mints includes a number of aromatic perennial herbs belonging to family Lamiaceae.
- *Mentha arvensis* var. *piperascense* (Japanese mint), *M. piperita* (Pepper mint), *M. spicata* (Spearmint) and *M. citrata* (bergot mint) are the common cultivated species of Mentha.
- It is a perennial herbaceous plant with the root-stock creeping along or just under the ground surface. It has a rigid branching, pubescent 60-90 cm tall stem bering aromatic leaves.
- Japanese mint has high percentage of menthol and is used in cold and cough drops, mouth washes, cosmetics, tobacco flavouring.
- Peppermint oil is carminative, antiseptic and gastro-stimulant properties and used in confectionaries, alcoholic drinks, dental creams and mouthwashes.
- Spearmint oil is rich in **carvone** and has digestive and gastro stimulant properties; It is used in confectionary, chewing gum and tooth pastes. Bergamot mint oil is rich in linalyl acetate and linalool and mainly used in cosmetic industry.
- Mints were first introduced to India in 1952 by the Regional Research Laboratory, Jammu.
- The herbage of the plants are economically useful which yields essential oils.
- The chief chemical constituents of the essential oils are menthol, carvone, linalyl acetate and linalool and they are highly demanded in pharmaceutical, food flavour, cosmetics, beverages and related industries.

Production technology

Climate	Sub-tropical areas receiving an annual rainfall of 100 - 150 cm are good.
Soil	Well drained loamy and acidic soil high in organic matter content is suitable.
Varieties	Japanese mint – Kalka, , Gombi, Himalaya, Sambhav, Saksham, MAS-1, MA-2, Hybrid-77,Shivalik and EC-41911Spear mint – Arka Neera, Ganga, Neerkalka, MSS-1, MSS-5 and Punjab Spearmint-1Bergamot mint – KiranPepper mint – Kukrail, Pramjal and Tushar
Propagation	It can be propagated through root suckers and terminal cuttings. Terminal cuttings are highly capable for rooting. The root suckers from the previous crop can be used for propagation.
Time of planting	June - July
Planting distance	40 x 10 cm
Manuring	Basal: Apply NPK at 80, 50 and 40 kg/ha.Top dressing: Apply 30 kg N in 2 splits at 60th and 120th day after planting.
Irrigation	Irrigation is given twice in a week.
Weeding	Frequent weeding is done up to 75 days of planting.
Crop maturity and harvesting	First cutting starts in about 5 months after planting and subsequently at 3 months interval. Under good management conditions, the crop will give economic yield for about four years.
Yield	Herbage 15 - 20 t/ha/year, Oil 50 - 100 kg/ha/year

Important cultivars and species

Cultivar/ and species	Description
Japanese Mint or Menthol Mint *(Mentha arvensis)*	
MAS-1	The plants are dwarf and early maturing. This variety gives 50% more oil yield with higher menthol content (80-85) and yields 200 q/ha of herbage and 125 kg oil / ha .It should not be used when ratooning is to be practiced.
Kalka (HY-77)	The plants are somewhat taller than MAS-1. This variety shares many of its characteristics with MAS-1 and is tolerant to several fungal diseases including leaf spot and rust diseases. The essential oil has more than 80% of menthol. Kalka matures early. The crop of Kalka yields about 250q/ha of herbage and 150 kg oil/ha. The variety Kalka is not recommended for ratooning. It is especially well adapted to the areas adjoining Lucknow, Sitapur and Barabanki where soils are sandy loam and climate is somewhat drier than that of the tarai region.
Shivalik	It is an introduction from China. This variety is highly suitable for obtaining second cut through ratooning. The herbage yield is 300q/ha

Contd.

Cultivar/ and species	Description
	while the essential oil yield amounts to about 180 kg/ha with a low menthol content (70%). Shivalik is highly sensitive to the fungal diseases and pests prevailing in the tarai area.
Himalaya	This variety yields high amount of essential oil and is not susceptible to pests and diseases. Like Shivalik, Himalaya can be ratooned for two cuts of herbage. Much like MAS-1 and Kalka, it can be used in areas and conditions where only one cut of herbage can be taken. The yield of herbage is 400q/ha and essential oil is 250 kg/ ha. Its oil is of high quality with menthol content of 80%.
Hybrid 77	It is a tall vigorous compact growing variety. It produces 762q/ha of fresh herb, 468 kg oil/ha containing 81.5% menthol (obtained from 3 cuts). It is highly resistant to rust diseases.
EC-41911	This is an erect type variety, which is less affected by rains. It produces 236.5 q/ha of herbage and 125.2 kg/ha of oil with 705 menthol. This oil is preferred in food flavours.
Spearmint *(Mentha spicata)*	
MSS-1	It is a selection from spearmint cultivars introduced from USA. It produces fresh herbage of 535q/ha and yield 253 kg of oil in 3 cuttings containing 60% carvone.
MSS-5	It is a selection from MSS-1, which yields 690q/ha of oil in 3 cuttings. It has 70% of carvone in the oil.
Punjab Spearmint-1	It is an erect growing variety producing 209q/ha of herb containing 0.57% oil. The oil yields 119.29 litres/ha containing 68% carvone.
Bergamot Mint (*Mentha citrata*)	
Kiran	The variety produces 239 kg of oil with 48% linalool, over 172 kg and 39% linalool over parent cultures.

Basil (Tulsi)

Botanical name	*Ocimum* spp.
Family	Lamiaceae
Centre of origin	Tropical and subtropical regions of Asia, Africa, and Central and South America
Chromosome number	Basic chromosome numbers X = 8, 10, 12, or 16
Economic part	Leaves and tender parts of the shoots
Main Phytochemical constituents	Cineol, linalool, citral, methyl chavicol (estragole), eugenol and methy \cinnamate
Uses	It is used to treat heart diseases, leucoderma, asthma, bronchitis and fever

- The genus ***Ocimum*** **L. (Lamiaceae)**, collectively called basil, comprises 30–160 annual and perennial herbs and shrubs native to the **tropical and subtropical regions of Asia, Africa, and Central and South America** (Darrah 1980; Paton et al. 1999).
- Ocimum species are characterized by the different basic chromosome numbers x = 8, 10, 12, or 16 (Morton 1962; Sobti and Pushpangadan 1979; Khosla and Sobti 1985).
- It is a source of essential oils and aroma compounds (Simon et al. 1990), a culinary herb, and an attractive, fragrant ornamental plant (Morales et al. 1993; Morales and Simon 1996).
- Sweet basil (***Ocimum basilicum***) is one of the leading herb crops, used fresh or dry.
- Two plant types are commonly available, one is with green leaves and the other one is with purple leaves.
- It is distributed through out India and is also under cultivation. Leaves, flowers and occasionally the whole plant are medicinally used to treat heart diseases, leucoderma, asthma, bronchitis and fever.
- The leaves and tender parts of the shoots are economically important and it yields essential oils.
- The essential oils obtained have immense value in aroma industry.
- The fragrance of the basil varies due to the varying types and quantities of essential oils contained in the plants.
- The most important are 1,8 cineol, linalool, citral, methyl chavicol (estragole), eugenol and methyl cinnamate, although hardly any basil contains all of these in any significant amount.

Production technology

Climate	Basil is a hardy, long day plant. It is adopted to a wide range of climatic conditions.
Soil	It grows luxuriantly over well-drained, loamy soils of medium fertile soils.
Varieties	Clo-ocimum, developed by recurrent selection by polycross hybridization has 70-80% of eugenol in the oil.
Propagation	Through seeds.
Seed rate	250g/ha.
Sowing	Sow in furrows at 60X45 cm or 45X45cm.
Manuring	40kg N and P each/ha.

Crop maturity and harvesting	It is harvested after 120-130 days of age at maximum flowering stage by sickle 20cm above the ground.
Yield	It gives an yield of 15-20 tonnes of herbage, which in turn produces 66-70kg of oil easily.

Patchouli

Botanical name	*Pogostemon patchouli* Pellet.= *Pogostemon cablin* (Blanco) Benth
Family	Lamiaceae
Centre of origin	Philippines
Chromosome number	2n=32
Type of Pollination	Self pollination
Economic part	Leaves
Main Phytochemical constituents	Patchouli alcohol, pogostol, seychellene, nor-patchoulinol, patchoulipyridine, methylchavicol, limonene, pinene, p- methoxy cinnamaldehyde etc.
Uses	The oil is used in scenting of soaps, cosmetics, after-shave lotions, detergents and many fancy products. The oil is also used in major food products as a flavouring agent.

- The plant belongs to family Lamiaceae is a perennial erect or ascending herb.
- It is a branched, erect or ascending pubescent herb or undershrub with quadrangular stem about 1 to 1.2 m tall.
- The leaves are simple, ovate to oblong ovate, pale to purplish green in colour.
- Commercial cultivation in India was first attempted by Tata Oil Mills in 1942.
- Oil distilled from shade dried leaves is used in perfumery to give a solid foundation and lasting character of other fragrances since it has a fixative property to prevent the rapid evaporation of perfumes.
- The patchauli oil is generally blended with other essential oils like geranium oil or clove oil before use.
- The oil possesses antibacterial activity and is used an ingredient in insect-repellent preparations.
- The oil is also used in major food products as a flavouring agent.

- Currently India is producing only a part of its annual requirement and about 20 tonnes of patchouli oil is importing to meet the requirements.
- It is propagated by shoot tip cuttings. Common strains available for cultivation are Singapore, Johore, Indonesian, Java and Malaysian strains.

Production technology

Climate	Thrives best in hot and humid conditions, under shade of rubber, coconut, coffee, etc.
Soil	Well drained loamy soil with slightly acidic condition is suitable.
Varieties	Johore, Singapore and Indonesia are commonly cultivated. Out of these, the first one yields the best quality oil where as the other two give high oil yield.
Propagation	Rooted stem cutting of 15-20 cm are used.
Planting	The field is prepared to a fine tilth and the rooted cuttings are planted at a spacing of 60 x 30 cm during April – May.
Manuring	Apply FYM 15t/ha and NPK 25:50:50 as basal. N 125 kg/ha is split into 5 equal doses and after each harvest i.e., at 4 months interval.
Irrigation	Grown as a rainfed crop in hills. In plains, irrigation is to be given at 7 - 10 days interval.
Crop maturity and harvesting	First harvest when fully grown (6 - 8 months). Subsequent harvest at 3 to 5 months interval. The leaves are dried in shade and distilled.
Yield	30 - 40 kg oil/ha/year.

Improved varieties

Variety	Salient features
Johore	This variety yields the best quality oil in terms of chemical composition and odour value.
Java	Java strain recorded the highest herbage yield but inferior quality oil.
Singapore	This strain has higher oil content and was thus superior to 'Java' in oil yield.

Jasmine

Botanical name	*Jasminum* spp.
Family	Oleaceae
chromosome number	2n = 26
Centre of origin	Tropical and sub-tropical regions
Economic part	Leaves and flowers
Main Phytochemical constituents	Fats, glycosides, carbohydrates, flavanoids, steroids, saponins, proteins and amino acids, tannins and phenolic compounds
Uses	The flowers are used in perfumery and leaves are used for treatment of fever, ulcer, diarrhea, diabetes and skin diseases like itching, leprosy etc.

- The species belongs to family Oleaceae and subfamily Jasminoideae and it is a more or less scandent shrub with highly fragrant flowers.
- Jasmines are native of tropical and sub-tropical regions. The Arabian or Tuscan jasmine (*J. sambac*) is considered as native of the East Indies.
- It is a climbing, trailing and erect flowering shrub. There are both deciduous and evergreen species in jasmine.
- Leaves are opposite and alternate, simple, trifoliate or pinnate; leaflets are entire. Flowers are white, yellow or rarely reddish, sometimes solitary, more often in cymose clusters of three to many, usually fragrant; 2 loculed with 1 to 4 erect ovules.
- Fruit is a berry, rarely with separate capsules and each having 2 seeds. The main beauty and uniqueness of jasmine is its fragrance, which cannot be imitated by any known synthetic aromatic chemical and has a unique status in the perfume world.
- The genus with basic chromosome number of n =13 generally consists of forms with 2n = 26.
- The essential oils obtained from the flowers are used in perfumery and are export oriented also. In India, organised cultivation of the species was started in Tamil Nadu.
- Plant starts blooming in the first year of planting and full yield will be in the third year and the crop sustains up to 15 years.

Production technology

Climate	A mild climate with well-distributed rainfall of 80-100 cm and sunshine around the year is desirable.
Soil	Jasmine is a hardy drought tolerant crop, growing luxuriantly in well-drained, loamy soils of 6-8 pH but it is highly susceptible to water logging.
Propagation	Plant is propagated by stem cuttings
Time of planting	Rainy season
Planting distance	1.8 X1.5 or 2X2 m
Manuring	The application of 100, 150, and 100g of N,P and K. bush annually produces maximum flower yield.
Harvesting	The flower crop (mature buds and open flowers in early stage) is picked in early morning (5-8AM). The open flower crop is processed for extraction of oil immediately.
Yield	On an average, 300-400kg of flowers are needed to produce 1kg of concrete; the commercial yield of concrete being 10kg/ha over the years.

Important species and cultivars

Species	Cultivars
Jasminum auriculatum	CO-1 Mullai, CO-2 Mullai, Pari Mullai, Long Point, Short Point.
J. grandiflorum	CO-1 Pitchi, CO-2 Pitchi, and Arka Surabhi.
J. sambac	Double Mohra, Gundumali, Iruvatchi, Kasturimalli, Madanbari, Oosimalli, Ramabanam, Single Mohra, Soojimalli, Khoya, Khoya Large, Arka Aradhana, Bale Japani, Butt Mohra.
J. multiflorum	Arka Arpan

Lemongrass

Botanical name	*Cymbopogon flexuosus* Nees ex. Steud Wats.
Family	Poaceae (Graminae)
Centre of origin	India
Chromosome number	2n=40, 60
Type of Pollination	Cross pollination
Economic part	Leaves
Main Phytochemical constituent	Citral
Uses	The oil is used in flavouring and medicinal industry and also for production of Vitamin A and synthetic violets.

- Lemongrass belongs to family Poaceae and there are three species of *Cymbopogon i.e., C. flexuosus, C. citratus and C. pendulus,* commonly known as lemongrass.
- The odour of the oil extracted from the leaves are lemon scented and hence the name 'lemongrass'. *C.flexuosus, C.citratus* are under cultivation.
- The species are perennial grass of about 2 m long with profuse tillering habit. *C. flexuosus* (East Indian lemongrass or true lemongrass) is more preferred by the industry due its superior oil quality.
- Lemongrass is distributed commonly in tropical and subtropical Asia, Africa and America. *C. flexuosus* is present in wild state in Western Ghats, Sikkim and Arunachal Pradesh.
- Commercial cultivation of the species is spread to Kerala, Assam, Maharashtra, Uttar Pradesh and Gujarat. *C. pendulus* is distributed naturally in West Bengal, Sikkim, Assam and Madhya Pradesh and cultivated in parts of Jammu.
- **Citral** extracted from the oil is of commercially importance. The oil is used in flavouring and medicinal industry and also for production of Vitamin A and synthetic violets.

- India is one of the major suppliers of lemongrass oil in the world market.

Production technology

Climate	It comes up well under tropical and sub-tropical conditions with a high rainfall 200 - 250 cm and humidity.
Soil	Sandy loam with abundant organic matter and pH of 6.0.
Varieties	Sugandhi, OD-19, OD-408, OD 440, RRL16, RRL-39, Pragathi, Praman (a tetraploid mutant), CKP-25, Krishna and Cauvery.
Propagation	Slips
Planting	55,600 slips/ha at a spacing of 60 x 30 cm in ridges in June –July.
Manuring	Apply FYM or compost at 20 - 25 t/ha as basal. Apply 50 kg N/ha annually, half at planting and half, one month after planting.
Irrigation	Light irrigation on third day and further irrigation at 7 - 15 days interval.
Weeding	Hand hoeing and weeding as and when required.
Crop maturity and harvesting	Harvest the leaves first at 90 days after planting and thereafter at 90 days interval. Cut the bush 10 - 15 cm above the ground level and herbage is withered under shade for one day prior to distillation. They are then chopped into small pieces and fed into steam distillation stills for distillation of essential oil. Oil recovery 0.3 - 0.5 %.
Yield	**Herbage:** 20 - 30 t/ha., **Oil:** First year : 50-100 kg oil/ha., Second year onwards : 200-300 kg oil/ha.

Improved varieties

Variety	Salient features
Sugandhi (OD 19)	It is adapted to a wide range of soil and climatic condition. A red stemmed variety with plant height 1 to 1.75 m and profuse tillering. The oil yield ranges from 80 to 100 kg per hectare with 85-88 per cent citral under rain-fed conditions. Aromatic and Medicinal Plants Research Station, Odakkali, Kerala released this variety.
Pragati (LS48)	It is a tall growing variety with dark purple leaf sheath suitable for north Indian Plains and Tarai belt of subtropical and tropical climate. Average oil content is 0.63 per cent with 86 per cent. This variety is a clonal selection from OD19 developed at Central Institute of Medicinal and Aromatic Plants, Lucknow.
Praman (Clone 29)	Evolved through clonal selection at Central Institute of Medicinal and Aromatic Plants, Lucknow and belong to species *C. pendulus*. It is a medium sized variety with erect leaves and profuse tillering. The oil yield is high with 82 per cent citral.
RRL 16	Average yield of this variety is 15 to 20 tonnes/hectare/annum giving 100 to 110 kg oil. Oil content varies from 0.6 to 0.8 per cent with 80 per cent citral. Evolved from *C. pendulus* and released for cultivation from Regional Research Laboratory , Jammu as Jammu Lemongrass.
CKP 25	A hybrid between *C.khasianum* X *C.pendulus.* Gives 60 t/ha herbage in North Indian plains under irrigation. Oil contains 82.85% citral.

Jama Rosa	Selection from interspecific cross between *C.jawarancusa*. very hardy with vigorous growth. The variety yields about 35.0tonnes/ha. of herbage containing 0.3-0.4 % oil(FWB). The variety yields upto 300kg oil in 4-5 cuts in 16-18 months growing period (RRL Jammu).
Other Varieties	Kaveri, Krishna(released by CIMAP regional station, Bangalore), OD-408 (released by AMPRS, Odakkali), GRL-1 and RRL-39.

Palmarosa

Botanical name	*Cymbopogon martini* Stap f. var. *Motia*
Family	Poaceae
Centre of origin	India
Chromosome number	2n=20
Type of Pollination	Cross pollination
Economic part	Leaves
Main Phytochemical constituent	Geraniol
Uses	The oil has high demand in perfumery, soap, cosmetics and blending tobacco products industries.

- Palmarosa or Rosha grass is a tall perennial herb belongs to family **Poaceae.**
- It is a drought hardy grass attaining a height of 1.5 to 2.5m having hairy and fibrous shallow root system with long linear lanceolate leaves.
- It produces large fawn coloured inflorescence containing while, hairy star like spiked flowers.
- It is distributed in most parts of subtropical India. Distillation of herbs with the flowering parts yields sweet, colourless to pale (greenish) yellow scented oil which is rich in geraniol.
- The oil has high demand in perfumery, soap, cosmetics and blending tobacco products industries.
- Owing to its low cost, in western countries palmarosa oil is preferred over sandalwood oil and true geranium oil, hence referred to as East Indian Geranium oil in commerce.
- The species is under cultivation in Central, western and southern states of India.

Production technology

Climate	Comes up well under tropical conditions with an annual rainfall of about 150 cm.
Soil	A well drained loamy soils with a pH of 6-8 produce luxuriant growth and high oil yield.
Varieties	Trishna, Tripta, PRC-1, IW 31245, IW 3629, IW 3244, OPD-1, OPD-2, Vaishnavi, and RRL(B)-77, RH 49
Propagation	Through seeds
Seed rate	2-2.5kg of seeds/ha., Sow in raised nursery beds in lines 15 - 20 cm apart.
Transplanting	Transplant at 3 - 4 weeks in ridges at a spacing of 60 x 60 cm during June – July.
Manuring	**Basal:** FYM 10 t/ha and NPK at 20:50:40 kg/ha.**Top dressing:** 15 kg/ha in 3 splits at 3, 6 and 9 months of planting.
Harvesting	The first harvest commences at 3 - 4 months of planting. Subsequent harvests at 3- 4 months interval.
Yield	**Herbage:** 20 - 30 t/ha/year., **Oil:** First year : 50 kg/ha., Second year : 200-250 kg/ha.

Improved varieties

Variety	Salient features
IW-31245	Profuse tillering. 20-25 t/ha of herbage per cutting under low irrigation. Oil yield 150 kg in 3 cuts during 16-18 months. Oil contains 90% total geraniol. It has superior rosaceous green odour characteristics of palmarosa.
Trishna	It possesses straw yellow clumps, long compact inflorescence and has profuse tillering. The oil contains 93% geraniol.
Jamrosa	It has vigorous growth, high yield of 35-45 t/ha of herbage, It yields between 0.3-0.4% of oil in the herb which is rich in geraniol content (54-83%) and geranil acetate (17.89%). The oil yield was estimated between 130-150 kg/ha in the second year.

Citronella

Botanical name	*Cymbopogon winterianus* Jowitt
Family	Poaceae (Graminae)
Centre of origin	Sri Lanka
Chromosome number	2n=20
Type of Pollination	Cross pollination

Contd.

Economic part	Leaves
Main Phytochemical constituents	Citronellal, geraniol and citronellol
Uses	Its oil from leaves has a strong characteristic lemonic odour extensively used in scenting of soaps, cosmetics, producing deodorants and mosquito repellent products.

- Citronella (*Cymbopogon winterianus* Jowitt) is an aromatic grass belonging to the **Poaceae** family which gives essential oils upon stem distillation.
- Citronella was first introduced in India in 1959 from Indonesia (Java Island, hence the name Java citronella) (Kaul *et al.* 1997).
- This is used extensively as a source of perfumery, soap, cosmetic and flavouring industry throughout the world.
- Citronella oil is commonly known for its natural insect repellent properties, though it has many other uses in aromatherapy.
- This oil is also known as Java citronella and has a fresh, powerful, lemon like scent.
- *C. winterianus* essential oil is rich in citronellal, geraniol and citronellol (Cessal and Vergas, 2006). There are other constituents like citronellyl acetate, L-limonene, ellemol and other sesquiterpene alcohols present in citronella.
- Citronella is an aromatic perennial herb with fibrous roots. It is erecting over 2 m tall with smooth and shiny leaves which are glabrous at the nodes.
- The leaf blades are linear, gradually tapering to a long, membranous, acuminate and upto 1 m long and 1.5 cm wide, drooping at two-third of its length.
- The leaf sheath is smooth and glabrous, yellow or purplish-red in colour. The inflorescence is very large, erect, finally drooping, consisting of a branched and rebranched axis.
- India has been a leading producer of essential oils including oil of citronella.
- At present, it is found in the states of U.P, M.P, Maharashtra and various parts of India.

Production technology

Climate	It comes up well under tropical and sub-tropical conditions with a high rainfall 200 - 250 cm and humidity.
Soil	The plant has been found to grow well under varying soil conditions but sandy loam soil with abundant organic matter is most suitable. Citronella thrives well in a range of soil pH ranging from 5.8-8.0.
Varieties	Java-2, Jorlab-2, Java citronella, Java II, Ceylon citronella and Bio-13, Mandakini andManjusha are the high yielding varieties.
Propagation	Through slips.
Planting	The slips are planted 8-12 cm deep into the soil maintaining a distance of 60×60 cm apart. As such, about 55,555 slips are required for planting in one hectare.
Manuring	Basal: FYM 30 t/ha and NPK at 25:40:40 kg/ha.Top dressing: 75 kg N/ha in 3 equal splits of 25 kg each at 3, 6 and 9 months after planting.
Irrigation	Light irrigation on third day and again at 10 – 15 days interval.
Crop maturity and harvesting	The crop gets ready for first harvest after about 9 months of planting. Harvesting is done by using an arider sickle, at about 20-45 cm above the ground. The crop is harvested in the month of March, June and September.
Yield	On an average, however, the oil content is about 1% on the basis of fresh rate of leaves. The yield of leaves may range from 15-20 tones per hectare in the first year and 20-25 tones per hectare in the second and third year. The yield of oil obtained during the first year is about 100-150 kg per hectare and, in subsequent years about 200 kg per hectare of oil of citronella may be obtained.

Improved varieties

Variety	Salient features
Mandakini	Clonal selection, gives a little less herb yield (35 t/ha) and oils (118 kg/ha). the variety is suitable for hills and Tarai tracts of Himalayas (CIMAP, Lucknow).
Manjusha	Clonal selection, gives a herbage yield of 43 tons/ha. and 150 kg/ha of oil per annum. The variety is suitable for indo-gangetic plains (CIMAP, Lucknow).
Manjiri	This variety has been released by University of Agriculture Sciences, Bangalore. It is an elite mutant clone of Manjusha M3-8. It has been found to possess 50-90% more oil, high cirtonellol and low elemol content. It has profuse tillering and rapid growing ability thus producing a high herb yields.
Other varieties	CIMAP Bio-13, Java-2, Jorhat-2.

Vetiver

Botanical name	*Vetiveria zizanioides (*L.) Nash.
Family	Poaceae
Centre of origin	Indian subcontinent
Chromosome number	2n=20
Type of Pollination	Cross pollination
Economic part	Roots
Main Phytochemical constituents	Vetiverol and vetiverone
Uses	The oil is used as a valuable fixative in blending of perfumes, cosmetics and scenting of soaps.

- It is a tall perennial grass of about 1.5 to 2.0 m height belongs to family Poaceae.
- It has a small stout rhizomatous stolon giving rise to spongy fibrous root system which grows 20 to 30 cm deep.
- The clumps form dense tuft bearing erect linear narrow leaves. The inflorescence is a long panicle.
- In India it is widely seen in wet and damp places, however, it tolerates extreme conditions.
- The wild variety grows especially in Rajasthan and Uttar Pradesh.
- The species is used against soil erosion because of its roots' soil binding property.
- The aromatic roots are traditionally used for making mats, fans, door-screens which produce fragrance imparting a cooling effect during extreme summer when sprinkled with water.
- The roots on distillation produce volatile oils. The oil finds extensive use in perfumery as a fixative and as an odour contributor in bases, and also as a raw material for the isolation of vetiverol and vetiverone.
- It is also medicinal and used against flatulence and colic pains.
- Sugandha and Nilambore are the two high yielding varieties available for cultivation.

Production technology

Climate	The crop grows luxuriantly with a rainfall of 100-200 cm and a temperature of 30-40°C.
Soil	Well drained, sandy loam and red laterite soils of pH 6-8 are considered ideal as the roots produced in such soils are thick and contain more essential oil.
Varieties	There, are basically two main types, viz. (1) Seedling type and (2) Non seedling type The one that grows wild in North India is mainly the seedling type while that of the South is the non-seedling type. Hybrid 8, sugandha, ODV-3 are the superior strains for cultivation. Nilambore is popular from Karnataka genepool.
Propagation	Vetiver can be propagated through slips. The top of the slips are cut down before planting to prune transpiration loss, thus giving a better chance for survival of the slips.
Planting	There are three different systems of planting adopted by different growers.System 1: Conical ridges, 30-38 cm high and 48 cm apart are made at the summit and the slips planted 23 cm apart on the summit.System 2: The land is laid out into beds of 30 cm high, 68 cm wide and 45 cm apart edge to edge and the slips are planted on these in two rows 22.5 cm apart, leaving 22.5 cm on either sides. System 3: The beds are made 45 cm high, 60 cm wide and 30 cm apart edge to edge and two rows, 30 cm apart, are planted on these leaving 15 cm on either side. The spacing within the row is also 30 cm in this system.
Time of planting	The best planting time to get higher oil yield under South Indian condition is June-July.
Method of planting	The slips are planted in pits, five to eight cm deep made with a pointed stick. One hectare requires 1,50,000 to 2,25,000 slips with 2 - 3 slips per pit in the commonly adopted system of planting (IInd method).
Manuring	Application of FYM at 10 ton/ha and 60 kg of N, 22.5 kg in each of P2O5 and K2O was found to be efficient in increasing the yield of vetiver.
Harvesting	The plants planted in July should be harvested after 18 months to get the maximum oil yield. Harvesting is usually done during dry season (December to February) by manually digging out the bush along with its roots. The roots are then separated from the leaves, washed and dried under the shade for 1-2 days before distillation.The roots that possess the following characteristics have good oil content. It should 1. Be slightly reddish brown 2. expose a hard surface when the skin is peeled off 3. be thick, hard, long and wiry and 4. give a very bitter taste when chewed.
Yield	On an average one hectare of vetiver plantation yields 3-4 -- tones of roots which on distillation yield 15 to 16 kg of oil. Roots yield 1.00 to 1.50 per cent of oil on dry weight basis.

Improved varieties

Variety	Salient features
Hybrid-8	It is a perennial herb reaching a height of 2 m with medium long roots having large number of rootlets per plant containing more essential oil. It possesses 70-85% vetiverol content. This variety produces profuse tillering under light soil texture of moderate fertility and gives a high root yield 12-15 q/ha.
Hybrid-26	This variety has been evolved by NBPGR, New Delhi. The variety gives higher yield of roots (14.58q/ha) with an oil content of 1.5%. It performs better in saline and alkaline soils.
Sugandha	This was isolated from a collection made from Bharatpur (North India) material. The cultivar shows higher root biomass and oil percentage. This new variety has a potential of producing 1.4 % oil in fresh root (46% free moisture) and 21.2 q/root/ha. This variety has been developed and released by Central Institute of Medicinal and Aromatic Plants Lucknow.
Nilambore	It is a popular South Indian variety that produces 3.8 t/ha of fresh roots and 15-22 kg. oil in well managed fields.

Scented rose

Botanical name	*Rosa damascena* Mill.
Family	Rosaceae
Centre of origin	Central Asia
Chromosome number	2n = 2x = 14
Centre of origin	Damascus
Economic part	Petals, fruits (hips) and buds
Main Phytochemical constituents	Rose oil , containing 34% 1- citronellol,, 35% geraniol and 12-13% nerol besides a large number of minor aroma compounds
Uses	Different perfumery products obtained from rose are rose attar, Gulkhand, Gul roghan, Punkhuri and Otto rose.

- Rose belongs to family Rosaceae and among flowers it has a reputed position because of its colour, fragrance, economic benefits and religious significance.
- Roses have gained the title of the king of flowers (Peter Bealis, 1990) and it is the most popular ornamental plant that has been cultivated systematically (Wylie, 1995).
- Genus Rosa consists of approximately 200 species and up to 18,000 cultivars (Gudin, 2000).

- Chromosome numbers in the genus Rosa are based on multiples of seven and range from 2n = 2x = 14 to 2n = 8x =56 (Darlington and Wylie, 1955). Most species are diploid, whereas most of the modern rose cultivars are tetraploid and usually interfertile (2n = 4x =28) (Smulders et al., 2011).
- Roses are grouped into classifications based on the botanical characteristics: hybrid teas, grand floras, polyanthas, floribundas, miniatures, climbing, shrub, old roses, but only a few species are scented (Gudin, 2000).
- These include *Rosa damascena* Mill., *R. gallica* Linn., *R. centifolia* Linn., *R. bourboniana* Desportes., *R. chinensis* Jacq., *R. moschata* Herrm. and *R. alba* Linn.
- Rosa × damascena, more commonly known as the Damask rose or simply as "Damask", or sometimes as the Rose of Castile, is a rose hybrid, derived from Rosa gallica and Rosa moschate (Huxley 1992).
- It is thought that the origin and centre of diversity of Damask roses could be found in Middle East especially, Damascus into Western Europe (Rusanov et al., 2005). Rosa damascena Mill. is a beautiful aromatic flower with immense horticultural importance, signification from economical and research point of view.
- Rosa × damascena, more commonly known as the Damask rose is a deciduous shrub growing to 2.2 metres (7 ft 3 in) tall, the stems densely armed with stout, curved prickles and stiff bristles.
- The leaves are pinnate, with five (rarely seven) leaflets.
- The roses are a light to moderate pink to light red. The relatively small flowers grow in groups.
- The bush has an informal shape. It is considered an important type of Old Rose, and also important for its prominent place in the pedigree of many other types.
- Developed from the hypanthium the fruit or hip is a reddish orange colour, oval in shape and contains many one seeded achene, which are brown in colour.
- The world main producers of rose oil are Bulgaria, Turkey and Iran and to a less extent France, India, China and Northern Africa.

Production technology

Climate	*Rosa damascena* grows best at moderate temperatures with a flow of humid air during flowering, it favours altitudes between 300-1800 m.
Soil	Rose is grown in Gangetic plains over well-drained alluvial soils. It withstands wide range of soil pH conditions from 6 to 8 and silty clay loam to sandy loam soils.
Varieties	Noorjehan, Jwala (IHBT Gulab I) and Himroz (IHBT Gulab II).
Propagation	Through semi-hard, woody stem cuttings.
Time of planting	October-December.
Planting distance	1x1m.
Harvesting	Rose bushes commence blooming in March of the second year after planting but commercial yields is obtained from third year onwards and continues till 12-15 years. Flowers are hand-picked when they begin to open in early morning hours before the sun rise.
Yield	Flower yield 35-40 q/ha, Oil recovery 0.025-0.03 %, Oil yield 0.75 – 1.0 kg.

□□□□

24

Recent Initiative in Crop Improvement

A high-yielding cultivar Vallabh Medha of mandukparni (*Centella asiatica*) was identified. Fresh herbage yield (12,331 kg/ha) and dry herbage yield (2,113 kg/ha) of this variety were far more than the local variety (2,050 and 392 kg/ha respectively), besides higher active ingredients. Among mucuna selections, IIHR PS 15 with long duration, IIHR PS 6 with medium duration and IIHR PS 14 with short duration recorded significantly higher yield and high L-dopa yield/plant. In Coleus, a promising hybrid, Hy 08-53, recorded significantly higher root yield (60.22g) and higher forskolin yield/plant (0.58 g). In ashwagandha, IIHR- WS 3 was found to be the best yielder (11.65 q/ha).

Source: DARE/ICAR Annual Report 2011–12

A molecular technique to detect common adulterant of Ashoka raw drug was developed. Species-specific RAPD bands of both species were cloned and sequenced. Primers were designed to amplify sequence characterized amplified region (SCAR) markers from individual species. The SCAR primer designed from *S. asoca* produced an amplified product of 193 bp from this species only. Similarly, species-specific amplification of a 395 bp product was achieved from *P.longifolia* using another set of primers.

Progeny rows of single plant selections of ashwagandha resulted in wide variability. The lines obtained included an extra dwarf plant (15–20 cm), plant with open type calyx, clustered flowering type, compact plant type with less fibrous roots and procumbent plant type. Plant types with four different berry colours, viz. red, yellow, orange and creamy white were also identified. A male sterile plant was identified, first time in this species.

Forty-three giloe accessions were characterized based on starch granule size (perimeter) and their frequency distribution. Based on the frequency distribution of starch granules of different classes, germplasm was also classified into five classes. Majority of them have small starch granules (46–75 μm, L1 10–25 μm and L2 of 10-20 μm). Fresh stem yield varied from 0.09 to 5.44 kg/plant and dry stem yield 0.03 to 1.43 kg/ plant. Highest dry stem yield was recorded in IC 283650 (1.43 kg/plant), followed by Kalyani (0.71 kg/plant), IC 310602 (0.64 kg/plant) and GUJ 4 (0.60 kg/plant). Stem starch content varied from 12 to 52% on dry- weight basis among the accessions.

In M3 generation progeny, rows of isabgol were grown from chemical mutagen (DES, EMS and Colchicine) treatments. Among these, an early-flowering(30-35 days after sowing) and-maturing (80–85 DAS) mutant, DPO 14, was identified. It had an average plant height of 37.6 cm, 60 leaves/plant, 37 spikes/ plant, 9.84 g biomass/plant and seed yield of 2.25 g/plant. From the colchicines treated plant progenies of GI 2 variety, a tetraploid plant, was isolated. Tetraploid plants were more vigorous than the diploids. Fifteen germplasm lines of asalio (*Lepidium sativum*) were evaluated. The highest seed yield was recorded in genotype HLS 5 (2,048.21 kg/ha), followed by HLS 7 (1,938.36 kg/ha). Out of 13 germplasm lines evaluated, highest mean seed yield was recorded in MLS-7 (1,941 kg/ha).

Source: DARE/ICAR Annual Report 2010–11

Mandookaparni: A distinct elite plant type of medicinal plant mandookaparni (*Centella asiatica*), collected from Faizabad, Uttar Pradesh, was characterised at DMAPR. The new plant type having accession No. IC 561247 was bigger in leaf size and superior to local plant type in yield and quality. Fresh herbage yield was about three times more the elite plant type compared to local plant type. Asiaticoside was 1.62% in registered material while local type had 1.47% active principle. Characterisation based on molecular markers also revealed the distinctness of IC 561247. It is registered as INGR No. 08105 as an elite germplasm with superior yield and quality.

Source: DARE/ICAR Annual Report 2009–10

25

Recent Initiative in Crop Production

Isabgol (*Plantago ovata* Forsk.)

The crop requires cool and dry climate during the growing season. Sowing of seeds at 4 kg/ha in 0.25-0.50 cm depth during November 20 to December 20 was recommended. In medium black cotton soil of Malwa region, second week of November was reported to be optimum time for sowing. Broadcasting of seeds followed by light sweeping with broom found to give uniform germination. A spacing of 30x 45 cm found to be most ideal to get higher seed yield in Isabgol under Madhya Pradesh situation. Response of chemical fertilisers was found low. However, a fertiliser dose of 25 kg/ha each of N and P2O5 as basal dose and 25 kg/ha N as top dressing at 30-42 DAS was recommended for commercial cultivation in Gujarat while 50 kg/ha N reported to increase the seed yield in Madsaur areas of Madhya Pradesh. At Anand three irrigations viz. first at the time of sowing and subsequently at 30 and 70 days after sowing (DAS) proved to be beneficial. However, four irrigations at sowing, 10,25, and 50 DAS at Mandsaur are recommended. Chemical weed control was found to be economical and a pre-sowing or pre-emergence application of Isoproturone (0.5 kg ai/ha) was recommended for weed control. One spray of Metalaxyl + two sprays of Mancozeb were found significantly superior in controlling downy mildew. In Gujarat , Pearl millet in Kharif and Isabgol in Rabi season are advocated as a suitable crop rotation. Groundnut-Isabgol rotation gave higher economic return followed by Soybean-Isabgol and Pigeon pea-Isabgol rotations in Maharashtra.

Opium poppy (*Papaver somniferum* Linn.)

The optimum time for sowing is the first fortnight of November. Delay in sowing causes poor growth. A seed rate of 6-7 kg /ha is recommended in case of broadcasting and 5-6 kg /ha for line sowing. Line sown crop yielded more latex/ha than broadcasting. Seed inoculation with Azotobacter culture (M-4/W-5) reduces the nitrogen requirement up to 40 kg /ha. A fertiliser dose of 90 kg/ha N, 50 kg /ha P2O5 and 40 kg /ha K2O was reported to maximise the latex and seed yield. However, N,P and K at 150:75:45 kg/ha was recommended for higher seed, husk and latex yield and morphine content of opium poppy at Mandsaur. For weed control, and integrated approach with Isoproturon (0.37 kg a.i./ha + hand weeding at 30 DAS) showed very good control without any phyto-toxic effects. Ten to fourteen light irrigations are required in sandy soils at an interval of 10 days. Lancing is usually started on developing capsules about 15 days of flowering. Maximum latex yield is harvested in first lancing which decreased at subsequent lancings. Early morning is the best time for collection of latex. Latex yield ranges between 35 and 55 kg/ha and seed yield between 8 and 12 q/ha. Crop rotations, maize-opium poppy; urd-opium poppy and groundnut-opium poppy are profitable. Intercropping with garlic gave higher profit compared to sole crop without affecting the latex yield.

Senna (*Cassia angustifolia* Vahl)

The crop prefers sandy loam to laterite soils. However, it is grown in marginal and sub marginal lands also with 7.0-8.5 pH range. The crop is sensitive to water logging condition, therefore, requires well drained soil. Sowing time recommended for rainfed crop in Western India is June-July. Line sowing at 45 x 30 cm gives higher yield under irrigated conditions. About 70-75 thousand/ha plant population is recommended for optimum yield. Application of FYM at 10 t/ha as basal and 60 kg/ha N in 3 split doses are recommended for better yield. Two hand weedings followed by hoeing at 25-30 and 50 DAS are essential. Harvesting is recommended in dry season to avoid spoilage of leaves due to fungal infection during storage. A well managed irrigated Senna crop yielded about 15-20 q/ha dry leaves and 7-10 q/ha dry pods. Whereas, a rainfed crop recorded on an average about 10 q/ha of dry leaves and 4-5 q /ha dry pods. Sun drying of leaves and pods is advisable. Crop rotation such as Senna-mustard and Senna-coriander were found profitable.

Periwinkle (*Catharanthus roseus* Linn. G. Don.)

Tropical and subtropical climates are found most suitable for its cultivation. However, water logged or highly alkaline soils may be avoided. A plant population of 75,000/ha is recommended to get higher yield. Nutrient dose of 15t/ha FYM + 80 kg N/ha under irrigated and 15t/ha FYM + 40 kg N/ha under rainfed conditions were recommended. Detopping of plants by 2 cm at 50% flowering improves the root yield and alkaloid content. Pre-emergence application of Fluchoraline 0.75 kg ai/ha was found to be very effective as weed control. The crop requires 4-5 irrigations. Intercropping with groundnut in 1:1 ratio was found to give highest monetary benefit. This crop is not suitable as inter-crop under shade condition because plant growth, root yield and alkaloid contents are observed to reduce under such condition. On an average, 1.8 t/ha dried leaves and 0.8 t/ha dried roots are harvested.

Safed musli (*Chlorophytum* spp Ker.)

The crop is grown in Kharif season in places having warm and humid climatic conditions and adequate soil moisture throughout the crop growth. Optimum time for its planting is middle of June under irrigated condition and onset of monsoon for rainfed condition. Fleshy roots at 2.5-3.0q /ha are planted in ridges at 30 cm row to row and 15 cm plant to plant distance. High density planting of 3.33 lakh/ha to 4.4 lakh/ha plant population was found the best in increasing root yield. Application of 20 -45 t/ha FYM showed significant yield increase. Fresh root inflorescence (detopping) improved the bulking of fleshy roots and increased root yield significantly. Separation of Safed musli fleshy roots in the month of April reduced the fleshy root damage as compared to March and February. Sprouting percentage and storability were also increased when separated with a major portion of stem disc. Paired fleshy roots used for planting recorded high degree of sprouting and survival in comparison to single root planting.

Liquorice (*Glycyrrhiza glabra* Linn.)

The performance of Liquorice was tested at Hisar condition it grows well in rich fertile sandy loam soil ranging between acidic and slight alkaline soils having pH 5.5-8.2. Best time for planting is in middle of November with a spacing of 90x45 cm. Underground stem cuttings of 15-25 cm having 2-3 buds are most suitable for planting. However, treatment of root cuttings with Seradix-B enhances sprouting. At the time of soil preparation 10 t/ha FYM and 40 kg/ha each of N and P2O5 added as basal dose. Thereafter, 20 kg/ha N as top dressing per year is recommended. The crop is harvested after 2 to 3 years. Root yield of 70-80 q/ha was recorded at Hisar.

Sarpagandha (*Rauvolfia serpentina* Beth. ex Kurz)

Frost free tropical to subtropical humid climate with irrigation facilities are found to be the most suitable for its cultivation. Root and stem cuttings have been recommended for vegetative propagation. For transplanting, nursery is raised in the end of April. A fertiliser dose of 30 kg/ha N and 60 kg /ha P2O5 was found to increase the total alkaloid yield. Water requirement is very high in this crop, about 15-16 irrigations are required to get good crop. Inter cropping of soybean (1:1) in kharif and garlic (1:3) in rabi were reported to be most suitable crop combinations.

Foxglove (*Digitalis purpurea* Linn. and *D. lanta* Ehrh)

The crop requires 20-300C temperature for seed germination and subsequent plant growth. Well drained loam to clay loam soils rich in organic matter are suitable for the crop. However, requirement of soil pH for higher yield of glycoside differs species to species. It was reported that Digitalis purpurea thrives well in acid soil whereas D. lanta in neutral soil. Most suitable time for transplanting of seedling is in April. Transplanting of seedlings at 45x30 cm proved to be better. Nitrogen requirement of foxglove is high. A fertiliser dose of 100 kg /ha N, 50 kg/ha P2O5 and 25 kg/ha K2O along with 30-40 t/ha FYM was found to be optimum for good crop. Five to six weedings followed by hoeings increased the foliage yield. Three to four irrigations are needed during April to June. One harvesting in first year at rosette stage and three harvestings in second year starting form August were recommended. However, harvest comes in the month of February-March contains almost double the glycoside than in August cutting. About 28 q/ha dry leaf yield is harvested. Sun drying of leaves at 30-400C is recommended to maintain quality.

Aswagandha (*Withania somnifera* Danunal)

It prefers well drained sandy-loam to red soil having pH 7.5-8.0. The crop is sown in late kharif in 2nd or 3rd week of August. On an average, 60 to 75 cm rainfall is best suited for rainfed crop. Broadcasting is practised by the farmers with high seed rate of 20-35 kg/ha. However, line sowing at 25 cm in rows facilitates better inter- cultural practices. One weeding and thinning at 25-30 days after sowing found to be sufficient in sub-marginal lands. Multi location trials showed that Isoproturon at 0.5 kg/ha, Glyphosate at 1.50 kg a.i./ha and Trifluralil (48% EC) @4.1 a.i/ha were found effective in controlling weeds and increasing root yield. Raised bed condition has been reported to yield higher quantity of root. Root size, root and shoot biomass and alkaloid content were found maximum in 180 DAS which should be considered as harvesting time. The whole plant is uprooted and roots are separated. About 3 to 4 q/ha dry root and 50-75 kg /ha seeds are harvested.

Khasi Kateri (*Solanum viarum* Dunal)

The crop is grown in various agroclimatic and soil types in India . However, commercial cultivation is confined to the Peninsular India and Diphu hills of Assam . Seeds are sown in nursery bed for raising seedlings for transplanting. Forty five-day old seedlings having 4-6 leaves give high success after transplanting. Although higher yield of dry berries and solasodine per hectare were obtained in close spacing (45x60 cm), a spacing of 90x150 cm is recommended to avoid difficulties in inter-culture and harvesting operations due to spiny stems. However, a less spiny variety, Arka Sanjeevini recorded higher berry yield per hectare under high density planting (10,989 - 13,889/ha). The variety showed about three fold increase in dry berry and solasodine yield. Top dressing of N delayed flowering whereas combination of N and CCC initiated early flowering. Application 250 ppm GA3 increases solasodine content in diploid and 250 and 1000 ppm in tetraploid. Further, foliar spraying with CCC at 1600 ppm was reported to increase solasodine content significantly. Turning yellow stage was found to be the most ideal for berry harvesting. Berry yield varies from 3.0-14.0 t/ha depending upon variety and soil status.

Long pepper (*Piper longum* Linn.)

Hot humid climate with 20-25% partial shade is ideal for its cultivation. The performance of long pepper was superior under partial shade of Cassava under Kerala condition. Well drained and nutrient rich soil is recommended. Three- to five-node rooted vine cuttings give cent percent field establishment. The best time for raising nursery is during March to April. Long pepper responded to summer irrigation. Integrated water management strategy involve in the application of irrigation water through the sprinkler system at IW/CPE ratio of one in combination with incorporation of coir pith for soil moisture conservation was found to enhance the annual spike yield in Long pepper. High density planting at 30 x30 cm combine with the organic farming technique, mulching and application of vermicompost resulted in higher spike yield under irrigated condition. Harvesting is done eight months after planting and 3-4 pickings are usually done in a year. About 400 kg/ha dried spikes in first year and 1000 kg/ha in second and third years are harvested. Harvested spikes are dried in sun for 4-5 days. Dried spikes are stored in moist proof containers. Besides spike, thicker stem and roots are cut and dried and used in Ayurvedic drug preparations. About 500 kg/ha roots are harvested.

Henbane (*Hyoscyamus niger* Linn.)

North western hilly tracts are most suitable for this crop. About 2-3 kg/ha seeds are recommended. Seeds are shown in first week of October 4 to 5 weeks old seedlings are transplanted in a spacing of 35 x 15 cm. A fertiliser dose of 80 kg/ha N, 40 kg/ha K2O and 15-20 t/ha FYM was recommended for optimum crop yield. Two-three weedings and 5-7 irrigations are needed. Crop rotations such as Senna - Henbane and Basil - Henbane were reported to be most profitable. The crop is harvested after 125 to 145 DAS. The older leaves at the base of the plant touching the ground are picked up first which is followed by picking of leaves from the upper portion of twigs and branches at 50% flowering. The bio-mass is thereafter thoroughly sun dried. This procedure of harvesting is quite economical and fetches higher market rate.

Palmarosa (*Cymbopogon martinii* var. *Motia*)

It is a crop of warm tropical climate. Sowing time for raising nursery is end of April to mid of May. Transplanting is done in the last week of June to mid August at 45x 30 or 60x60 cm spacing, depending upon soil fertility and climatic conditions. Optimum dose of fertilisers was 75 kg/ha N and 40 kg each of P_2O_5 and K_2O to get higher herbage and oil yield. Recently an integrated nutrient management - trial was conducted in Palmarosa. It was found that FYM @ 10 t/ha, N&P @ 20 kg/ha each and Azospirillum or Azotobacter favoured higher productivity and oil yield. Frequent light irrigations are required during rain free period. Inflorescence is ready to harvest at 7-10 days after opening of flowers. Crop is harvested 10-15 cm above the ground level. First year 2-3 harvest and in subsequent years 3-4 harvests are taken. On an average 80 kg/ha oil yield from rainfed crop and upto 220 -250 kg/ha oil yield from irrigated crop were achieved from the second year.

Vetiver (*Vetiveria zizanioides* (l.) Nash)

As a commercial crop vetiver flourishes over rich sandy - loam soils having 6-8 pH under warm and humid weather conditions. Earthing up operation increases the root yield. Irrigation at 0.4 IW/CPE ratio showed maximum root yield (14.2 q/ha). However, 8 irrigations are required within 15 months. A fertiliser dose of 80 kg N and 30 kg each of P2O5 and K2O/ha increase the root yield without affecting the oil quality. Combined application of farm yard manure @10 t/ha, N,P @37.5 & 20 kg/ha respectively and Azospirillium or Azotobacter enhanced oil yield. Cowpea, cluster bean and black gram as intercrops added the income of the cropping system. Best time for harvesting of roots is 15 months after planting.

Mints (*Mentha* spp)

The crop prefers well drained soil rich in organic matter and 6-7.5 pH. This group of plants needs very high amount of nutrients particularly nitrogen. Application of FYM at 25-30 t/ha at land preparation, 30 kg/ha N and 40-60 kg/ha each of P2O5 and K2O as basal dose followed by 75 kg/ha N at 40 days after sprouting and 75 kg/ha N after first harvest is recommended. Application of 120 kg N/ha and harvesting of spearmint shoots in June showed maximum herbage and oil yield. Zinc deficiency is common in Indo-Gangetic plains. The crop responses maximum at 20 kg/ha Zn application at planting. Iron and Boron deficiencies are also reported. Two to three weedings and hoeings are essential in mint farming. Pre-emergence Terbacil (2.0 kg a.i./ha) application found to be effective in controlling of weeds. Six to nine irrigations are required during dry seasons. First crop is harvested at 105-110 DAS and subsequent harvest takes about 90 days thereafter. On an average 30 t/ha of herbage yield in Japanese mint and 20-25 t/ha of Bergamot mint are harvested which yield about 150 and 100 kg/ha oil respectively.

Babchi (*Psoralea corylifolia* Linn.)

Maximum seed yield was obtained when 40 kg N, 20 kg P2O5 per ha were applied at 30x30 cm spacing. Psoraline content was also found high in this treatment.

Kalmegh (*Andrographis paniculata* Nees)

Optimum transplanting and harvesting time for high herbage yield was obtained in 16th July transplanting and 16th November harvesting at Anand and Indore ; 1st July transplanting and 16th November harvesting at Akola and Udaipur and 1st July transplanting and 1st November harvesting at Faizabad.

Lavender (*Lavender officinalis* Chaix)

Sowing of lavender seeds during February and harvesting of the flowers for oil extraction at 50% flower initiation stage proved beneficial to get higher oil yield (32.41 kg/ha). The optimum spacing in lavender was found to be 45x60cm for higher foliage yield.

Melisa (*Melissa officinalis* Linn.)

Fresh herbage yield (141.4 q/ha) and oil recovery (29.60 kg/ha) were found to be maximum with GA3 application at 200 ppm concentration.

Salvia (*Salvia scalarea* Linn.)

Transplanting of Salvia in February and application of 90+30+30 kg/ha NPK recorded highest oil yield of 35.24 kg/ha.

Source: Directorate of Medicinal and Aromatic Plants Research, Anand.

□□□□

26

Recent Initiatives in Crop Protection

Azadiractin (1%) was found effective for the control of Hadda beetle on *Ashwagandha*. Nineteen arthropods were recorded on isabgol. Among them, 13 were phytophagous species belonging to order Lepidopteran (09), Hemipteran (03) and Homopteran and six predatory species belonging to order Coleoptera (03), Neuroptera (01) and Odanata (02). The sequential occurrence of arthropods revealed the presence of aphid (*Aphis gossypii*) from second week of January to fourth week of February, whereas Lepidopteran (*Helicoverpa armigera, Trichoplusia ni, Thysanoplusia oricalchae, Spliarictia* sp. *Hyposidra successari* and *Olene mendosa* and true bugs (*Graptostethus servus* and *Spilostethus pandurus*) were observed associated during early crop growth stages (i.e. from second week of December). Infestation of aphid was uniformly severe, whereas that of Lepidopterans and true bugs were sporadic and less severe. However, presence of these pest arthropods caused significant reduction in seed yield.

Source: DARE/ICAR Annual Report 2012–13

In isabgol, aphid infestation was positively influenced by N doses. Hence, irrespective of irrigation schedules, maximum aphid infestation (33.78) was highest at N/60 kg/ha. Downy mildew disease severity was also significantly influenced by fertilizer doses but not due to irrigation schedules. Maximum disease was recorded at 60 kg N/ha (14.23 PDI) which was at par with 45 kg N/ha (13.29 PDI). Both sowing time and seed rate had significant effects on *Alternaria* leaf blight disease development in Asalio (*Lepidium satibum*).

Delay in sowing increased the disease severity till 10 November and it reduced thereafter. Increasing seed rate resulted in increase in blight severity. Hence, 6–8 kg/ha seed sown on 10 October had minimum (10.8–13.2 PDI) disease severity, while maximum (38.9 PDI) was recorded at 10 November sowing with highest seed rate. At Faizabad, *Alternaria* leaf blight could be best managed with foliar application of Mancozeb thrice (40, 55 and 70 DAS) which resulted in maximum diseases suppression and consequently increased seed yield (1,880 kg/ha). Systemic fungicide (Propaconazole) was less effective.

Source: DARE/ICAR Annual Report 2010–11

Recommendations for management of Major diseases

Crop	Disease	Causal Organism	Findings & Recommendations
Medicinal Crops			
Opium poppy (*Papaver somniferum*)	Downy mildew	*Peronospora arborescens*	Seed treatment with Metalaxyl and four sprays of Metalaxyl @ 0.2% at 35, 55, 75 and 95 days after sowing found most effective in reducing primary and secondary disease incidence. It also increased latex yield (26.7%) and seed yield (24.7%) over control. Jawahar Opium poppy - 540 having some resistance against the disease has been recommended for M.P.
Senna (*Cassia angustifolia*)	Leaf spot	*Alternaria alternata*	Spraying of Bordeaux mixture, Benomyl, and Copper oxy chloride is recommended.
Isabgol (*Plantago ovata*)	Downy mildew	*Peronospora alta P. plantaginis Psendoperonospora plantaginis*	Seed treatment with Metalaxyl at 4g/kg seed and 3 sprays of Metalaxyl (0.2%) at 15 days interval starting from first appearance is best for disease control. December sown crops are mostly disease free.
Sarpagandha (*Rauvolfia serpentina*)	Leaf blight and Bud rot	*Alternaria tenuis*	Spraying of Mancozeb (0.2%) or Rovaral (0.2%). Seed treatment with contact fungicides.
Liquorice (*Glycyrrhiza glabnra*)	Leaf spot	*Cercospora cavarae*	Mancozeb spray at 0.2%.
Long pepper (*Piper longum*)	Leaf spot & rotting	*Colletotrichum and Cercospora spp.*	Spraying of Bordeaux mixture - one during May and 2-3 sprays during rainy season.
Khasi kateri (*Solanum viarum*)	Wilt	*Fusarium oxysporum*	Direct sowing of seeds instead of transplanting of seedlings can reduce the disease incidence
Aromatic Crops			
Palmarosa & Lemongrass (*Cymbopogon* spp.)	Red leaf spot	*Collectorichum graminicola*	Two sprays of Carbendazim (0.1%) or 3 sprays of Mancozeb (0-.3%) starting from first appearance of disease at 20 days or 10-12 days intervals, respectively.
Rose (*Rosa damascena*)	Die back	*Diplodia rosarum*	Pruned ends of the shoots should be treated with Copper fungicides.
Ambrette seeds (*Abelmoschus moschatus*)	Anthracnose	*Collectotrichum hibiscicum*	Seed treatment with contact fungicides and spray of Bordeaux mixture are recommended.

Source: Directorate of Medicinal and Aromatic Plants Research, Anand

□□□□

27

Diseases of Major Medicinal and Aromatic Plants

Common name, symptoms of damage and management of diseases of major medicinal and aromatic plants

Name of disease and its causal organism	Symptoms of damage	Management
Coleus		
Downy mildew: *Peronospora sp.*	• This mildew appears on the leaves making the plant look dirty because it is brown in colour. And can also result in curled leaves. • The leaves also twist. Sometimes symptoms are not found on leaves which make the disease harder to control.	• Spray 0.3 % Copper oxychloride.
Impatiens Necrotic Spot Virus	• Causes brown or yellow spots on leaves, rings, black or brown stem discoloration, and brown leaf veins. It is a virus that causes plants to die. • The disease is spread by insect thrips. It only takes a few of these insects to infect a whole greenhouse.	• Spray Dimethoate 2ml/litre.
Wilt:*Fusarium chlamydosporum*	• Slow wilting of seedlings and drooping of leaves.	• Inoculation with *Trichoderma viride* + *Glomus mosseae* gave the best result in controlling the disease. • The same treatment also resulted in maximum growth, yield and root forskolin concentration of coleus.
Isabgol		
Powdery mildew: *Podosphaera leucotricha*	• White powdery growth on the upper surface of leaves.	• Powdery mildew sometimes attacks the crop and is brought under control by • Spraying the crop with wettable sulphur two or three times at 15-day intervals after the appearance

Name of disease and its causal organism	Symptoms of damage	Management
		of infection.
Opium poppy		
Downy mildew : *Peronospora arborescens*	• On the upper surface of leaves yellow patches were seen and on the lower side. Sometimes causes serious damage to the crops at the time of capsule formation.	• It is controlled by the application of 0.2% Mancozeb.
Citronella		
Leaf Blight: *Curvularia andropogonis*	• The fungus affects the crop in the beginning of monsoon. Small brownish spots appear initially during the month of July, which enlarge into long patches along the tips and margins of the leaves. • In case of serious infections, the entire leaves dry up resulting in considerable decrease in leaf and oil production. • Black dots in from of sporulation of the fungus have been seen on dead spots of the leaves.	• The disease can be controlled by prophylactic spraying Mancozeb@ 20g/10 litres of water at an interval of 10-15 days during the growing period.
Geranium		
Alternaria Leaf Spot: *Alternaria tenius*	• Small brown spots with alternating light and dark bands are most evident on lower surface of older leaves. • It occurs on garden geraniums during cool, wet weather. • The fungus rarely sporulates on attached leaves but does so readily on fallen leaves.	• Removal and destruction of infected foliage and application of foliar fungicides immediately upon the appearance of disease helps to control the disease.
Root rot/wilt: *Rhizoctonia solani*	• Root rot/wilt of geranium is observed very often on the crop.	• At the time of planting, fresh cuttings in nursery beds are dipped in 0.2% Mancozeb.

Name of disease and its causal organism	Symptoms of damage	Management
	• In the beginning, only a few plants exhibit the typical wilting find drooping symptoms, which soon attain alarming rate if ignored.	• The nursery bed is also drenched with the same fungicide. • Rooted cuttings are dipped in 0.2% Mancozeb before transplanting. • The soil around the plant is also drenched with the same fungicide. • After every harvest or after every heavy rain the soil around the plant is drenched with 0.2% Mancozeb. • Over watering increases the incidence of wilt, and hence should be avoided.Sharp sickles should be used for harvesting to avoid mechanical damage to the plant.
Lemongrass		
Leaf Blight: Curvularia Andropogonis	• The disease appears in the form of minute, circular, reddish brown spots mostly on the margins and tip of the leaves which merge to form elongated reddish brown necrotic lesions resulting in premature drying of leaves, older leaves are more susceptible to infection.	• Spray Mancozeb (0.2%) with an interval of 10-12 days or 0.3% Copper oxychloride.
Long Smut:*Tolyposporium Christensenii*	• All the flowers in the inflorescence are transformed into, slender, tubular to conical cream-coloured sori. • These sori flake off at maturity, starting from the tip and hang in shreds.	• Prolific spraying of Mancozeb @0.2% just before flower initiation and treating the seeds with Carbendazim 2g/kg of seed before sowing effectively controls the disease.
Little Leaf or Grassy Shoot: *Balansia Sclerotica*)	• The characteristic symptoms are in the form of stunting and little leaf formation in place of normal inflorescence.	• Spraying Mancozeb (0.2%) just before flowering stage at an interval of 10-12 days helps to reduce the disease incidence.

Name of disease and its causal organism	Symptoms of damage	Management
		• Fresh seedlings should be used for plantation and crop rotation.
Red Leaf Spot: *Colletotrichum Graminicola*	• The symptoms are in the form of brown spots with concentric rings in the centre appearing on the lower surface of the leaves. • The spots may be formed on leaf sheaths and midrib. Later the spot merge to form bigger patches and the affected leaves dry away.	• Two sprays of Carbendazim 0.1% just after the appearance of the disease at an interval of 20 days or three sprays of Mancozeb (0.2%) at an interval of 10-12 days.
Rust: *Puccinia Nakarishikii*	• The symptoms are in the form of Discrete, linear brown uredinia appear on the lower surface of the leaves associated with chlorotic streaks.	• The disease is controlled by spraying Mancozeb (0.2%) or Copper oxychloride (0.3%) or Plantavax (0.1%) at an interval of10-12 days.
Mint Carbendazim		
Fusarium Wilt: *Fusarium Oxysporum*	• The leaves of the affected plants turn yellow, curled and finally dry.	• Benlate, Carbendazim and Topsin can be used to control the disease.
Leaf Spot: *Corynespora cassiicola; Curvularia lunata; Alternaria spp*	• Small dirty brownish necrotic spots are observed on the leaves and twigs leading to defoliation.	• Treating the stolons with 0.1% Benlate or spraying with 0.3% Copper Oxychloride is recommended.
Powdery Mildew: *Erysiphae cichoracearum*	• Powdery white patches on the upper and lower surfaces of the leaves and stems are observed. • Under favourable conditions, the disease causes severe loss due to defoliation. It also reduces the oil content up to 50%.	• Spraying Sulphur formulation or fungicide like Karathen-WD is recommended. • Spraying should commence as soon as the disease appears in the field and should be repeated every 10-15 days.
Rust: *Puccinia menthae*	• Golden brown spots are observed on the leaves and twigs. The leaves turn yellow and fall off.	• Spraying the crop with 0.3% Copper fungicides immediately on the appearance of the disease is recommended. • Early harvest of the crop during May is suggested so as to decrease the losses due to falling of leaves.

Name of disease and its causal organism	Symptoms of damage	Management
Stolon Rot: *Macrophomina phaseoli*	• The disease causes rotting of the underground stem and death of the entire plant. • One of the main damages caused by the disease is the reduction in the production and rotting of the stolons by the end of the cropping season.	• The disease can be controlled by proper crop rotation. • Mint should not be planted continuously in those areas where attack of stolon-rot is observed. It is better to follow 3-year-crop rotation with rice, wheat and mint. • It is safer to discard all the diseased stolons while planting and treat the stolons with solution of 0.1% Benlate.
Palmarosa		
Little Leaf or Grassy Shoot: *Balansia Sclerotia*	• The disease is attacks the floral parts and consequently the oil yield and seed production is adversely affected. • The characteristic symptoms are stunting and formation of little leaf on the plants in place of normal inflorescence.	• Spray Mancozeb (0.2%) before flowering stage at an interval of 10-12 days.
Red Leaf Spot: *Colletotrichum graminicola*	• The disease is much prevalent in all the Palmarosa growing regions and appears during rainy season. • Under warm and humid conditions, the disease spreads rapidly and may cause up to 35 % reduction in the oil production. • The disease perpetuates on the left-out stumps and stubbles on the field. • Thesymptoms can be seen as small brownish spots, scattered all over the ventral surface of the leaves. • The disease also infects the leaf sheaths and midrib.	• Two Sprays of Carbendazim (0.1%) just after the first appearance of the disease at an interval of 20 days or three sprays of Mancozeb (0.2%) at an interval of 10-20 days control the disease.
Smut: *Tolyposprium christensenii*	• This is an inflorescence borne disease adversely affects the oil yield and seed production.	• Seeds should be treated with organomercurial fungicides before sowing in the nursery.

Name of disease and its causal organism	Symptoms of damage	Management
	• The symptoms are characterized by the transformation of seed into the smut sori, which are slightly long covered by dull yellow membrane.	• Spraying Mancozeb (0.2%) just at the time of flower initiation is effective in controlling the disease.
Patchouli		
Leaf Blight: *Cercospora* sp	• Mostly when the plants are almost one year old brown spots are formed near the margin or at the apical region, which enlarge irregularly, coalesce, cover the entire lamina and create drying. • It is, however, not of common occurrence and less serious as compared to root-knot.	• The disease can be controlled with two sprays of Mancozeb 0.3%, at one-month interval. • This, however, depends upon the severity of the disease.
Vetiver		
Leaf Blight: *Curvularia trifolii*	• The infected leaves bear dark spots, which turn black with age. • The roots of affected plants become yellow and gradually dry out.	• Two to three sprays of Copper fungicide (0.3%) containing 50 % Metallic Copper is recommended as a control measure.

□□□□

28

Insect Pests of Major Medicinal and Aromatic Plants

Common name, symptoms of damage and management of insect pests of major medicinal and aromatic plants

Common and Scientific name of pest	Symptoms of damage	Management
Aswagandha		
Cut worm: *Agrotis* sp.	• Larva attacks the tender seedlings.	• Hand picks the caterpillars and destroys. • Spray NSKE 5%. • Use poison bait to lure the larvae.
Epilachna Beetle: *Henosepilachna vigintioctopunctata*	• Skeletonisation of leaves.	• Collect the beetles, grubs, pupae and destroy. • Spray Carbaryll 50 WP 2 g/lit mixed with wettable sulphur 50 WP 2g/lit of water.
Fruit borer: *Helicoverpa armigera*	• Neonates after hatching feed on tender leaves. • Later stages of larvae bore into fruit.	• Field release of *T. chilonis* at 5cc/ha. • *Bacillus thuringiensis* 0.5 kg/ha. • Spray application of NPV @ 250 LE/ha. • Set up pheromone trap at 12/ha.
Glory lily		
Leaf Webber:*Archips micaceana*	• Larva green- feeds under thin webbing on the epidermis of leaves, main stalks, bunches and on berries.	• Spray NSKE 5% or Neem oil 0.5%.
Lily caterpillar: *Polytela gloriosae*	• Early instars feed on chlorophyll of the leaves. • Later instars feed voraciously leaving only the hard stem of the plant. • Complete devastation.	• Spray application of Neem oil 0.5% or NSKE 5% . • Collection of growing stages and destruction.
Red Hairy Caterpillar: *Amsacta lactinea*	• Larvae feed voraciously on leaves. • Similar to grazing of cattle. • Moth emerge after monsoon.	• Hand picking of larvae. • Collection of egg masses. • Summer ploughing. • Bon fire. • Spray 2.5 kg of Carbaryll in 625 lit of water/ha. • Poison baiting.

Common and Scientific name of pest	Symptoms of damage	Management
Semilooper: *Plusia signata*	• Caterpillar feed on the leaves. • Defoliation	• Collect and destroy the infested plant parts. • Hand picking and destruction of the caterpillars.Spray NSKE 5% or Neem oil 0.5%.
Tobacco Cutworm: *Spodoptera litura*	• Caterpillar cut and feed on the leaves. • Defoliation	• Hand pick the caterpillars and destroyed. • Spray NSKE 5% or Neem oil 0.5%.
Long pepper		
Mealy bug:*Dysmicoccus* sp.	• The root is infested by the mealy bug and the affected plant shows yellowing. • Stunted growth of the plant.	• Soil application of Neem cake at 60 kg/ac. • Soil drenching with Dimethoate 2 ml / lit near the root zone of affected plants.
Tea mosquito bug: *Helopeltis theivora*	• Nymphs and adults severely attack the plant by feeding on the tender foliage. • Necrotic lesions develop around the feeding puncture leaving shot holes on the lamina.	• Spray Monocrotophos 2ml / lit of water in the evening hours for better control.
Opium poppy		
Aphid: *Myzus persicae*	• Nymphs and adult suck the sap on leaves.	• Spray Neem oil 0.5% (or) NSKE 5%. • Encourage natural enemies like coccinellid beetle.
Capsule borer: *Helicoverpa armigera*	• It feeds on flower heads and seeds of flower heads.	• Hand picking of the larvae. • Use of pheromone traps at 12/ha. • Spray NPV 250 LE / ha. • Use of Bt spray formulation at 2g/2ml per litre of water. • Use of egg parasitoids *Trichogramma chilonis* at 5cc/ha.
Cutworm: *Agrotis spp*	• Larva cut younger plant at the base above the ground level. • Larva remains concealed in cracks in the ground and active at night.	• Hand pick the caterpillars and destroy. • Spray NSKE 5%. • Poison bait with rice bran, jiggery and Carbaryll can be used.

Common and Scientific name of pest	Symptoms of damage	Management
Root weevil: *Sternocarus fuliginosus*	• The grubs feed on roots.	• Apply Neem cake 5q/ha in the soil before planting.
Senna		
Green leaf eating caterpillar: *Catopsilia pyranthi, Eurema hacabe*	• Feed on young leaves.	• Hand picking. • Destruction of infested pods along larva. • Spray application of NSKE 5% or Bt formulation of 2g/lit of water.
Pod borer: *Etiella zinckenella*	• Make minute holes on the pods and attack the seeds.	• Hand picking. • Destruction of infested pods along larva. • Spray BT – 0.1% @ 50% flowering stage. • Field release of *Trichogramma chilonis* @5 cc/ha. • Use light trap at 1 /ha.
Sarpagandha		
Ash weevil: *Myllocerus viridanus*	• Adults feeding on the leaves from the edges in a serrated manner. • Notching of leaf margins. Grubs feed on roots and cause wilting and drying of plants.	• Collection and destruction of adults. • Spraying of Carbaryll 0.2% or Malathion 0.1%. • Incorporation of soil with dust or drenching of the soil with 0.1% Chlorpyriphos.
Cut worm: *Agrotis sp.*	• Larva attack the tender seedlings. • Caterpillars are nocturnal in habit and found under debris around plants. • Cuts seedlings near the ground level and eat the tender parts. • Damage is more in low lying water logged areas.	• Fork the soil (summer months) – to expose larva and pupa for avian predators. • Install light traps (summer) to attract adult moths. • Install Pheromone traps at 20/ha to monitor and attract male moths. • Irrigate the field in day time to expose larva for predation by birds. • Drench collar region of the plants in evening hours with Chlorpyriphos 2ml/lit.a day after planting.

Common and Scientific name of pest	Symptoms of damage	Management
Epilachna beetle: *Henosepilachna vigintioctopunctata*	• Grubs and adults feed by scrapping chlorophy ll from epidermal layers of leaves. • Skeletnized and gradually dry away.	• Collect and destroy grubs and adults. • Spray Carbaryll 50 WP 3g/lt + wettable sulphur2g/lt.
Mealy bug:*Coccidohystrix insolita*	• Nymphs and adults present on under surface of leaves and suck the sap. • Affected plants turn yellow, wilt and dry. • Honeydew secretion causes sooty mould.	• Removal of affected parts in the early stage of attack. • Spray application of Dimethoate 0.05 % (or) malathion 50 EC 0.05% • Field release of coccinellid predators. • In the later stage of the crop Neem oil 5ml or fish oil rosin soap 25 g per litre of water can be sprayed.
Sphingid caterpillar: *Deilephila nerii*	• Caterpillar feed on the leaves. • Defoliation	• Soil digging to expose the pupae for predation by birds. • Handpicking and destruction of caterpillars. • Use light trap 1/ha. • Plant neerium around the bunds and take up plant protection on neerium with chemicals.

□□□□

29

Glossary

Abortifacient	:	A drug that induces foetus expulsion
Achene	:	Single seeded, unicellular, dry, indehiscent fruit also called caryopsis
Acuminate	:	Long, pointed, gradually tapering towards apex
Alexiteric	:	Developing resistance against infectious diseases
Alterative	:	A drug that alters body condition by improving metabolism; used against long effect of a medicine
Amenorrhoea	:	Failure of menstruation
Amplexicaul	:	Encircling of the node by leaf bases
Analgesic	:	Pain killer
Anodyne	:	A drug used to allay pain
Antacid	:	To neutralize acidic effect in abdomen
Anthelmintic	:	A drug used to expel or destroy intestinal worms
Antibronchial	:	Working against respiratory track infection and congestion
Antiemetic	:	A drug used to control vomiting
Antihistaminic	:	A drug used for controlling skin irritation and itching caused due to increase of blood histamine
Anti-inflammatory	:	A drug used to cure swellings
Antiperiodic	:	A drug that prevents recurrence of a disease. Used against malarial fever
Antiphlogestic	:	An agent used for reducing or subsiding inflammations

Antipyretic	:	A drug or a medicine used to lower body temperature in fever
Antirhinitis	:	Clearing of nasal mucous by subsiding nasal membrane inflammation
Antiseptic	:	Prevention of putrefaction or sepsis of wounds and cuts
Antispasmodic	:	A medicine that releases nervous irritability and reduces spasm or convulsion
Antitussive	:	A drug controlling cough
Aperient	:	Mild laxative/cathartic
Apex	:	Tip, uppermost part
Aphrodisiac	:	Drug increasing the sexual desire and longevity
Apices (apex)	:	Top
Appetizer	:	Increasing digestion and hunger
Aristate	:	Ending in bristle or awn
Aromatic	:	An agent that emits sweet smell
Articulate	:	Jointed
Astringent	:	A drug that contracts the muscular membrane
Bracteole	:	A small bract or leaf structures below perianth in a flower
Bronchodiltaor	:	A drug that widens the trachea, thus easing congestion
Bulbils	:	Vegetative propagative spherical structure arising at the leaf base (as in *Dioscorea bulbifera*)
Caducous	:	Falling soon
Campanulate	:	Bell-shaped
Capitulum	:	Head-shaped inflorescence, as in Asteraceae
Carminative	:	A drug that releases intestinal gases or flatulence
Catarrhal	:	Mucous membrane inflammation with excessive secretion of mucous
Cathartic	:	Drastic purgative, totally expelling rectal stool
Cauline	:	Arising from stem
Cholagogue	:	A drug inducing excessive secretion of bile juice
Cladodes	:	Modification of dwarf branches into leaf-like structure
Clasping	:	Wrapping
Colic	:	Severe spasmodic and gripping pain in colon region
Comose	:	With long, white bunch of hair
Cordate	:	Heart-shaped
Coriaceous	:	Thick, stiff
Corm	:	A stem modification, underground spherical in shape with reserve food material

Corona	:	Bundle of hair between corolla and stamens arising from base of the corolla
Cuneate	:	Wedge-shaped, tapering towards base
Cuspidate	:	Tapering to long point at tip
Cyme	:	Arrangement of flowers with older flower on top and younger flower towards base
Demulcent	:	Soothing medicine for digestive function
Depurant	:	Purifier
Diadelphous	:	Stamen divided into two groups 9+1 as in Fabaceae
Diaphoretic	:	Drug inducing perspiration
Didymous	:	Two-sized (filament of stamens)
Dioceous	:	Male and female flowers on different plants
Discoid	:	Disc-shaped
Diuretic	:	Increasing urination frequency
Dysmenorrhoea	:	Painful menstrual flow
Dyspepsia	:	Indigestion with gastric pain
Ellipsoid	:	Eclipse-shaped
Emarginate	:	Deeply and irregularly notched at apex
Emmenagogue	:	A drug that restores regularity in menstrual cycle
Entemophylous	:	Pollinated by insects
Entire	:	Even margin, complete margin, no cut or lobation on margin (of leaves)
Epigynous	:	Ovary seated above perianth
Epilepsy	:	An affectation of the nervous system resulting from excessive or disordered discharge of cerebral neurons
Expectorant	:	A drug expelling phleg from trachea
Exstipulate	:	Without stipule, a leafy structure at the base of leaf
Extrose	:	Facing outward
Fascicled	:	Clustered at one point
Febrifuge	:	A drug used to cure fever
Fluted	:	Hollow
Follicle	:	Dry dehiscent fruit opening only by ventral suture
Fragrant	:	Emitting sweet smell
Galactagogue	:	Increasing and activating mammary gland
Gamo petalous	:	Petals united with each other
Gamo sepalous	:	Sepals united with each other
Gamo tepallus	:	Perianth united with each other
Geriatric	:	Pertaining to old age
Glabrous	:	Without any hairy structure
Gleet	:	Chronic discharge from vagina

Gonorrhoea	:	Inflammation of the gentio-urinary passage with pain and discharges
Gout	:	A purine metabolic disease with raised level of serum uric acid (blood urea)
Gregarious	:	Very long – robust, profuge
Gynaecium	:	Female part of flower having ovary, style, and stiga
Gynaecological	:	Pertaining to female genital organ
Haemophilic	:	Loss of blood coagulation property in which blood continues to flow on cuts
Haemoptysis	:	Spitting of blood
Haemorrhage	:	Bleeding piles
Haemostatic	:	Blood coagulant, preventing bleeding
Heart wood	:	Central hardest part of wood/trunk
Hydrogogue	:	Promoting expulsion of water or serum
Hyper lipidemia	:	Reducing fat on joints
Hypoglycaemic	:	Lowering blood sugar
Hypogynous	:	Ovary inferior, sepals, petals, and stamen above the ovary
Hypotensive	:	Lowering blood pressure
Hysteria	:	Neurotic attack with unusual activities and symptoms
Imbricate	:	Arrangement of corolla with two outer, one inner, and two with one side outer other side inner
Imparipinnate	:	Leaflet in odd number on top
Insomnia	:	Sleeplessness
Lactagogue	:	Increasing milk secretion
Lanceolate	:	Shape of convex lens
Latex	:	Oozing milky sap
Laxative	:	Smoothening rectal wall and loosening the stool
Lenticellate	:	Slit-like raised cortical structure on the branches
Leucorrhoea	:	White fluid discharge from vagina
Linear	:	Very narrow, like a line
Lomentum	:	Single seeded cell of pod, septate, and constricted between two seeds
Menorrhagia	:	Excessive menstrual flow
Moniliform	:	Beaded in a row like a garland
Monoecious	:	Unisexual, male and female flowers on the same plant
Mucronate	:	Small projection at the apex (acume)
Neuropathy	:	Diseases related to CNS (central nervous system)
Oblong	:	Longer than broad with narrowing margin towards base

Obpyramidal	:	Inverted pyramid shaped
Obsolete	:	Minute or wanting
Obtuse	:	Blunt top (apex)
Oedima	:	Inflammation
Ophthalmic	:	Pertaining to eye diseases like conjunctivitis
Orbicular	:	Almost circular
Orchitis	:	Inflammation of testis with hypertrophy and pain
Ovate	:	Egg shaped
Paraplegia	:	Paralysis, loss of ability to move or feel in the lower part of body
Pedicel	:	Stalk of flower
Pedicillate	:	Stalked flowers
Peduncle	:	Stalk of inflorescence
Perianth	:	Vegetative covering of sexual organ in flower, sometime differentiated into calyx and corolla
Peripinnate	:	Leaflet in even number
Pesticide	:	An agent that kills unwanted plants and insects
Phleg	:	Mucous secretion in respiratory track
Pinnate	:	Compound leaf with leaflets arranged on same rachis at length
Polyhederal	:	Many faced, many angled
Post-natal	:	After child birth
Procumbent	:	Creeping on ground then rising up
Pruritus	:	Skin itching
Pubescent	:	Carpeting of small soft hair
Purgative	:	Loosening stool to help exersion, thus curing constipation
Radical	:	Arising from stem base (leaves or branches)
Refrigerant	:	Cooling effect
Rejuvinative	:	Antiageing, prolonging life
Reniform	:	Almost kidney-shaped
Resolvant	:	Causing resolution of a tumor or swelling
Reticulate	:	Weaved
Rhizome	:	Subterranean part between stem and root bearing buds that may be used as a propagative part
Rhomboid	:	Quadrihedral with only opposite angles equal
Rubifacient	:	Producing counter effect on external application
Sarmentose	:	Growing among bushes, with long flexuous runners
Scandent	:	Weak plants that need support; climbing without any climbing organ, and so on

Scrofula	:	Tubercular cervical adenitis, with or without ulceration
Sedative	:	Central nervous system depressant in which a person is made calm or asleep
Serrate	:	Margin of leaf cut into saw-shaped structure, pointing upwards
Sessile	:	Without any stalk
Sinuate	:	Wavy margins
Spathulate	:	Service spoon shaped
Spike	:	Sessile flowers arranged on peduncle
Staminode	:	Barren stamen (infertile anthers)
Stellate	:	Star-shaped arrangement of short stiff hair (trichomes)
Stipitate	:	Stalked
Stomachic	:	A drug used for improving digestion
Striate	:	Marked with vertical lines
Styptic	:	Blood purifier
Succulent	:	Thick, soft, and juicy
Suffruticose	:	A herb becoming perennial at base and herbaceous at apices
Syncarpous	:	Fused carpels
Terete	:	Lined
Terrestrial	:	Growing in soil
Thermogenic	:	Producing heat offer metabolism
Thrombosis	:	A blockage preventing the flow of blood in the body caused by clot
Tomentose	:	Dense, soft, layer of hair or cotton easily scraped off
Tranquilizer	:	A drug used to calm a person and reduce mental activity
Truncate	:	Flat topped
Tuber	:	A swollen, subterranean root containing reserve food material
Turbinate	:	Tube shaped
Urticaria	:	Nettle rashes on skin
Variegated	:	Spotted with various colour
Vermifuge	:	Expelling or destroying intestinal worms
Vertigo	:	Dizziness, a feeling of spinning
Villous	:	Long soft shaggy hair
Whorl	:	Arising more than two from one node (leaf or branches)
Zygomorphic	:	Asymmetrical plain of flowers not divisible into equal halves

□□□□

Appendices

Appendix-I: List of Spices under the Purview of the Spices Board

S. No.	Common Name	Botanical Name	Family	Part used as spice
1.	Garlic	*Allium sativum* L.	Alliaceae	Bulb
2.	Coriander	*Coriandrum sativum* L.	Apiaceae	Leaf & Fruit
3.	Cumin	*Cuminum cyminum* L.	Apiaceae	Fruit
4.	Fennel	*Foeniculum vulgare* Mill.	Apiaceae	Fruit
5.	Celery	*Apium graveolens* L.	Apiaceae	Leaf, Fruit & Stem
6.	Aniseed	*Pimpinella anisum* L.	Apiaceae	Fruit
7.	Ajowan	*Trachyspermum ammi* L.	Apiaceae	Fruit
8.	Caraway	*Carum carvi* L.	Apiaceae	Fruit
9.	Dill	*Anethum graveolens* L.	Apiaceae	Fruit
10.	Parsley	*Petroselinum crispum* Mill.	Apiaceae	Leaf
11.	Asafoetida	*Ferula asafoetida* L	Apiaceae	Oleogum resin from rhizome and thickened root
12.	Lovage	*Levisticum officinale* Koth.	Apiaceae	Leaf & Stem
13.	Sweet flag	*Acorus calamus* L.	Araceae	Rhizome
14.	Tarragon	*Artemisia dracunculus* L.	Asteraceae	Leaf
15.	Mustard	*Brassica juncea* L.Czern	Brassicaceae	Seed
16.	Horse Radish	*Armoracia rusticana* Gaertn.	Brassicaceae	Root
17.	Tamarind	*Tamarindus indica* L.	Caesalpiniaceae	Fruit
18.	Caper	*Capparis spinosa* L.	Capparidaceae	Flower buds
19.	Kokam	*Garcinia indica* Choisy	Clusiaceae	Rind
20.	Camboge Desr	*Garcinia cambogia (*Gaertn).	Clusiaceae	Rind

Contd.

S. No.	Common Name	Botanical Name	Family	Part used as spice
21.	Juniper berry	*Juniperus communis* L.	Cupressaceae	Berry
22.	Fenugreek	*Trigonella foenum-graecum* L.	Fabaceae	Seed
23.	Star Anise	*Illicium verum* Hook.	Illiciaceae	Fruit
24.	Saffron	*Crocus sativus* L.	Iridaceae	Stiga
25.	Mint	*Mentha piperita* L.	Lamiaceae	Leaf
26.	Hyssop	*Hyssopus officinalis* L.	Lamiaceae	Leaf
27.	Marjoram	*Marjorana hortensis* Moench.	Lamiaceae	Leaf
28.	Basil	*Ocimum basilicum* L.	Lamiaceae	Leaf
29.	Rosemary	*Rosmarinus officinalis* L.	Lamiaceae	Leaf
30.	Sage	*Salvia officinalis* L.	Lamiaceae	Leaf
31.	Savory	*Satureja hortensis* L.	Lamiaceae	Leaf
32.		*Thymus vulgaris* L.	Lamiaceae	Leaf
33.	Oregano	*Origanum vulgare* L.	Lamiaceae	Leaf
34.	Cinnamon	*Cinnamomum zeylanicum* Breyn	Lauraceae	Bark
35.	Cassia	*Cinnamomum cassia.* Blume	Lauraceae	Bark
36.	Tejpat	*Cinnamomum tamal* (Buch Ham) Nees & Eberum	Lauraceae	Bark & Leaf
37.	Bay Leaf	*Laurus nobilis* L.	Lauraceae	Leaf
38.	Nutmeg	*Myristica fragrans* Houtt.	Myristicaceae	Seed
39.	Mace	*Myristica fragrans* Houtt.	Myristicaceae	Aril
40.	Clove	*Syzygium aromaticum(L)* Merr.& Perry	Myrtaceae	Unopened Flower bud
41.	Allspice	*Pimenta dioica* (L) Merr.	Myrtaceae	Fruit & Leaf
42.	Vanilla	*Vanilla planifolia* Andr.	Orchidaceae	Pod
43.	Poppy seed	*Papaver somniferum* L.	Papaveraceae	Seed
44.	Pepper	*Piper nigrum* L.	Piperaceae	Fruit
45.	Pepper Long	*Piper longum* L.	Piperaceae	Fruit
46.	Pomegranate	*Punica granatum* L.	Punicaceae	Seed
47.	Curry leaf	*Murraya koenigii(L)* Sprengel	Rutaceae	Leaf
48.	Chilli			
	Bird's Eye	*Capsicum frutescens* L.	Solanaceae	Fruit
	Capsicum	*Capsicum annuum* L.	Solanaceae	Fruit
	Chilli	*Capsicum annuum* L.	Solanaceae	Fruit
	Paprika	*Capsicum annuum* L.	Solanaceae	Fruit
49.	Cardamom (Small)	*Elettaria cardamomum* Maton	Zingiberaceae	Fruit, Seed
	Cardamom (Large)	*Amomum subulatum* Roxb.	Zingiberaceae	Fruit, Seed
50.	Ginger	*Zingiber officinale* Rosc.	Zingiberaceae	Rhizome
51.	Turmeric	*Curcuma longa* L.	Zingiberaceae	Rhizome
52.	Greater Galanga	*Alpinia galanga* Willd.	Zingiberaceae	Rhizome

Source: Spices Board of India Website-November 2014

Appendix-II: Item-Wise Export of Spices from India

Item	2011-12		2012-13		2013-14 (Est)	
	Qty	Value	Qty	Value	Qty	Value
Pepper	26,700	87,813.45	15,363	63,810.29	21,250	94,002.34
Cardamom (s)	4,650	36,322.28	2,372	21,215.04	3,600	28,380.88
Cardamom (l)	935	6,830.00	1,217	6,254.59	1,110	7,961.15
Chilli	241,000	214,408.00	301,000	238,060.90	312,500	272,227.20
Ginger	21,550	20,420.02	22,207	18,725.14	23,300	25,614.27
Turmeric	79,500	73,434.40	88,513	55,487.70	77,500	66,675.85
Coriander	28,100	16,401.85	35,902	20,182.59	45,750	37,185.65
Cumin	45,500	64,442.05	85,602	115,306.61	121,500	160,006.00
Celery	3,650	2,340.05	5,171	2,977.26	5,600	3,661.48
Fennel	8,100	7,209.20	13,811	10,466.12	17,300	16,001.42
Fenugreek	21,800	7,275.20	29,622	10,488.12	35,575	13,378.37
Other seeds (1)	13,050	5,881.25	18,442	11,178.60	27,800	15,425.65
Garlic	2,200	1,415.70	22,872	6,868.14	25,650	8,387.05
Tamarind	21,395	12,364.00	17,950	10,753.15	16,000	9,440.00
Nutmeg & mace	3,620	24,097.51	3,231	22,591.87	4,450	26,285.62
Other spices (2)	14,505	19,669.00	16,348	19,455.88	18,700	32,406.80
Curry powder/paste	17,000	25,208.25	17,436	27,515.66	23,750	40,132.03
Mint products (3)	14,750	222,372.00	20,039	394,049.95	24,500	343,042.20
Spice oils and oleoresins	7,265	130,438.28	9,515	155,888.19	11,415	173,324.85
Total	575,270	978,342.48	726,613	1,211,275.80	817,250	1,373,539.26
Value in million US $		**2,037.76**		**2,212.13**		**2,267.67**

(Est): Estimate

1. Include Bishops Weed(Ajwanseed), Dill Seed, Poppy Seed, Aniseed, Mustard Etc
2. Include Asafoetida, Cinnamon, Cassia, Cambodge, Saffron, Spices (Nes) Etc.
3. Include Menthol, Menthol Crystals And Mint Oils.

Source: Spice Board of India

Appendix-III: Major Item/Country-Wise Export of Spices from India

Major Item/Country	2010-11		2011-12		2012-13(P)	
	Qty	Value	Qty	Value	Qty	Value
Pepper						
U.S.A	6920.95	13883.41	9383.35	30979.23	7177.77	29589.56
U.K	1476.97	3273.56	1443.90	4972.93	1309.75	5658.19
Germany	715.36	1333.37	1181.30	3842.25	747.96	3358.90
Japan	592.96	1318.51	690.29	2387.88	727.74	3233.34
Canada	1065.56	2111.25	897.11	2785.94	538.72	2061.95
Australia	592.01	1369.12	991.60	3464.30	468.74	2027.61
Netherlands	403.22	804.15	475.06	1802.73	375.60	1905.13
South Africa	382.54	734.45	666.98	2081.17	389.65	1666.90
Italy	899.94	1807.05	1021.20	3543.95	406.03	1574.36
U.A.E	327.61	650.52	927.19	2810.55	399.60	1513.07
France	369.93	596.87	569.72	1967.46	218.73	1062.57
Belgium	378.79	833.71	445.17	1450.18	233.78	929.63
Iran	126.15	288.16	234.43	839.04	188.83	872.26
Vietnam	779.50	1354.21	2216.31	6943.71	200.59	810.48
Spain	393.84	649.46	436.55	1272.96	251.96	762.53
Saudi Arabia	177.38	385.22	232.04	779.35	127.87	556.07
Singapore	203.74	379.81	344.18	1063.45	138.00	473.63
Malaysia	131.71	265.82	119.82	400.27	110.55	463.19
Philippines	113.80	297.71	149.08	616.79	81.13	440.15
Poland	314.86	666.71	242.25	759.81	188.45	427.34
Denmark	168.02	268.67	79.79	244.26	79.25	340.04
Norway	102.94	275.43	121.54	475.46	69.50	328.55
Sweden	556.05	1173.20	721.08	2244.37	63.97	284.94
Taiwan	101.73	200.82	191.98	536.15	76.23	279.87
Turkey	34.85	72.77	73.50	278.12	76.20	272.58
New Zealand	43.40	103.20	80.09	297.07	55.38	268.12
Israel	108.72	219.35	224.40	720.49	62.20	263.59
Russia	141.90	281.21	550.73	1923.82	38.50	233.26
Greece	37.37	85.02	238.11	812.15	97.16	212.54
Thailand	40.07	101.30	28.00	93.56	50.25	207.18
Bahrain	23.94	57.32	59.18	197.43	44.99	196.26
Kuwait	65.75	133.22	138.27	392.73	32.57	149.56
China	71.00	179.20	96.45	339.67	28.70	134.12
Total (Incl. Others)	18850.08	38318.50	26699.73	87813.42	15362.96	63810.28

Contd.

Major Item/Country	2010-11		2011-12		2012-13(P)	
	Qty	Value	Qty	Value	Qty	Value
Cardamom (Small)						
Saudi Arabia	723.50	8155.96	3097.51	23828.52	1777.36	16082.36
U.A.E	61.61	641.73	555.49	4349.71	148.44	1380.32
Kuwait	101.96	1109.60	116.62	910.67	112.44	1038.67
U.S.A	40.40	627.47	39.14	455.11	54.51	429.12
U.K	18.30	238.14	142.96	1008.34	40.17	342.49
Oman	3.08	46.14	84.75	659.72	34.51	314.07
Australia	15.85	185.24	35.95	383.69	33.94	303.61
Malaysia	20.79	229.06	78.16	664.21	35.59	186.46
Canada	10.96	141.83	16.55	222.85	18.99	163.56
Qatar	3.02	34.88	39.53	287.60	13.23	146.83
Pakistan	32.82	288.34	136.65	962.42	17.90	139.17
South Africa	9.08	105.47	25.34	205.77	14.00	110.63
Japan	26.26	376.52	46.24	525.00	8.22	76.68
Total (Incl. Others)	1175.09	13216.24	4649.71	36322.30	2371.59	21215.00
Cardamom (Large)						
Pakistan	581.18	3231.58	719.52	4859.99	905.11	4489.86
U.S.A	32.97	119.29	30.84	286.93	69.51	425.51
U.K	39.15	247.60	35.12	382.89	52.34	223.89
Oman	1.92	16.02	5.54	43.58	37.82	211.62
U.A.E	46.22	379.88	87.30	726.30	29.27	162.06
Canada	6.88	42.34	10.52	97.05	25.84	146.61
Total (Incl. Others)	775.04	4462.89	934.91	6829.99	1217.22	6254.63
Chilli						
Vietnam	3382.62	2399.94	3751.02	3132.50	43658.64	36201.06
Malaysia	48248.35	35641.96	35447.39	37544.85	38187.62	31579.24
U.S.A	17362.49	13801.24	13528.00	17626.34	25405.88	27728.64
Thailand	2600.96	1850.79	11974.78	10781.95	33591.81	26234.74
Sri Lanka	34072.20	19728.61	44666.20	31955.73	38635.45	22479.86
Mexico	8499.62	7627.51	5300.01	6166.27	12869.30	14472.87
Bangladesh	32741.61	18207.91	17365.98	14176.92	19392.92	13212.92
U.A.E	20702.72	8478.70	21040.89	11708.36	19545.43	9459.26
China	6771.36	4699.28	13633.21	11062.58	9537.81	7824.71
Indonesia	10242.14	6035.16	21297.31	17748.93	11730.85	7527.29

Contd.

Major Item/Country	2010-11		2011-12		2012-13(P)	
	Qty	Value	Qty	Value	Qty	Value
Pakistan	25712.02	13491.59	24137.75	18998.32	8544.15	4867.37
U.K	3612.45	3744.56	2810.34	4346.88	3901.39	4697.69
Saudi Arabia	1725.51	1365.46	3052.58	3153.53	2917.85	2794.46
Singapore	1744.90	1350.55	2353.67	2600.37	2979.78	2481.49
Nepal	2196.99	860.97	290.15	161.29	4832.11	2454.05
South Africa	2469.39	1843.59	2135.62	2401.90	2591.99	2312.29
Canada	1087.10	1023.14	617.21	826.95	1703.12	2006.56
Australia	858.79	887.87	999.68	1261.25	1420.95	1708.74
Italy	792.88	847.28	841.46	1350.35	1235.78	1648.06
Egypt(A.R.E)	1465.10	938.82	525.40	459.69	2149.62	1610.92
Iran	356.98	249.91	642.48	556.21	1605.50	1219.68
Tunisia	293.53	229.38	992.50	1104.42	1154.75	1111.16
Russia	1297.55	538.36	844.68	670.89	1542.04	1028.79
Oman	623.38	520.52	768.07	858.02	933.89	849.32
Kuwait	827.05	558.19	629.13	740.52	888.35	821.77
Netherlands	681.92	677.51	726.17	1008.45	676.70	779.77
Turkey	89.90	56.94	159.00	183.11	866.87	777.33
Qatar	597.80	409.63	680.24	615.24	917.03	764.31
Spain	349.20	318.04	311.30	425.74	558.63	703.69
France	421.25	454.41	520.62	829.58	407.25	623.39
Brazil	633.97	454.11	452.47	571.95	645.86	590.91
Germany	413.18	444.40	195.86	303.34	332.49	522.21
Israel	355.63	238.53	590.68	573.53	827.19	508.37
Philippines	380.47	300.53	261.39	312.63	420.00	447.93
Bahrain	555.22	322.61	478.28	373.15	483.88	350.68
Greece	122.95	101.23	82.20	95.93	297.14	308.00
Maldives	229.21	151.06	108.32	134.44	308.23	247.96
Hong Kong	150.39	153.85	154.31	210.88	277.52	245.95
Mauritius	219.80	167.16	3737.70	3648.75	267.08	231.89
Sweden	136.86	162.67	172.23	304.99	129.28	219.62
Algeria			14.00	8.28	258.60	207.43
Taiwan	205.47	149.53	116.82	114.53	192.02	177.04
New Zealand	92.45	107.52	92.67	130.96	132.84	172.85
Lebanon	81.60	35.03	59.63	41.03	213.27	141.94
Total (Incl. Others)	240000.04	153553.96	240999.81	214408.01	301000.21	238060.90

Contd.

Major Item/Country	2010-11		2011-12		2012-13(P)	
	Qty	Value	Qty	Value	Qty	Value
Ginger						
U.S.A	991.28	1726.24	1854.39	3198.91	2239.70	2611.03
Spain	231.63	419.62	719.53	963.29	1973.10	2143.64
Bangladesh	8275.38	1735.22	6583.10	1215.48	6612.67	1762.78
U.K	532.91	939.75	453.16	923.48	793.22	1187.07
Morocco	750.89	1331.64	1220.50	1578.38	1051.06	1007.31
Germany	67.90	123.59	169.81	357.70	403.46	913.46
Saudi Arabia	741.78	1378.42	1602.17	1946.07	767.85	783.68
Egypt(A.R.E)	134.38	186.36	603.48	709.82	787.99	723.23
Vietnam			410.95	541.58	659.06	616.32
U.A.E	434.16	478.77	2189.39	2695.46	644.56	563.89
Australia	67.49	144.85	164.11	352.54	326.43	479.89
Netherlands	187.36	404.77	300.69	595.21	336.58	454.91
Iran	138.13	218.83	92.13	114.21	608.10	433.11
South Africa	126.28	214.47	235.79	435.78	338.02	416.63
Nepal			165.40	184.58	381.71	406.61
Canada	37.77	65.47	64.64	95.62	413.80	361.38
Y.A.R	350.70	661.26	558.39	614.48	366.35	349.15
Malaysia	85.13	74.45	197.45	166.25	218.84	242.42
Belgium	16.23	27.98	30.44	54.15	140.50	235.10
Total (Incl. Others)	15750.06	12131.23	21550.24	20419.92	22207.52	18725.17
Turmeric						
U.S.A	2664.02	3916.33	3323.54	5216.66	6274.92	5410.16
U.A.E	8988.10	11184.49	16726.31	12781.73	10881.53	5281.14
Iran	2563.43	3724.07	3810.23	3117.28	7638.03	3946.52
Malaysia	4330.36	6700.41	4298.98	3807.32	5928.47	3786.37
Bangladesh	2545.24	3281.59	8566.95	7540.94	4176.96	3021.31
Japan	3066.03	5799.29	3214.13	4450.26	3332.61	2791.27
U.K	2091.29	3060.82	2082.22	2537.37	3437.08	2480.20
Sri Lanka	1913.67	1933.38	2811.35	1336.89	3920.10	2166.08
Saudi Arabia	1334.35	1952.05	2946.83	2468.52	3707.78	2154.67
South Africa	1891.67	2907.78	2133.16	2533.95	2864.66	1866.68
Tunisia	1347.00	2062.13	3408.68	2980.29	3336.47	1732.08
Germany	1695.13	2828.47	1860.49	2187.22	2195.27	1723.06

Contd.

Major Item/Country	2010-11		2011-12		2012-13(P)	
	Qty	Value	Qty	Value	Qty	Value
Netherlands	1509.53	2500.64	1766.92	2340.18	2265.47	1600.46
Spain	719.99	898.34	1534.01	1376.73	2601.81	1528.25
Egypt(A.R.E)	1656.63	1805.24	2583.85	1647.74	3253.69	1488.80
Korea(South)	198.88	350.63	322.47	468.61	1042.01	1010.20
Morocco	1583.23	1981.54	2704.92	2419.89	1833.00	1002.12
France	677.41	1192.84	851.80	1121.05	1029.39	986.34
Jordan	452.87	479.46	157.48	149.32	1676.27	753.38
Vietnam			135.28	115.06	1353.04	741.16
Canada	359.95	524.33	442.84	529.35	802.32	655.75
Australia	296.14	522.75	437.49	624.71	668.19	547.98
Qatar	186.55	257.74	188.29	246.51	1026.81	538.93
China	68.50	123.67	572.23	440.47	562.54	509.74
Kuwait	501.86	680.90	726.34	771.16	1049.80	503.16
Oman	405.62	724.66	582.20	600.96	753.09	479.17
Russia	520.65	752.89	662.95	581.22	677.01	422.73
Israel	674.39	951.69	2527.66	1775.38	923.31	400.81
Singapore	532.10	772.53	432.50	345.23	624.98	380.50
Italy	53.09	112.80	98.52	97.70	256.73	367.18
Brazil	48.80	53.35	343.75	295.28	582.20	357.72
Y.A.R	287.30	343.93	795.51	549.85	712.68	343.42
Bahrain	362.26	602.43	310.33	370.13	581.35	291.05
Turkey	363.25	522.39	493.03	413.35	437.20	289.31
Algeria	137.00	130.11	685.00	475.06	722.54	284.70
Poland	285.00	469.96	315.00	349.76	318.86	271.61
Hong Kong	153.39	218.59	186.93	183.10	388.91	215.75
Belgium	285.54	483.97	175.56	225.69	229.97	202.81
Libya	34.00	50.90	195.15	119.09	386.60	179.32
Taiwan	183.31	245.71	176.10	193.17	277.64	155.80
Syria	169.41	229.60	602.30	451.49	258.00	128.23
Sweden	103.84	198.80	115.95	181.18	152.87	128.06
Pakistan	115.28	152.67	711.73	506.97	236.97	127.89
Iraq	305.12	347.06	327.04	190.49	234.44	114.63
Indonesia	48.95	73.29	93.87	72.11	132.17	114.23
Trinidad	144.50	242.44	164.99	176.08	176.24	112.16
Ukraine	248.00	229.36	73.50	60.20	187.81	104.71
New Zealand	44.18	81.42	58.28	86.61	153.44	100.18
Mauritius	125.07	212.03	478.82	366.97	181.70	100.05
Total (Incl. Others)	49250.03	70285.14	79499.80	73434.36	88513.03	55487.74

Contd.

Major Item/Country	2010-11		2011-12		2012-13(P)	
	Qty	Value	Qty	Value	Qty	Value
Coriander						
Malaysia	8449.81	3450.01	8856.19	5376.46	9252.17	5180.47
U.A.E	5473.13	2141.95	4451.22	2197.51	5858.57	2891.81
Saudi Arabia	2570.88	1105.28	2918.47	1653.06	3041.72	1771.92
South Africa	2391.10	1148.26	1536.77	910.36	2126.20	1381.47
U.K	2635.12	1427.28	1907.43	1249.50	2005.95	1264.21
Nepal	1156.49	357.92	355.16	164.90	2485.98	1109.72
U.S.A	625.59	399.67	556.88	435.98	1085.58	1002.44
Pakistan	3455.18	977.76	1618.75	765.49	1658.92	713.66
Egypt(A.R.E)	1757.24	521.87	490.00	193.60	1346.79	534.25
Singapore	1162.70	450.09	894.72	517.05	822.42	490.42
Y.A.R	595.32	188.00	797.48	338.59	1028.35	489.32
Algeria	194.52	60.19	142.00	59.45	857.03	476.76
Oman	705.35	373.71	715.23	459.72	563.83	407.55
Kuwait	403.92	220.23	412.00	317.79	466.49	342.97
Australia	214.79	154.86	296.59	276.96	344.97	331.28
Qatar	430.57	192.94	235.47	174.35	432.01	264.38
Bahrain	450.26	241.62	343.04	204.23	334.58	220.32
Canada	284.17	139.63	135.27	117.29	171.24	148.96
Sri Lanka	2217.25	650.28	95.58	42.84	335.07	138.55
Total (Incl. Others)	40500.06	16663.26	28100.03	16401.83	35902.10	20182.59
Cumin						
Vietnam	5.85	8.25	6662.40	9460.08	10933.90	15169.64
U.S.A	2214.21	2979.93	4387.25	7108.27	9559.76	14185.70
U.A.E	1887.97	2150.71	6261.43	8277.25	8440.89	10832.52
Egypt(A.R.E)	326.55	349.29	1798.24	2318.19	5493.32	7077.36
Nepal	1897.84	2260.20	230.23	285.85	4332.07	5771.27
Spain	230.21	272.99	1641.48	2324.95	4033.66	5402.25
Brazil	2552.23	2909.48	3765.50	4999.75	3859.12	5016.78
U.K	2457.16	3242.77	2150.88	3432.72	3065.69	4430.16
Malaysia	1537.07	2061.52	2131.88	3257.42	2902.21	4162.88
Saudi Arabia	209.52	247.67	963.10	1224.83	3223.58	4136.87
Pakistan	1661.60	1737.74	519.34	568.48	3232.57	3972.17
Bangladesh	388.99	357.73	1416.00	1727.70	2110.19	2626.94

Contd.

Major Item/Country	2010-11		2011-12		2012-13(P)	
	Qty	Value	Qty	Value	Qty	Value
Mexico	381.53	462.09	1801.80	2733.77	1943.22	2610.82
Turkey	91.00	100.62	569.88	615.81	2008.25	2138.29
Y.A.R	434.31	460.14	732.27	883.39	1754.00	2132.99
Algeria	34.47	28.07	588.90	632.36	1652.02	2001.33
Sri Lanka	270.82	334.81	382.62	505.03	1221.87	1639.73
Iran	173.00	190.05	660.20	768.49	1381.13	1593.01
South Africa	787.68	1032.64	733.36	1122.57	1017.54	1476.45
Japan	488.17	832.03	693.99	1408.81	565.32	1049.85
Ecuador	317.50	385.90	650.21	922.90	745.50	1013.03
Indonesia	445.43	493.29	539.81	664.82	788.60	959.56
Israel	176.95	255.84	320.94	448.83	727.67	952.42
Tunisia			85.00	120.77	727.75	930.52
Canada	420.27	539.63	267.82	471.73	585.72	882.67
Australia	258.43	383.32	281.26	478.45	602.58	850.37
Morocco					592.50	757.21
Philippines	19.37	33.34	46.38	83.24	435.95	698.21
Netherlands	96.67	110.59	119.49	164.08	423.47	649.99
Venezuela	361.50	437.73	359.50	519.44	423.75	617.93
Chile	213.40	273.44	333.00	482.97	440.52	599.86
Oman	50.94	73.41	293.18	439.20	419.36	572.33
Singapore	232.65	304.18	897.67	1285.17	392.12	551.57
Bulgaria	106.50	141.10	161.00	215.86	391.00	528.54
Kuwait	169.76	218.85	182.10	282.29	360.60	497.67
Burma (Myanmar)	151.65	183.27	237.94	340.10	350.69	485.81
France	38.71	46.99	81.95	114.97	304.07	482.58
Mauritius	135.98	190.81	269.20	400.69	293.80	442.57
Russia	60.00	85.47	136.61	198.25	273.33	396.28
China			6.80	3.85	292.57	372.61
Honduras	288.70	310.59	127.50	183.78	211.00	302.29
Thailand	74.20	102.44	135.43	204.12	183.76	268.81
Trinidad	147.66	204.20	196.02	294.88	181.97	267.34
Germany	97.11	114.87	74.59	89.20	194.56	249.25
Bahrain	78.06	134.37	209.41	299.93	162.46	244.01
Argentina	13.00	17.07	79.00	119.12	164.50	239.28
Greece	65.15	85.99	63.38	99.29	157.37	236.38
Kenya	82.68	86.23	88.46	130.50	137.83	217.75
Korea(South)	11.02	22.10	12.38	27.11	127.57	214.98
Columbia	396.00	428.48	125.00	188.16	148.00	209.93
Total (Incl. Others)	32500.08	39597.74	45499.68	64442.04	85601.89	115306.56

Contd.

Major Item/Country	2010-11		2011-12		2012-13(P)	
	Qty	Value	Qty	Value	Qty	Value
Celery						
U.S.A	1783.47	1167.21	2165.26	1292.34	3074.07	1719.82
South Africa	223.25	145.92	265.48	191.33	253.97	163.27
Netherlands	169.72	148.63	193.63	153.44	176.18	134.20
Canada	65.43	46.62	53.57	32.58	171.44	92.23
Australia	61.49	45.72	98.78	69.78	97.74	75.54
U.K	73.82	56.92	71.46	58.42	104.34	72.66
Japan	135.92	97.49	63.30	37.16	118.30	71.51
Germany	60.28	43.68	96.61	55.01	124.16	64.68
China	0.65	0.58	47.18	32.77	134.34	62.98
Egypt (A.R.E)	19.50	9.81	33.00	18.12	100.03	49.41
Total (Incl. Others)	3750.05	2585.92	3650.44	2340.06	5171.06	2977.31
Fennel						
U.S.A	814.26	768.14	1111.80	1039.64	3238.94	2237.62
Malaysia	652.91	595.66	1145.43	1049.28	1760.81	1369.34
Pakistan	1873.24	1410.21	448.00	340.69	1748.44	1216.99
Saudi Arabia	561.24	484.92	819.48	627.92	861.97	574.34
U.K	406.76	402.37	371.79	470.55	475.67	568.24
U.A.E	488.57	392.46	551.61	423.71	793.71	542.43
Sri Lanka	138.58	127.98	473.07	326.06	697.25	428.49
Morocco	101.50	88.40	275.20	318.40	339.28	386.14
Bangladesh	-	-	518.50	280.24	611.00	319.19
Brazil	-	-	72.12	52.24	470.00	313.13
Tunisia	75.03	65.93	274.33	287.23	291.00	296.55
South Africa	131.77	129.99	264.88	277.34	258.93	249.35
Canada	132.58	140.81	126.83	166.45	210.59	247.18
Australia	101.32	109.42	103.30	129.98	152.41	175.63
Indonesia	98.10	65.45	253.35	142.07	292.96	153.77
Netherlands	7.35	10.16	67.21	47.32	173.60	126.32
Algeria	-	-	91.00	71.65	145.49	114.32
Vietnam	-	-	39.00	28.37	160.79	104.14
Singapore	112.12	100.54	159.95	156.54	125.26	101.26
Korea (South)	74.05	85.64	92.03	96.87	94.04	88.74
Total (Incl. Others)	7250.03	6588.25	8099.99	7209.22	13810.92	10466.14

Contd.

Major Item/Country	2010-11		2011-12		2012-13(P)	
	Qty	Value	Qty	Value	Qty	Value
Fenugreek						
Y.A.R	2211.42	617.08	3737.68	999.76	3600.77	1107.93
Egypt(A.R.E)	681.00	194.82	2181.33	575.62	3710.20	1082.92
China	130.20	41.46	990.90	282.52	3373.59	1073.37
U.S.A	477.80	199.57	661.21	321.34	1741.96	822.41
Japan	1354.38	804.77	1126.31	574.95	1001.67	695.68
U.A.E	2047.87	662.01	1830.86	549.49	2028.94	689.89
Sri Lanka	337.83	96.13	473.00	127.89	1830.67	533.60
Nepal	471.14	132.08	159.64	63.57	1179.71	371.03
South Africa	915.82	315.56	740.97	242.34	1032.66	364.23
U.K	656.34	275.65	628.98	347.73	594.12	337.55
Saudi Arabia	597.88	173.63	934.36	267.55	896.61	274.88
Israel	350.40	118.56	710.79	210.00	742.73	234.18
Korea(South)	70.04	27.81	324.23	389.84	191.23	211.12
France	413.49	148.58	447.48	160.04	524.60	202.20
Malaysia	440.54	140.69	429.52	130.74	586.05	194.62
Germany	305.44	98.44	531.15	160.68	561.68	194.35
Vietnam	21.60	7.96	97.56	39.54	557.14	179.66
Morocco	165.00	63.03	214.63	66.34	514.00	165.30
Turkey	-	-	96.00	26.36	505.09	149.96
Bangladesh	657.78	188.39	665.00	199.44	508.00	149.76
Sudan	395.47	130.50	567.05	151.69	447.00	144.24
Iraq	269.87	78.69	624.51	174.89	404.92	133.37
Kuwait	158.57	54.23	80.79	32.21	355.02	123.23
Algeria	48.00	13.06	517.00	143.37	375.00	117.63
Netherlands	340.42	111.33	403.97	120.57	333.73	112.13
Total (Incl. Others)	18500.02	6548.09	21800.41	7275.18	29622.20	10488.07
Other seeds						
Nepal	2541.53	692.90	5905.93	1838.76	10247.04	5858.28
U.S.A	625.28	533.91	522.40	474.29	844.78	855.04
Sri Lanka	1688.00	440.28	1600.12	550.34	1559.04	737.69
U.A.E	1037.55	353.18	822.07	304.44	1141.81	627.59
Saudi Arabia	533.68	390.19	598.96	586.31	699.15	507.60
Malaysia	326.43	194.02	295.96	212.31	502.06	367.69

Contd.

Major Item/Country	2010-11		2011-12		2012-13(P)	
	Qty	Value	Qty	Value	Qty	Value
U.K	568.05	357.56	388.29	311.66	352.98	307.14
Singapore	174.50	72.16	208.55	189.22	257.21	248.93
Vietnam	14.10	9.68	45.10	21.18	484.05	220.10
Canada	135.89	129.69	134.07	155.78	124.92	165.98
Russia	79.74	36.24	87.98	41.64	202.80	113.15
Australia	57.74	40.27	66.01	48.92	142.58	106.39
Total (Incl. Others)	12500.08	5558.07	13049.84	5881.44	18441.70	11178.62

Garlic

Major Item/Country	Qty	Value	Qty	Value	Qty	Value
Indonesia	848.30	330.01	533.40	103.21	7689.44	1598.84
Thailand	4971.48	2107.16	719.40	260.31	3842.60	1039.41
Malaysia	710.42	323.68	72.95	37.74	5207.37	845.49
U.S.A	111.09	140.01	177.04	291.36	360.89	375.61
U.K	131.74	104.47	95.69	75.45	374.26	324.40
U.A.E	38.73	49.46	105.73	136.47	382.40	290.54
Spain			18.00	9.50	330.93	211.31
Russia	13.05	9.46	10.07	6.09	318.80	198.86
Vietnam	136.05	64.21	10.00	3.93	818.40	192.60
Canada	2.52	2.67	13.45	16.71	162.27	139.69
Total (Incl. Others)	17300.04	6977.31	2199.77	1415.76	22872.35	6868.12

Nutmeg & Mace

Major Item/Country	Qty	Value	Qty	Value	Qty	Value
Vietnam	267.00	1178.94	631.19	5032.37	914.86	6510.27
U.A.E	651.57	2913.76	987.65	6175.39	802.58	5651.45
Russia			166.02	1182.96	185.39	1707.50
U.S.A	164.22	657.86	289.10	2149.05	172.42	982.99
Germany	24.47	119.60	85.89	554.59	133.49	840.17
Brazil	25.25	140.21	133.23	960.76	92.25	669.22
Argentina	33.65	122.02	72.18	443.92	89.28	471.21
Pakistan	16.10	54.96	61.67	275.26	68.92	438.64
South Africa	169.58	812.55	99.16	667.33	52.96	389.47
Israel	36.63	160.94	60.73	497.97	48.76	384.15
Australia	37.07	131.21	57.46	490.90	49.80	375.38
Netherlands	0.12	1.53	18.40	224.51	41.03	353.41

Contd.

Major Item/Country	2010-11		2011-12		2012-13(P)	
	Qty	Value	Qty	Value	Qty	Value
Singapore	135.39	671.57	235.73	1739.67	47.15	350.10
Ghana	0.44	2.28	48.60	326.77	47.41	325.51
Nigeria	126.90	1065.64			91.81	269.23
Canada	13.81	60.97	16.44	133.53	33.44	236.87
Y.A.R	12.60	28.35	2.03	12.98	29.30	193.73
Dominican Republic	48.00	205.07	52.00	377.32	33.00	192.96
Egypt(A.R.E)	0.04	0.78	64.52	535.85	18.24	151.50
Spain	25.63	117.32	32.88	210.69	18.53	151.33
Iran	10.00	49.01	11.50	80.60	18.30	149.09
Norway	9.58	81.54	9.40	153.69	8.75	148.52
U.K	1.24	5.41	0.05	0.72	8.72	118.99
Qatar	2.48	8.26	34.84	66.48	19.20	118.15
Mexico	6.03	23.49	6.62	56.75	14.70	105.83
Guatemala	13.93	75.02	5.80	42.07	16.30	104.94
Total (Incl. Others)	2100.05	9776.81	3620.41	24097.51	3230.88	22591.88
Other Spices						
U.S.A	877.81	1265.20	1444.79	2517.94	1939.96	3403.36
U.K	876.53	876.79	504.67	804.68	1773.44	2069.06
Nepal	0.19	0.69	0.05	0.08	1943.24	2054.45
U.A.E	1105.73	1201.77	1115.79	1166.95	927.97	1469.33
Russia	1316.61	211.36	1620.81	472.84	1993.06	897.74
Vietnam	424.00	659.85	869.33	894.64	797.07	780.97
South Africa	388.91	378.91	357.80	489.74	383.50	606.44
Australia	216.99	315.28	361.68	605.43	249.64	574.21
Germany	237.16	135.09	215.29	141.56	456.53	567.36
Canada	524.23	498.19	637.77	788.23	429.74	546.22
Spain	444.93	96.16	399.39	145.82	778.44	439.02
France	278.46	154.53	124.98	305.16	246.15	432.62
Saudi Arabia	284.49	282.19	394.08	386.88	372.35	425.68
Pakistan	537.56	372.02	453.73	374.01	402.06	312.56
Ukraine	519.50	86.64	677.61	164.83	741.79	312.49
Singapore	128.85	304.03	957.95	6276.42	126.36	305.50
Malaysia	71.91	128.62	188.92	396.66	95.65	247.54
Netherlands	543.72	191.41	522.44	245.76	388.02	230.90
Italy	196.93	118.95	53.50	18.58	176.47	222.63
Nigeria	243.65	398.47	80.35	189.96	71.64	213.18
Total (Incl. Others)	7575.06	6140.82	14474.05	19382.35	16292.30	18766.75

Contd.

Major Item/Country	2010-11		2011-12		2012-13(P)	
	Qty	Value	Qty	Value	Qty	Value
Curry Powder/Mixture						
Saudi Arabia	1510.88	2348.90	2525.39	4110.84	2271.64	3899.07
U.A.E	1241.39	2255.81	1510.34	2726.09	1860.20	3299.19
U.S.A	1601.49	2288.45	1342.89	2410.85	1711.20	3223.62
U.K	2282.57	2328.57	2214.39	2954.06	2434.85	3131.92
Nigeria	1954.87	2935.54	1578.61	2307.78	1822.36	2434.56
Hong Kong	424.37	488.54	522.87	629.38	940.07	1468.17
Kuwait	438.88	662.24	676.68	1140.38	891.36	1465.26
Qatar	448.51	830.52	636.76	1247.36	555.59	1120.16
Australia	552.79	691.66	447.20	726.26	495.94	975.01
Oman	238.44	473.78	337.06	634.34	409.19	832.85
Canada	264.97	420.89	245.87	482.24	265.35	517.76
Nepal	0.91	1.56			402.41	489.47
Jordan	157.58	193.05	59.34	57.85	332.35	430.95
China	49.65	91.68	93.75	120.78	222.00	379.36
Bahrain	180.63	301.75	219.61	335.78	190.30	347.12
Japan	197.12	277.89	133.96	234.17	104.05	342.31
Singapore	161.36	259.64	126.15	214.21	154.15	293.45
South Africa	333.03	331.18	188.46	187.48	229.62	251.83
France	253.24	223.32	170.94	179.71	206.43	202.53
Total (Incl. Others)	15250.03	21050.50	17000.31	25208.29	17436.46	27515.65
Spice Oils & Oleoresins						
U.S.A	1776.21	22420.68	1679.02	31941.09	1988.83	36816.42
China	337.63	3991.31	363.73	4191.25	1946.41	21036.31
Germany	670.18	9098.01	765.07	14410.35	811.10	15602.03
U.K	555.24	6425.09	451.46	8784.18	443.20	8086.88
Korea(South)	366.94	3668.80	309.02	4879.77	343.46	4874.94
Japan	223.58	5398.35	128.47	6121.64	120.57	4851.04
Indonesia	201.29	1625.93	190.34	1909.70	467.73	4426.74
South Africa	206.14	2686.11	254.80	4160.31	247.10	4106.69
France	142.85	2624.05	166.73	3063.20	168.88	3527.71
Spain	287.84	1883.23	251.70	2515.19	202.00	3451.53
Netherlands	119.40	2116.41	129.98	2883.90	110.80	3048.43
Thailand	288.81	2001.79	312.88	3055.56	205.38	2963.11

Contd.

Major Item/Country	2010-11		2011-12		2012-13(P)	
	Qty	Value	Qty	Value	Qty	Value
Russia	129.23	2440.07	96.93	2239.22	96.87	2722.23
Brazil	88.03	1364.25	132.15	2847.73	141.18	2717.82
Mexico	151.22	1491.25	138.82	1926.26	177.76	2662.78
Canada	158.60	1760.01	140.41	2143.56	160.50	2459.66
Poland	103.24	1197.34	151.63	2571.95	112.22	2183.49
Australia	133.87	1391.50	167.86	2165.23	151.36	2173.95
Hong Kong	141.99	1585.70	103.35	1847.42	146.30	2129.82
Egypt(A.R.E)	75.10	1042.05	87.25	1372.79	135.25	2056.37
Philippines	101.77	1048.52	115.92	1577.63	114.94	1994.24
Denmark	66.24	1709.60	78.66	3162.90	68.37	1960.23
Cayman Island	-	-	71.74	1069.25	91.83	1745.72
Austria	54.54	456.39	15.68	210.84	67.57	1194.09
Singapore	75.84	762.03	92.43	1295.00	85.78	1182.59
Estonia	31.47	302.83	53.31	912.67	57.15	1180.34
Turkey	88.29	1036.43	126.01	1635.39	88.75	1147.11
Switzerland	20.72	608.09	29.63	725.67	33.01	1091.55
Italy	33.10	420.56	23.92	565.71	30.72	983.85
Israel	53.49	691.28	55.40	1034.35	48.06	857.25
Belgium	95.77	934.93	63.32	658.14	69.50	683.77
Taiwan	47.15	347.67	71.25	596.06	69.80	672.08
Sweden	31.93	348.13	35.57	598.87	27.34	596.36
Total (Incl. Others)	7599.99	91062.41	7265.00	130438.10	9515.44	155888.22
Mint Products						
China	5473.36	54559.27	4989.88	70176.82	9525.52	170298.22
U.S.A	3445.80	31444.03	3198.55	52416.19	3695.00	78918.72
Singapore	1414.25	13695.10	1418.37	19498.29	1446.23	29235.25
Netherlands	501.26	5265.60	951.89	16091.22	795.93	19934.07
Germany	1249.38	12078.24	910.10	14210.66	795.23	15209.10
U.K	606.09	6167.34	462.77	7109.35	531.67	10955.22
France	247.81	2218.05	153.48	2321.89	459.30	9707.82
Brazil	579.02	5553.25	398.50	5880.43	355.37	8409.20
Hong Kong	360.68	3918.98	187.50	3147.29	185.53	4137.37
Japan	482.00	5421.53	469.88	7107.59	176.22	3966.98
Philippines	99.60	1237.93	131.40	2156.41	169.24	3561.47
Nigeria	94.23	1149.26	52.45	741.34	181.39	3541.05

Contd.

Major Item/Country	2010-11		2011-12		2012-13(P)	
	Qty	Value	Qty	Value	Qty	Value
Indonesia	78.68	852.15	16.37	240.40	154.21	3089.94
Russia	107.84	1056.20	136.80	2248.82	126.71	2875.91
Switzerland	1.26	9.87	8.00	106.50	98.54	2487.34
Spain	133.42	1359.75	105.94	1695.03	133.90	2407.32
Belgium	171.68	1639.22	196.30	2792.05	111.48	2308.34
Mexico	89.18	982.52	62.08	711.02	102.63	2048.26
Paraguay	206.52	1798.66	266.96	3710.10	82.15	1598.31
Turkey	40.00	405.52	72.79	1165.81	58.69	1551.90
Vietnam	126.16	1390.58	39.90	605.13	71.15	1424.77
Poland	170.50	1317.89	0.90	19.14	63.27	1248.63
Ukraine	18.40	233.32	43.40	739.57	50.74	1242.09
Columbia	19.38	215.21	14.94	213.26	40.00	1018.29
Korea(South)	30.56	395.35	35.03	624.95	49.26	911.64
Burma(Myanmar)	1.05	15.82	1.01	13.78	40.18	842.16
Taiwan	51.75	604.56	64.65	1059.65	45.99	814.41
Thailand	75.20	701.44	31.77	432.90	34.94	809.39
Canada	14.64	143.35	23.42	405.30	28.16	678.82
Australia	18.27	183.75	15.68	231.72	36.50	663.29
Total (Incl. Others)	17450.06	169678.99	14749.91	222372.03	20039.24	394049.54
Grand Total	525750	684070.70	575270	978342.48	726613.13	1211275.80
(P) Provisional						

Source: Spices Board of India Website-November 2014

Appendix -IV: List of 178 Medicinal Plant Species in High Volume Trade

S. No	Name of the Species	Trade Name	API names	Major Supply Source	Estimated Annual Trade (MT)	Price Range (Rs. per Kg)
1.	*Abelmoschus moschatus* Medik. [=*Hibiscus abelmoschus* L.]	Mushakdana	Kasturilatika	C	200 - 500	60 - 100
2.	*Abies spectabilis* (D.Don) Spach	Talispatra	Talisa	HF	500 - 1000	30 - 50
3.	*Abrus precatorius* L.	Gunja	Gunja	W	200 - 500	10 - 15
4.	*Acacia catechu* (L.f.) Willd.	Katha		TF	200 - 500	10 - 15
5.	*Acacia nilotica* (L.) Willd. *ex* Del.	Babul	Babbula	TF	200 - 500	150 - 250
6.	*Acacia sinuata* (Lour.) Merr. [=*A. concinna* (Willd.) DC.]	Shikakai		TF	200 - 500	20 - 30
7.	*Achyranthes aspera* L.	Apamarga	Apamarga	W	200 - 500	10 - 15
8.	*Aconitum ferox* Wall. *ex* Seringe[1]	Vachnag		HF	100 - 200	150 - 250
9.	*Aconitum heterophyllum* Wall. *ex* Royle[2]	Atis	Ativisa	HF	200 - 500	2000 - 4000
10.	*Acorus calamus* L.	Vach	Vaca	C	500 - 1000	30 - 35
11.	*Adhatoda zeylanica* Medic. [=*A. vasica* Nees; *Justicia adhatoda* L.]	Adusa	Vasa	C	2000 - 5000	10 - 15
12.	*Aegle marmelos* (L.) Corr. *ex* Schultz	Bael	Bilva	TF	2000 - 5000	10 - 25
13.	*Aerva lanata* (L.) Juss. *ex* Schult.	Cheroola	Pattura	W	100 - 200	10 - 15
14.	*Albizia amara* (Roxb.) Boivin	Krishna shirish		TF	200 - 500	5 - 10
15.	*Aloe barbadensis* Mill.[=*A. vera* L.]	Kumari	Kanyasara	C	1000 - 2000	130 - 180
16.	*Alpinia calcarata* Roscoe	Chittartha		C	100 - 200	40 - 50
17.	*Alstonia scholaris* R.Br.	Saptaparni	Saptaparna	TF	100 - 200	15 - 25
18.	*Andrographis paniculata* (Burm.f.) Wall. *ex* Nees	Kalmegh		W	2000 - 5000	17 - 30
19	*Anogeissus latifolia* Wall. *ex* Guill. & Perr.	Dhawada		TF	100 - 200	50 - 100

Contd.

S. No	Name of the Species	Trade Name	API names	Major Supply Source	Estimated Annual Trade (MT)	Price Range (Rs. per Kg)
20.	*Aquilaria agallocha* Roxb.	Agar kala	Agaru	I	100 - 200	15 - 25
21.	*Asparagus racemosus* Willd.	Shatavari	Satavari	TF	2000 - 5000	40 - 70
22.	*Azadirachta indica* A.Juss.	Neem	Nimba	C	2000 - 5000	10 - 15
23.	*Bacopa monnieri* (L.) Pennell	Brahmi	Brahmi	W	2000 - 5000	30 - 35
24.	*Baliospermum montanum* (Willd.)Muell.-Arg. [=*B. solanifolium* (J.Burm.) Suresh]	Dantimool	Danti	TF	100 - 200	20 - 30
25.	*Berberis aristata* DC.[3]	Daruhaldi	Daruharidra	HF	500 - 1000	15 - 35
26.	*Bergenia ciliata* (How.) Stern.	Pashanabheda	Pasana-bhedha	HF	200 - 500	15 - 20
27.	*Boerhavia diffusa* L.	Punarnava	Punarnava rakta, Rakta-punarnava	W	2000 - 50000	30 - 40
28.	*Bombax ceiba* L.[=*B. malabaricum* DC.]	Mochrus	Salmali	TF	100 - 200	50 - 60
29	*Boswellia serrata* Roxb.	Salai guggul	Kunduru	TF	500 - 1000	40 - 55
30.	*Buchanania lanzan* Spr.	Chironji	Priyala	TF	5000 - 10000	5 - 10
31.	*Butea monosperma* (Lam.) Taub. [=*B. frondosa* Koen.]	Tesu phool	Palasa, Palasah	TF	200 - 500	15 - 20
32.	*Caesalpinia sappan* L.	Pathimugam	Pattanga	C	200 - 500	125 - 175
33.	*Cardiospermum halicacabum* L.	Mudakkathan	Karnasphota	W	200 - 500	10 - 20
34.	*Careya arborea* Roxb.	Vaari kumbha	Kumbhikah	TF	100 - 200	10 - 20
35.	*Cassia absus* L.	Chaksoo		W	100 - 200	40 - 60
36.	*Cassia angustifolia* Vahl[=*C. senna* L.; *Senna alexandrina* Gars. ex Mill.]	Sonamukhi	Svarnapatri	C	2000 - 5000	33 - 50
37.	*Cassia fistula* L.	Amaltas	Aragvadha	TF	200 - 500	30 - 40
38.	*Cassia tora* L. [=*Senna tora* (L.) Roxb.]	Chakoda beeja	Prapunnada	W	5000 - 10000	5 - 10

Contd.

S. No	Name of the Species	Trade Name	API names	Major Supply Source	Estimated Annual Trade (MT)	Price Range (Rs. per Kg)
39.	*Catharanthus roseus* (L.) G.Don[=*Vinca rosea* L.]	Sadabahar		C	200 - 500	40 - 55
40.	*Cedrus deodara* (Roxb.) Loud.	Devdar	Devadaru	HF	500 - 1000	25 - 35
41.	*Celastrus paniculatus* Willd.	Malkangani	Jyotismati	TF	200 - 500	48 -55
42.	*Centella asiatica* (L.) Urban	Brahmi booti	Mandukaparni	W	500 - 1000	30 - 35
43.	*Centratherum anthelminticum* (L.) O.Kuntze [=*Vernonia anthelmintica* (L.) Willd.]	Kali zeeri	Vanyajiraka	W	500 - 1000	70 - 75
44.	*Chlorophytum tuberosum* Baker[4]	Safed musli		TF	100 - 200	325 - 425
45.	*Cichorium intybus* L.	Kasani		C	500 - 1000	20 - 25
46.	*Cinnamomum sulphuratum* Nees[5]	Dalchini		TF	100 - 200	30 - 35
47.	*Cinnamomum tamala* Nees & Eberm.[6]	Tejpatta	Tvakapatra	HF	500 - 1000	15 - 35
48.	*Citrullus colocynthis* (L.) Schrader	Indrayan	Indravaruni	W	2000 - 5000	15 - 20
49.	*Clerodendrum phlomides* L.f.	Arnimool	Agnimantha	TF	200 - 500	15 - 20
50.	*Commiphora wightii* (Arn.) Bhandari [= *C. mukul* (Hook. ex Stocks) Engl.]	Guggul	Guggulu	I	500 - 1000	140 - 160
51.	*Convolvulus microphyllus* Sieb. *ex* Spreng.[7] [=*C. prostratus* Forssk.; *C. pluricaulis* Chois.]	Shankhapushpi	Sankapuspi	W	1000 - 2000	15 - 20
52.	*Coscinium fenestratum* (Gaertn.) Coleb.	Maramanjal	Kaliyaka	TF	100 - 200	30 - 35
53.	*Croton tiglium* L.	Jamalghota	Jayapala	C	100 - 200	90 - 100
54.	*Curculigo orchioides* Gaertn.	Kali musli	Talamuli	W	200 - 500	35 - 45
55.	*Curcuma angustifolia* Roxb.	Thikhur		C	200 - 500	90 - 100
56.	*Curcuma zerumbet* Roxb. [=*C. zedoaria* (Christm.) Rosc.]	Kachur	Karcura	C	100 - 200	15 - 25
57.	*Cyclea peltata* (Lam.) Hook.f. & Thoms.	Paadu kizhangu		TF	100 - 200	120 - 135
58.	*Cynodon dactylon* (L.) Pers.	Durva	Durva	W	100 - 200	10 - 25

Contd.

S. No	Name of the Species	Trade Name	API names	Major Supply Source	Estimated Annual Trade (MT)	Price Range (Rs. per Kg)
59.	*Cyperus esculentus* L.	Musta		W	1000 - 2000	10 - 15
60.	*Cyperus rotundus* L.	Nagarmotha	Musta	W	2000 - 5000	15 - 30
61.	*Datura metel* L.	Duttura	Dhattura	W	200 - 500	10 - 15
62.	*Decalepis hamiltonii* Wight & Arn.	Magali		TF	100 - 200	40 - 45
63.	*Desmodium gangeticum* (L.) DC.	Salparni	Salaparni	TF	1000 - 2000	5 - 10
64.	*Eclipta prostrata* (L.) L.	Bhringraj	Bhrngaraja	W	2000 - 5000	15 - 20
65.	*Embelia tsjeriam-cottam* A.DC.[8][=*E. basaal* sensu (Roem. & Schult.) A.DC., non Mez	Vaividang		TF	1000 - 2000	40 - 45
66.	*Emblica officinalis* Gaertn. [=*Phyllanthus emblica* L.]	Aonla	Amalaki	TF	16000	30 - 35
67.	*Ephedra gerardiana* Wall.	Somlata		HF	200 - 500	25 - 35
68.	*Ficus benghalensis* L. Nayagrodha	Vada chhal	Nayagrodha jata,	C	200 - 500	10 - 15
69.	*Ficus religiosa* L.	Lakh pippal	Asvattha	C	200 - 500	5 - 10
70.	*Fumaria indica* Pugsley	Shahtara	Parpata	W	100 - 200	10 - 15
71.	*Garcinia indica* (Dup.) Choisy	Kokam		TF	1000 - 2000	20 - 25
72.	*Gardenia resinifera* Roth	Dikamali		TF	100 - 200	20 - 25
73.	*Gloriosa superba* L.	Kalihari	Langali	C	100 - 200	600 - 750
74.	*Glycyrrhiza glabra* L.	Mulethi	Yasti	I	1000 - 2000	35 - 50
75.	*Gelina arborea* L.	Gambar chhal	Gambhari	TF	1000 - 2000	10 - 20
76.	*Gymnema sylvestre* R.Br.	Gudmar	Mesarngi	TF	500 - 1000	40 - 50
77.	*Hedyotis corymbosa* (L.) Lam. [=*Oldenlandia corymbosa* L.]	Pitpapra		W	200 - 500	10 - 20
78.	*Helicteres isora* L.	Marodphali		TF	200 - 500	20 - 25

Contd.

S. No	Name of the Species	Trade Name	API names	Major Supply Source	Estimated Annual Trade (MT)	Price Range (Rs. per Kg)
79.	*Hemidesmus indicus* (L.) Schult.	Anatmool	Sveta sariva	W	500 - 1000	60 - 65
80.	*Holarrhena pubescens* (Buch.-Ham.) Wall. *ex* G.Don[=*H. antidysenterica* (Roxb. ex Fleming) Wall.]	Kutja	Indrayava, Kutaja	TF	1000 - 2000	75 - 100
81.	*Holoptelea integrifolia* (Roxb.) Planch.	Aavithali	Cirabilva	TF	100 - 200	10 - 15
82.	*Holostemma ada-kodien* Schult. [=*H. annulare* (Roxb.) K.Schum.; *H. rheedii* Wall.]	Jeevanti		TF	100 - 200	250 - 270
83.	*Hygrophila schulli* (Buch.-Ham.) M.R. & S.M.Almeida[=*H. auriculata* (Schum.) Heine; *H. spinosa* T.And.]	Tal makhana	Kokilaksa	W	200 - 500	10 - 20
84.	*Indigofera tinctoria* L.	Akika	Nili	C	100 - 200	30 - 35
85.	*Inula racemosa* Hook.f.	Pushkarmool	Puskara	C	200 - 500	150 - 160
86.	*Ipomoea mauritiana* Jacq.	Palmudhakkan	Kshiravidari	TF	500 - 1000	10 - 15
87.	*Ipomoea nil* (L.) Roth	Kaladana		W	100 - 200	10 - 15
88.	*Ixora coccinea* L.	Thechippoovu		TF	100 - 200	10 - 15
89.	*Jatropha curcas* L.	Nepalam seed		C	200 - 500	15 - 30
90.	*Juniperus communis* L.	Hauber	Hapusa	HF	500 - 1000	35 - 45
91.	*Jurinea macrocephala* (Royle) C.B.Clarke [=*J. dolomiaea* Boiss.]	Dhoop		HF	1000 - 2000	60 - 150
92.	*Kaempferia galanga* L.	Kachora		C	100 - 200	80 - 90
93.	*Lannea coromandelica* (Houtt.) Merr.	Jingini		TF	100 - 200	10 - 20
94.	*Lawsonia inermis* L.	Mehndi	Madaynati	C	200 - 500	30 - 35
95.	*Lepidium sativum* L.	Kurassani	Candrasura	C	100 - 200	30 - 50
96.	*Litsea glutinosa* (Lour.) Robinson	Maida chhal	Medasakah	TF	100 - 200	20 - 35

Contd.

S. No	Name of the Species	Trade Name	API names	Major Supply Source	Estimated Annual Trade (MT)	Price Range (Rs. per Kg)
97.	*Lobelia nicotianaefolia* Roth *ex* Roem. & Schult.	Lobelia leaves		TF	100 - 200	80 - 100
98.	*Madhuca indica* Gel.[=*Bassia latifolia* Roxb.]	Madhuka	Madhuka	TF	>10000	10 - 20
99.	*Merremia tridentata* (L.) Hall.	Prasarani		W	100 - 200	45 - 55
100.	*Mesua ferrea* L.[=*M. nagassarium* (Burm.f.) Kosterm.]	Nagekesar	Nagakesar	TF	200 - 500	50 - 60
101.	*Mimusops elengi* L.	Bakul		TF	100 - 200	30 - 50
102.	*Morinda pubescens* J.E.Sm.	Manjanathi		TF	100 - 200	15 - 20
103.	*Mucuna pruriens* (L.) DC.	Kaunch beej	Atmagupta	TF	1000 - 2000	30 - 35
104.	*Nardostachys grandiflora* DC.	Jatamansi	Jatamansi	HF	200 - 500	110 - 150
105.	*Nilgirianthus ciliatus* (Nees) Bremek. [=*Strobilanthes ciliatus* Nees]	Kurinji		TF	200 - 500	15 - 20
106.	*Ocimum americanum* L.	Ban tulasi		W	500 - 1000	20 - 25
107.	*Ocimum basilicum* L.	Kali tulasi		C	1000 - 2000	20 - 25
108.	*Ocimum tenuiflorum* L. [=*O. sanctum* L.]	Tulasi	Tulasi	C	2000 - 5000	20 - 25
109.	*Onosma hispidum* Wall. *ex* G.Don	Ratanjot		HF	500 - 1000	50 - 60
110.	*Operculina turpethum* (L.) S.Manso [=*Merremia turpethum* (L.) Shah & Bhat]	Nishoth	Trivrta	TF	100 - 200	40 - 45
111.	*Oroxylum indicum* (L.) Vent.	Tetu chhal	Syonaka	TF	1000 - 2000	20 - 30
112.	*Parmelia perlata* (Huds.) Ach.	Chadila	Saileya	HF	1000 - 2000	80 - 90
113.	*Peganum harmala* L.	Harmal		W	200 - 500	40 - 45
114.	*Phyllanthus amarus* Schum. & Thenn.[9]	BhumiAonla	Tamalaki	W	2000 - 5000	20 - 25
115.	*Picrorhiza kurroa* Royle *ex* Benth.[10]	Kutaki	Katuka	HF	200 - 500	220 - 230
116.	*Piper chaba* Hunter	Kabab chini	Cavya	I	500 - 1000	30 - 35
117.	*Piper longum* L.	Pippali	Pippalimula, Pippali	C	1000 - 2000	90 - 100

Contd.

S. No	Name of the Species	Trade Name	API names	Major Supply Source	Estimated Annual Trade (MT)	Price Range (Rs. per Kg)
118.	*Pistacia integerrima*	Kakar singi		HF	150 - 200	90 - 110
119.	*Plantago ovata* Forssk.	Isabgol		C	1000 - 2000	70 - 80
120.	*Plectranthus barbatus* Andr. [=*Coleus forskohlii* (Willd.) Briq.]	Gandhira	Gandira	C	1000 - 2000	65 - 75
121	*Pluchea lanceolata* Oliver & Hiern.	Rasna	Rasna	W	100 - 200	10 - 15
122	*Plumbago zeylanica* L.	Chitrak	Citraka	W	2000 - 5000	20 - 25
123	*Pongamia pinnata* (L.) Pierre [=*P. glabra* Vent.; *Derris indica* (Lam.) Bennett]	Karanji	Karanja	C	2000 - 5000	20 - 25
124	*Premna integrifolia* L.	Arnimool		TF	500 - 1000	10 - 15
125	*Prunus armeniaca* L.	Chuli		C	200 - 500	15 - 30
126	*Pseudarthria viscida* Wight & Arn.	Moorva		W	200 - 500	30 - 40
127	*Psoralea corylifolia* L.	Bawachi	Bakuchi	W	200 - 500	10 - 20
128	*Pterocarpus marsupium* Roxb.	Damulakhwain	Asana	TF	500 - 1000	15 - 25
129	*Pterocarpus santalinus* L.f.	Rakatachandan	Raktacandana	TF	200 - 500	65 - 80
130	*Quercus infectoria* Oliv.	Majuphal	Mayakku	I	500 - 1000	70 - 80
131	*Rauvolfia serpentina* (L.) Benth. *ex* Kurz	Sarpagandha	Sarpagandha	TF	200 - 500	110 - 125
132	*Rheum australe* D.Don[11] [=*R. emodi* Wall. ex Meissn.]	Revan chini		HF	500 - 1000	25 - 30
133	*Rhododendron anthopogon* D.Don	Talispatra		HF	100 - 200	15 - 30
134	*Rubia cordifolia* L.	Manjistha	Manjistha	TF	500 - 1000	60 - 70
135	*Santalum album* L.	Chandan	Svetacandana	TF	200 - 500	700 - 850
136	*Sapindus mukorossi* Gaertn.[12]	Reetha		TF	200 - 500	30 - 40
137	*Saraca asoca* (Roxb.) de Wilde	Ashoka chhal	Ashoka	TF	2000 - 5000	20 - 25
138	*Saussurea costus* (Falc.) Lipsch. [=*S. lappa* C.B.Clarke]	Kuth	Kustha	C	100 - 200	200 -225

Contd.

S. No	Name of the Species	Trade Name	API names	Major Supply Source	Estimated Annual Trade (MT)	Price Range (Rs. per Kg)
139	*Schrebera swietenioides* Roxb.	Ghanti phool		TF	100 - 200	5 - 10
140	*Semecarpus anacardium* L.f.	Balave	Bhallataka	TF	1000 - 2000	5 - 10
141	*Shorea robusta* Gaertn.f.	Raal		TF	100 - 200	40 - 60
142	*Sida rhombifolia* L.[13]	Bala	Mahabala	W	5000 - 10000	5 - 10
143	*Silybum marianum* (L.) Gaertn.	Milk Thistle		C	500	
144	*Simmondsia chinensis* (Link) C.K. Schneid.	Jojoba		C	900	
145	*Sisymbrium irio* L.	Khubkalan	Khubkalan	W	200 - 500	70 - 80
146	*Smilax glabra* Roxb.	Chopchini		TF	100 - 200	150 - 200
147	*Solanum anguivi* Lam.[=*S. indicum* auct. non L.; *S. violaceum* Ortega]	Katheli badi	Brhati	W	1000 - 2000	5 - 10
148	*Solanum nigrum* L.	Makoi	Kakamaci	W	2000 - 5000	40 - 45
149	*Solanum virginianum* L.[=*S. surattense* Burm. f.; *S. xanthocarpum* Schr. & Wendl.]	Kateli	Kantakari	W	1000 - 2000	10 - 15
150	*Soymida febrifuga* (Roxb.) A.Juss.	Rohan		TF	100 - 200	10 - 15
151	*Sphaeranthus indicus* L.	Gorakh mundi	Munditika	W	200 - 500	15 - 20
152	*Sterculia urens* Roxb.	Karaya		TF	500 - 1000	80 - 100
153	*Stereospermum chelonoides* (L.f.) DC.[14] [=*S. suaveolens* (Roxb.) DC.]	Patala	Patalai	TF	1000 - 2000	20 - 30
154	*Strychnos nux-vomica* L.	Kuchla	Visamusti	TF	2000 - 5000	40 - 50
155	*Strychnos potatorum* L.	Nirmali	Kataka	TF	200 - 500	40 - 55
156	*Swertia chirayita* (Roxb. *ex* Fleming) Karst.[15]	Chiraiyata	Kiratatikta	HF	500 - 1000	200 - 225
157	*Symplocos racemosa* Roxb.[16]	Pathani Lodh	Lodhra	TF	500 - 1000	20 - 25
158	*Taxus wallichiana* Zucc.	Talispatra	Sthauneya	HF	100-200	75 - 90
159	*Tephrosia purpurea* Pers.	Sarpankha		W	500 - 1000	5 - 10

Contd.

S. No	Name of the Species	Trade Name	API names	Major Supply Source	Estimated Annual Trade (MT)	Price Range (Rs. per Kg)
160	*Terminalia arjuna* (Roxb.) Wight & Arn.	Arjun	Arjuna	TF	2000 - 5000	10 - 15
161	*Terminalia bellirica* Roxb.	Behra	Bibhitaka	TF	2000 - 5000	10 - 15
162	*Terminalia chebula* Retz. & Willd.	Harda	Haritaki	TF	5000 - 10000	10 - 15
163	*Tinospora cordifolia* (Willd.) Hook.f. & Thoms.[17]	Giloy	Guduci	W	2000 - 5000	10 - 15
164	*Trachyspermum ammi* (L.) Sprague	Ajwain	Yavani	C	1000 - 2000	45 - 55
165	*Tragia involucrata* L.	Barhanta	Vrscikalli	W	100 - 200	10 - 20
166	*Tribulus terrestris* L.[18]	Gokhru	Goksura	W	2000 - 5000	10 - 20
167	*Trichosanthes cucumerina* L.	Patol panchang		W	500 - 1000	15 - 20
168	*Valeriana jatamansi*.[19]	Musakbala	Tagara	HF	100 - 200	95 - 100
169	*Vateria indica* L.	Manda dhupa	Saraja	TF	500 - 1000	40 - 60
170	*Vetiveria zizanioides* (L.) Nash	Lavancha	Usira	W	200 - 500	30 - 35
171	*Viola pilosa* Bi.[20]	Banafsha		HF	200 - 500	300 - 350
172	*Vitex negundo* L.	Neergundi	Nirgundi, Renuka	C	200 - 500	10 - 15
173	*Withania coagulens* Dunal	Paneerdodi		W	100-200	20 - 35
174	*Withania somnifera* Dunal	Ashwagandha	Asvagandha	C	2000 - 5000	60 - 70
175	*Woodfordia fruticosa* (L.) Kurz	Dhai phool	Dhataki	W	2000 - 5000	10 - 15
176	*Wrightia tinctoria* R.Br.	Inderjau		TF	200 - 500	100 - 120
177	*Ziziphus jujuba* Lam.[=*Z. mauritiana* Lam.]	Ber	Kola	C	200 - 500	60 - 70
178	*Ziziphus xylopyrus* (Retz.) Willd.	Ghonta phala	Ghonta	TF	100 - 200	5 - 10

TF = Tropical Forest; HF = Himalayan Forest; W = Wastelands; C = Largely Cultivated; I = Imported

Notes

1. Includes *Aconitum chasmanthum*
2. Includes *Aconitum kashmericum, Delphinium denudatum, Chaerophyllum villosum*, the probable 'patis' adulterants.
3. Includes other species of *Berberis* viz. *B. lycium, B. asiatica, B. chitria*, being traded as 'daruhaldi'.
4. Includes *Chlorophytum borivilianum* and *C.arundinaceum*
5. Leaves of other species of *Cinnamomum* viz. *C. zeylanica*, *C. malabathrum* are also traded as 'tejpatta'.
6. Bark of other species of *Cinnamomum* viz. *C. cassia*, *C. zeylanica*, *C. malabathrum* is also traded as 'dalchini'
7. Includes *Evolvulus alsinoides, Clitoria ternatea, Canscora decussata* traded as 'shankhapushpi'
8. Includes the fruits of *Embelia ribes*, the most accepted candidate for 'Vaividang'.
9. Includes other herbaceous species of *Phyllanthus* viz. *P. urinaria, P.reticulatus, P.virgatus, P.debilis* & *P. madraspatensis* .
10. Also includes *Picrorhiza scrophulariiflora*.
11. Includes other species of *Rheum* viz. *R. moorcroftianum* & *R. webbianum* , being traded as 'revandchini'.
12. Includes *Sapindus emarginatus* and *Sapindus laurifolius*
13. Includes *Sida acuta, Sida cordifolia, Sida cordata*, etc., being traded as 'bala'
14. Includes *Stereospermum colais*.
15. Also includes other species of *Swertia* viz. *S. angustifolia*, *S. alata*, etc.
16. Includes *Symplocos cochinchinensis* and *Symplocos paniculata*.
17. Also includes *Tinospora sinensis*
18. Includes other species of *Tribulus* viz. *T. lanuginosus*, *T. subramanyamii*, *T. alatus*, being traded as 'gokhru'.
19. Includes *Valeriana hardwickii*
20. Includes other species of *Viola* viz. *V. odorata, V. canescens, V. biflora, V. betonicifolia* etc. traded as 'banafsha'.

This indicative list only provides an idea of accepted equivalents and substitutes and is not exhaustive.
Source: National Medicinal Plants Board (NMPB) Website-November 2014

Appendix-V: ICAR Institutions, Deemed Universities, National Research Centres, National Bureaux, Directorate/Project Directorates

S. No.	Institute	Full name	Location	State/UT
A. Deemed Universities – 4				
1.	NDRI	National Dairy Research Institute	Karnal	Haryana
2.	CIFE	Central Institute on Fisheries Education	Mumbai	Maharashtra
3.	IARI	Indian Agricultural Research Institute	New Delhi	New Delhi
4.	IVRI	Indian Veterinary Research Institute	Izatnagar	Uttar Pradesh
B. Institutions – 48				
1.	CARI	Central Agricultural Research Institute	Port Blair	Andaman & Nicobar Is.
2.	CRIDA	Central Research Institute of Dry land Agriculture	Hyderabad	Andhra Pradesh
3.	CTRI	Central Tobacco Research Institute	Rajahmundry	Andhra Pradesh
4.	NAARM	National Academy of Agricultural Research & Management	Hyderabad	Andhra Pradesh
5.	ICARRCER	ICAR Research Complex for Eastern Region including Centre of Makhana	Patna	Bihar
6.	NBSMI	National Biotic Stress Management Institute	Raipur	Chhattisgarh
7.	ICARRCG	ICAR Research Complex Goa	Goa	Goa
8.	CIRB	Central Institute for Research on Buffaloes	Hissar	Haryana
9.	CSSRI	Central Soil Salinity Research Institute	Karnal	Haryana
10.	CPRI	Central Potato Research Institute	Shimla	Himachal Pradesh
11.	CITH	Central Institute of Temperate Horticulture	Srinagar	Jammu & Kashmir
12.	IIAB	Indian Institute of Agricultural Biotechnology	Ranchi	Jharkhand
13.	IINRG	Indian Institute of Natural Resins and Gums	Ranchi	Jharkhand
14.	IIHR	Indian Institute of Horticultural Research	Bengaluru	Karnataka
15.	NIANP	National Institute of Animal Nutrition and Physiology	Bengaluru	Karnataka
16.	NIVEDI	National Institute of Veterinary Epidemiology and Disease Informatics	Hebbal, Bengaluru	Karnataka
17.	CIFT	Central Institute of Fisheries Technology	Cochin	Kerala
18.	CMFRI	Central Marine Fisheries Research Institute	Kochi	Kerala
19.	CPCRI	Central Plantation Crops Research Institute	Kasargod	Kerala

Contd.

S. No.	Institute	Full name	Location	State/UT
20.	CTCRI	Central Tuber Crops Research Institute	Thiruvananthapuram	Kerala
21.	IISR	Indian Institute of Spices Research	Kozhikode (Calicut)	Kerala
22.	CIAE	Central Institute of Agricultural Engineering	Bhopal	Madhya Pradesh
23.	IISS	Indian Institute of Soil Sciences	Bhopal	Madhya Pradesh
24.	CICR	Central Institute of Cotton Research	Nagpur	Maharashtra
25.	CIRCOT	Central Institute of Research on Cotton Technology	Mumbai	Maharashtra
26.	NIASM	National Institute of Abiotic Stress Management	Malegaon	Maharashtra
27.	ICARRCNEHR	ICAR Research Complex for NEH Region	Barapani	Meghalaya
28.	IASRI	Indian Agricultural Statistical Research Institute	New Delhi	New Delhi
29.	CIFA	Central Institute of Freshwater Aquaculture	Bhubneshwar	Odisha
30.	CRRI	Central Rice Research Institute	Cuttack	Odisha
31.	CIPHET	Central Institute on Post harvest Engineering and Technology	Ludhiana	Punjab
32.	CAZRI	Central Arid Zone Research Institute	Jodhpur	Rajasthan
33.	CIAH	Central Institute of Arid Horticulture	Bikaner	Rajasthan
34.	CSWRI	Central Sheep and Wool Research Institute	Avikanagar	Rajasthan
35.	CIBA	Central Institute Brackish water Aquaculture	Chennai	Tamil Nadu
36.	SBI	Sugarcane Breeding Institute	Coimbatore	Tamil Nadu
37.	CARI	Central Avian Research Institute	Izatnagar	Uttar Pradesh
38.	CIRG	Central Institute for Research on Goats	Mathura	Uttar Pradesh
39.	CISH	Central Institute of Sub Tropical Horticulture	Lucknow	Uttar Pradesh
40.	IGFRI	Indian Grassland and Fodder Research Institute	Jhansi	Uttar Pradesh
41.	IIPR	Indian Institute of Pulses Research	Kanpur	Uttar Pradesh
42.	IISR	Indian Institute of Sugarcane Research	Lucknow	Uttar Pradesh
43.	IIVR	Indian Institute of Vegetable Research	Varanasi	Uttar Pradesh
44.	VPKAS	Vivekananda Parvatiya Krishi Anusandhan Sansthan	Almora	Uttarakhand
45.	CSWCRTI	Central Soil and Water Conservation Research & Training Institute	Dehradun	Uttrakhand
46.	CIFRI	Central Inland Fisheries Research Institute	Barrackpore	West Bengal
47.	CRIJAF	Central Research Institute for Jute and Allied Fibres	Barrackpore	West Bengal
48.	NIRJAFT	National Institute of Research on Jute & Allied Fibre Technology	Kolkata	West Bengal

Contd.

S. No.	Institute	Full name	Location	State/UT
C. National Research Centres – 17				
1.	NRCM	National Research Centre on Meat	Hyderabad	Andhra Pradesh
2.	NRCY	National Research Centre on Yak	West Kemang	Arunachal Pradesh
3.	NRCP	National Research Centre on Pig	Guwahati	Assam
4.	NRCL	National Research Centre for Litchi	Muzaffarpur	Bihar
5.	NRCE	National Research Centre on Equines	Hisar	Haryana
6.	NRCC	National Research Centre for Citrus	Nagpur	Maharashtra
7.	NRCG	National Research Centre for Grapes	Pune	Maharashtra
8.	NRCP	National Research Centre for Pomegranate	Solapur	Maharashtra
9.	NRC Mithun	National Research Centre on Mithun	Medziphema	Nagaland
10.	NCAP	National Centre for Agril. Economics & Policy Research	New Delhi	New Delhi
11.	NCIPM	National Centre for Integrated Pest Management	New Delhi	New Delhi
12.	NRCPB	National Research Centre on Plant Biotechnology	New Delhi	New Delhi
13.	NRC Camel	National Research Centre on Camel	Bikaner	Rajasthan
14.	NRCSS	National Research Centre on Seed Spices	Ajmer	Rajasthan
15.	NRCO	National Research Centre on Orchids	Pakyong	Sikkim
16.	NRCB	National Research Centre for Banana	Trichi	Tamil Nadu
17.	NRCAF	National Research Centre for Agro forestry	Jhansi	Uttar Pradesh
D. National Bureaux – 6				
1.	NBAGR	National Bureau of Animal Genetic Resources	Karnal	Haryana
2.	NBAII	National Bureau of Agriculturally Important Insects	Bangalore	Karnataka
3.	NBSSLUP	National Bureau of Soil Survey and Land Use Planning	Nagpur	Maharashtra
4.	NBPGR	National Bureau of Plant Genetics Resources	New Delhi	New Delhi
5.	NBAIM	National Bureau of Agriculturally Important Micro-organisms	Mau, Pradesh	Uttar Pradesh
6.	NBFGR	National Bureau of Fish Genetic Resources	Lucknow	Uttar Pradesh

Contd.

S. No.	Institute	Full name	Location	State/UT
E. Directorates/Project Directorates – 25				
1.	DOPR	Directorate of Oil Palm Research	Pedavegi, West Godawari	Andhra Pradesh
2.	DOR	Directorate of Oilseed Research	Hyderabad	Andhra Pradesh
3.	DRR	Directorate of Rice Research	Hyderabad	Andhra Pradesh
4.	DSR	Directorate of Sorghum Research	Hyderabad	Andhra Pradesh
5.	PDP	Project Directorate on Poultry	Hyderabad	Andhra Pradesh
6.	DGR	Directorate of Groundnut Research	Junagarh	Gujarat
7.	DMAPR	Directorate of Medicinal and Aromatic Plants Research	Anand	Gujarat
8.	DWR	Directorate of Wheat Research,	Karnal	Haryana
9.	DMR	Directorate of Mushroom Research	Solan	Himachal Pradesh
10.	DCR	Directorate of Cashew Research	Puttur	Karnataka
11.	PDADMAS	Project Directorate on Animal Disease Monitoring and Surveillance	Hebbal, Bangalore	Karnataka
12.	DSR	Directorate of Soybean Research	Indore	Madhya Pradesh
13.	DWSR	Directorate of Weed Science Research	Jabalpur	Madhya Pradesh
14.	DOGR	Directorate on Onion and Garlic Research	Pune	Maharashtra
15.	DFR	Directorate of Floriculture Research	New Delhi	New Delhi
16.	DKMA	Directorate of Knowledge Management in Agriculture	New Delhi	New Delhi
17.	DMR	Directorate of Maize Research	New Delhi	New Delhi
18.	DRWA	Directorate of Research on Women in Agriculture	Bhubaneshwar	Odisha
19.	DWMR	Directorate of Water Management Research	Bhubaneshwar	Odisha
20.	PDFMD	Project Directorate on Foot & Mouth Disease	Mukteshwar	Punjab
21	DRMR	Directorate of Rapeseed & Mustard Research	Bharatpur	Rajasthan
22.	DSR	Directorate of Seed Research	Mau	Uttar Pradesh
23.	PDFSR	Project Directorate for Farming Systems Research	Modipuram, Meerut	Uttar Pradesh
24.	PDC	Project Directorate on Cattle	Meerut	Uttar Pradesh
25.	DCWFR	Directorate of Cold Water Fisheries Research	Bhimtal, Nainital	Uttarakhand

***Source*:** Indian Council of Agricultural Research, New Delhi

Appendix-VI: List of Authorities/Agencies/Institutions/Functionaries

1. National Horticulture Board (NHB), Gurgaon
2. Coconut Development Board, Kochi
3. Small Farmer's Agri-Business Consortium (SFAC), New Delhi
4. Directorate of Cashew and Cocoa Development (DCCD), Kochi
5. Directorate of Arecanut and Spices Development (DASD), Calicut
6. National Committee on Plasticulture Applications in Agriculture & Horticulture (NCPAAH), New Delhi
7. National Horticulture Research & Development Foundation, Nashik
8. National Seeds Corporation, New Delhi
9. State Farm Corporation of India, New Delhi
10. National Institute for Agricultural Extension Management (MANAGE), Hyderabad
11. Fresh & Healthy Enterprises Ltd. (FHEL), New Delhi
12. Agricultural and Processed Food Products Export Development Authority (APEDA), New Delhi
13. Directorate of Marketing & Inspection (DMI), New Delhi
14. Ministry of Food Processing Industries (MFPI), New Delhi
15. National Medicinal Plants Board (NMPB), New Delhi
16. Central Institute of Horticulture (CIH), Nagaland
17. National Centre for Cold Chain Development (NCCD)
18. National Institute of Food Technology Entrepreneurship and Management (NIFTEM), Sonipat, Haryana
19. United Planters' Association of Southern India (UPASI) Tea Research Foundation (TRF), Coimbatore District, Tamil Nadu, India.

Appendix-VII: International Agricultural Organizations

Name, Year of Establishment and Location of International Agricultural Organizations

S. No.	Institutes	Full Name	Year	Location
1.	CIMMYT	International Maize and Wheat Improvement Center	1943	El Batán, Edo Mex, Mexico
2.	ISHS	International Society for Horticultural Science	1959	Belgium
3.	IRRI	International Rice Research Institute (IRRI)	1960	Los Baños, Laguna, Philippines
4.	CIAT	International Center for Tropical Agriculture	1967	Cali, Colombia.
5.	IITA	International Institute of Tropical Agriculture	1967	Ibadan, Nigeria
6.	Africa Rice Center	Formerly known as the West Africa Rice Development Association (WARDA)	1971	Cotonou, Benin
7.	CIP	International Potato Center	1971	Lima, Peru
8.	ICRISAT	International Crops Research Institute for the Semi-Arid Tropics (ICRISAT)	1972	Hyderabad, India
9.	Bioversity International	Bioversity International was originally established by the CGIAR as the International Board for Plant Genetic Resources (IBPGR) in 1974.	1974	Maccarese, Rome
10.	ICARDA	International Center for Agricultural Research in the Dry Area (ICARDA)	1975	Beirut, Lebanon
11.	IFPRI	International Food Policy Research Institute	1975	Washington, D.C., USA
12.	WFC	World Fish Centre	1975	Penang, Malaysia
13.	World Agroforestry Centre	known as the International Centre for Research in Agroforestry, ICRAF before 2002	1978	Nairobi, Kenya,
14.	IWMI	International Water Management Institute	1985	Colombo Sri Lanka

Contd.

S. No.	Institute	Full name	Year	Location
15.	INIBAP	International Network for the Improvement of Banana and Plantain.	1985	Montpellier, France
16.	IPGRI	International Plant Genetic Resources Institute	1991	Rome, Italy
17.	CGIAR	Consultative Group on International Agricultural Research	1991	Montpellier, France
18.	CIFOR	Center for International Forestry Research	1993	Bogor, Indonesia
19.	ILRI	International Livestock Research Institute	1994	Nairobi, Kenya
20.	BISA	Borlaug Institute for South Asia	2011	Ludhiana in Punjab, Pusa in Bihar and

International Organizations

Name, Year of Establishment, HQ of International Organizations

S. No.	Abbreviation	Full name	Establishment year	HQ
21.	ILO	International Labour Organization	1919	Geneva, Switzerland
22.	IBRD	International Bank for Reconstruction and Development	1944	Washington, D.C., United States
23.	FAO	Food and Agriculture Organization	16 October 1945	Rome, Italy
24.	UN	United Nations	24 October 1945	New York, USA
25.	UNESCO	United Nations Educational, Scientific and Cultural Organization	16 November 1945	Paris, France
26.	IMF	International Monetary Fund	27th December 1945	Washington, D.C., United States
27.	UNICEF	United Nations International Children's Emergency Fund	December 1946	New York City, USA
28.	ISO	International Organization for Standardization	23 February 1947	Geneva, Switzerland
29.	ADB	Asian Development Bank	22 August 1966	Manila, Philippines
30.	ASEAN	Association of Southeast Asian Nations	8 August 1967	Jakarta, Indonesia
31.	SAARC	South Asian Association for Regional Cooperation	8 December 1985	Kathmandu, Nepal
32.	WTO	World Trade Organization	1 January 1995	Geneva, Switzerland

Appendix-VIII: List of Agricultural Universities

S.No.	Name of the University	Location	State
1.	Acharya NG Ranga Agricultural University	Hyderabad	Andhra Pradesh
2.	Dr YSR Horticultural University	West Godavari	Andhra Pradesh
3.	Sri Venkateswara Veterinary University	Tirupati	Andhra Pradesh
4.	Assam Agricultural University	Jorhat	Assam
5.	Bihar Agricultural University	Sabour, Bhagalpur	Bihar
6.	Rajendra Agricultural University	Samastipur	Bihar
7.	Chhattisgarh Kamdhenu Vishwavidyalaya	Anjora, Durg	Chhattisgarh
8.	Indira Gandhi Krishi Vishwavidyalaya	Raipur	Chhattisgarh
9.	Anand Agricultural University	Anand	Gujarat
10.	Junagadh Agricultural University	Junagadh	Gujarat
11.	Navsari Agricultural University	Navsari	Gujarat
12.	Sardarkrushinagar-Dantiwada Agricultural University	Banaskantha	Gujarat
13.	Chaudhary Charan Singh Haryana Agricultural University	Hisar	Haryana
14.	Lala Lajpat Rai University of Veterinary & Animal Sciences	Hisar	Haryana
15.	CSK Himachal Pradesh Krishi Vishvavidyalaya	Palampur	Himachal Pradesh
16.	Dr Yashwant Singh Parmar, Univ of Horticulture & Forestry	Solan, Nauni	Himachal Pradesh
17.	Sher-E-Kashmir Univ of Agricultural Sciences & Technology	Jammu	Jammu & Kashmir
18.	Sher-E-Kashmir Univ of Agricultural Sciences & Technology of Kashmir	Shrinagar	Jammu & Kashmir

Contd.

S.No.	Name of the University	Location	State
19.	Birsa Agricultural University	Ranchi	Jharkhand
20.	Karnataka Veterinary, Animal and Fisheries Sciences University	Bidar	Karnataka
21.	University of Agricultural Sciences, Bangalore	Bengaluru	Karnataka
22.	University of Agricultural Sciences, Dharwad	Dharwad	Karnataka
23.	University of Agricultural Sciences, Shimoga	Shimoga	Karnataka
24.	University of Horticultural Sciences	Bagalkot	Karnataka
25.	University of Agricultural Sciences	Raichur	Karnataka
26.	Kerala Agricultural University	Vellanikara, Trichur	Kerala
27.	Kerala University of Fisheries & Ocean Studies	Papangad, Kochi	Kerala
28.	Kerala Veterinary and Animal Sciences University	Thiruvananthapuram	Kerala
29.	Jawaharlal Nehru Krishi Viswavidyalaya	Jabalpur	Madhya Pradesh
30.	Nanaji Deshmukh Veterinary Science University	Jabalpur	Madhya Pradesh
31.	Rajmata Vijayraje Sciendia Krishi Vishwa Vidyalaya	Gwalior	Madhya Pradesh
32.	Dr Balasaheb Sawant Konkan Krishi Vidyapeeth	Dapoli Distt, Ratnagiri	Maharashtra
33.	Dr Panjabrao Deshmukh Krishi Vidyapeeth	Krishinagar, Akola	Maharashtra
34.	Maharashtra Animal Science & Fishery University	Nagpur	Maharashtra
35.	Mahatma Phule Krishi Vidyapeeth	Rahuri	Maharashtra
36.	Marathwada Agricultural University	Parbhani	Maharashtra
37.	Central Agricultural University	Imphal	Manipur
38.	Odisha Univ. of Agriculture & Technology	Bhubaneshwar	Odisha
39.	Guru Angad Dev Veterinary and Animal Science University	Ludhiana	Punjab
40.	Punjab Agricultural University	Ludhiana	Punjab

Contd.

S.No.	Name of the University	Location	State
41.	Maharana Pratap Univ. of Agriculture & Technology	Udaipur	Rajasthan
42.	Rajasthan University of Veterinary and Animal Sciences	Bikaner	Rajasthan
43.	Swami Keshwanand Rajasthan Agricultural University	Bikaner	Rajasthan
44.	Tamil Nadu Agricultural University	Coimbatore	Tamil Nadu
45.	Tamil Nadu Fisheries University	Nagapattinam	Tamil Nadu
46.	Tamil Nadu Veterinary & Animal Science University	Chennai	Tamil Nadu
47.	Chandra Shekar Azad University of Agriculture & Technology	Kanpur	Uttar Pradesh
48.	Manyavar Shri Kanshiram Ji University of Agriculture and Technology	Banda	Uttar Pradesh
49.	Narendra Deva University of Agriculture & Technology	Faizabad	Uttar Pradesh
50.	Sardar Vallabhbhai Patel University of Agriculture and Technology	Meerut	Uttar Pradesh
51.	UP Pandit Deen Dayal Upadhaya Pashu Chikitsa Vigyan Vishwa Vidhyalaya evam Go Anusandhan Sansthan	Mathura	Uttar Pradesh
52.	Uttarakhand University of Horticulture and Forestry	Pauri Garhwal	Uttarakhand
53.	Govind Ballabh Pant University of Agriculture & Technology	Pantnagar	Uttarakhand
54.	Bidhan Chandra Krishi Viswavidyalaya	Mohanpur, Nadia	West Bengal
55.	Uttar Banga Krishi Viswavidyalaya	Coach Bihar	West Bengal
56.	West Bengal University of Animal & Fishery Sciences	Kolkata	West Bengal

Source: Indian Council of Agricultural Research, Website, New Delhi

Appendix-IX: List of Various Boards/ Organizations

Name, Establishment year and HQ of Various Boards/ Organizations Related to Agriculture

S. No.	Name of board/ organization	Establishment year	HQ	Remarks
1.	Coffee Board of India	1942	Bangalore, Karnataka, India.	Functions under the Ministry of Commerce and Industry, Government of India.
2.	Indian Rubber Board	1947	Kottayam, Kerala, India	Government of India
3.	Tea Board of India	1st April 1954	Kolkata (formerly Calcutta)	Under the control of the Union Govt.
4.	Central Insecticide Board and Registration Committee	1968	Faridabad	Ministry of Agriculture, Government of India
5.	Coconut Development Board	1979	Kochi, Kerala, India	Functions under Ministry of Agriculture, Government of India
6.	National Horticulture Board	1984	Gurgaon (Haryana).	Functions under Department of Agriculture & Co-operation, Ministry of Agriculture, Govt. of India.
7.	Spices Board of India	26th February 1986	Cochin,Kerala, India	Functions under Ministry of Commerce, Government of India
8.	National Medicinal Plants Board	2000	New Delhi	Department of AYUSH, Ministry of Health & Family Welfare, Government of India
9.	National Development Council	6th August 1952	New Delhi	Government of India
10.	National Agricultural Cooperative Marketing Federation of India Ltd	October 2, 1958	New Delhi	
11.	Agricultural and Processed Food Products Export Development Authority (APEDA)	December, 1985	New Delhi	Established by the Government of India under the Agricultural and Processed Food Products Export Development Authority Act passed by the Parliament in December, 1985. The Act (2 of 1986) came into effect from 13th February, 1986 by a notification issued in the Gazette of India

Contd.

S. No.	Name of board/ organization	Establishment year	HQ	Remarks
12.	Food Corporation of India (FCI)	1964	New Delhi, India	
13.	Agricultural Prices Commission	January, 1965		Since March 1985, the Commission has been known as Commission for Agricultural Costs and Prices. (CACP)
14.	Indian Farmers Fertiliser Co-operative Limited (IFFCO)	November 3, 1967	New Delhi	
15	National Fertilizers Limited (NFL)	23rd August, 1974	Noida, India	
16.	National Commission on Agriculture	1976		Government of India
17.	National Environment Tribunal	1995		Function under ministry of environment and forest, Government of India
18.	National Environment Appellate Authority (NEAA)	1997		Function under ministry of environment and forest, Government of India
19.	National Biodiversity Authority	2003	Chennai, Tamil Nadu	The National Biodiversity Authority (NBA) was established in 2003 to implement India's Biological Diversity Act (2002)
20.	National Green Tribunal (NGT)	October 18, 2010	New Delhi	Function under ministry of environment and forest, Government of India
21.	National Commission on Farmers (NCF)	November 18, 2004		
22.	Food Safety and Standards Authority of India (FSSAI)	2006	New Delhi	Ministry of Health & Family Welfare, Government of India
23.	International Flower Auction Centre	2007	Bangalore, Karnataka	Government of India
24.	Directorate of Plant Protection, Quarantine and Storage (DPPQS)		Faridabad	Functions under Ministry of Agriculture
25.	Genetic Engineering Approval Committee (GEAC)			Function under ministry of environment and forest, Government of India

Appendix-X: List of National and International Horticulture's Journals and Magazines

Name of Journal/Magazine, Category, Publication Frequency, Language and Publisher of Agriculture Science Related Journals and Magazines

S.No.	Name of journal/magazine	Category	Publication frequency	Language	Publisher
1.	Current Science	Journal	Fortnight	English	Current Science Association in collaboration with the Indian Academy of Sciences (India)
2.	Indian Journal of Horticulture	Journal	Quarterly	English	The Horticultural Society of India
3.	Journal of Horticultural Sciences	Journal	Half-yearly	English	Society for Promotion of Horticulture (SPH)
4.	Journal of Medicinal and Aromatic Plants	Journal	Half-yearly	English	Medicinal and Aromatic Plants Association of India (MAPAI)
5.	Krishika	Journal	Half-yearly	Hindi	Indian Council of Agricultural Research (ICAR)
6.	Nature	Journal	Weekly	English	Nature Publishing Group (NPG)
7.	Potato Journal	Journal	Half-yearly	English	Potato Journal
8.	The Indian Journal of Agricultural Sciences	Journal	Monthly	English	Indian Council of Agricultural Research (ICAR)
9.	Vegetable Science	Journal	Half-yearly	English	Indian Society of Vegetable Science(ISVS)
10.	Indian Farming	Magazine	Bi-monthly	English	Indian Council of Agricultural Research (ICAR)
11.	Indian Horticulture	Magazine	Bi-monthly	English	Indian Council of Agricultural Research (ICAR)
12.	Kheti	Magazine	Monthly	Hindi	Indian Council of Agricultural Research (ICAR)
13.	Phal Phool	Magazine	Monthly	Hindi	Indian Council of Agricultural Research (ICAR)
14.	Kurukshetra	Magazine	Monthly	Hindi and English	Publications Division, Ministry of Information and Broadcasting, Government of India
15.	Yojna	Magazine	Monthly	13 languages	Publications Division, Ministry of Information and Broadcasting, Government of India

Appendix-XI: List of Acts and Rules

S. No.	Acts/Rules	Year	Remarks
1.	Land Acquisition Act	1894	The Land Acquisition (Amendment) Bill, 2007, was Passed in the Lok Sabha in 2009
2.	Livestock Importation Act	1898	Amended in 2001
3.	Destructive Insects and Pests (DIP) Act	1914	
4.	Agricultural Produce Act (Grading and Marking)	1937	(Amended in 1986), Act is popularly known as 'AGMARK' Standards.
5.	Central Committee for Food Standards	1941	
6.	The Rubber Act	1947	
7.	Vegetable Oil Products Order	1947	
8.	Vegetable Oil Products Order	1998	
9.	Minimum Wages Act	1948	
10.	Industries (Development and Regulation) Act (IDRA)	1951	
11.	Prevention of Food Adulteration (PFA) Act	1954	
12.	Food Product Order	1955	
13.	Sugar Control Order	1956	
14.	Solvent Extracted Oil, De-oiled Meal and Edible Flour (Control) Order	1967	
15.	Meat Food Products Order	1973	
16.	Milk and Milk Product Order (MMPO)	1992	
17.	Milk & Milk Product Amendment Regulations	2009	
18.	Edible Oils Packaging (Regulation) Order	1998	

Contd.

S. No.	Acts/rules	Year	Remarks
19.	Essential Commodities Act	1955	
20.	Fertilizer (Control) Order	1957/1985	
21.	Fertilizer (Movement Control) Order	1960/1973/2001	
22.	Prevention of Cruelty to Animals Act	1960	
23.	National Co-operative Development Corporation (NCDC) Act	1962	
24.	Seeds Act	1966	
25.	Seeds Rules	1968	
26.	Seeds (Control) Order	1983	
27.	National Seed Policy	2002	
28.	Seeds Bill	2004	
29.	Insecticide Act	1968	
30.	Insecticides Rules	1971	
31.	Wildlife Protection Act	1972	The Act was amended in January 2003
32.	Water (Prevention and Control of Pollution) Act	1974	The Act was last amended in 2003
33.	Standards for Weight and Measures Act	1976	
34.	Forest Conservation Act	1980	
35.	The Consumer Protection Act	1986	
36.	Environment Protection Act	1986	The Act was last amended in 1991
37.	Bureau of Indian Standards (BIS) Act	1986	
38.	The Bureau of Indian Standards Rules	1987	
39.	The Bureau of Indian Standards (Certification) Regulations	1988	
40.	Recombinant DNA Safety Guidelines	1990	

Contd.

S. No.	Acts/rules	Year	Remarks
41.	Revised Recombinant DNA Safety Guidelines	1994	
42.	Revised Guidelines for Research in Transgenic Plants & Guidelines for Toxicity and Allergenicity Evaluation of Transgenic Seeds, Plants and Plant Parts	1998	
43.	Protection of Plant Varieties and Farmers' Rights Act (PPVFRA)	2001	
44.	Multi-State Co-operative Societies (MSCS) Act	2002	
45.	Biological Diversity Act	2002	
46.	Plant Quarantine (Regulation of Import into India) Order	2003	Came into force on January 1, 2004
47.	Biological Diversity Rules	2004	
48.	National Biotechnology Strategy	2005	
49.	Environment Policy	2006	
50.	Mahatma Gandhi National Rural Employment Guarantee Act	2005	The law was initially called the National Rural Employment Guarantee Act (NREGA) and was renamed with the prefix "Mahatma Gandhi" on 2 October 2009, Gandhi's birth anniversary
51.	Foreign Trade Policy	2006	
52.	Food Safety and Standards Act	2006	Amended in 2008
53.	National Food Security Act	2013	
54.	National Policy on Skills Development	2009	

Appendix-XII: Agricultural Revolution

Name, Crop/Entrepreneur, Associated Person and Year of Starting of Revolution in Agriculture

S. No.	Name of Revolution	Crop/Entrepreneur	Person associated with	Year of starting
1.	Green revolution	Food grain production	Dr. N.E. Borlaug	1960
2.	Green revolution in India	Food grain production	Dr. M.S. Swaminathan	1966-67
3.	White revolution (In India it is known as Operation Flood)	Milk production	Varghese Kurien	1970
4.	Blue revolution	Aquaculture		
5.	Evergreen revolution	Overall development of agriculture	Dr. M.S. Swaminathan	1995
6.	Yellow revolution	Oilseed production		
7.	Grey revolution	Fertilizers		
8.	Red revolution	Meat/tomato production		
9.	Round revolution	Potato production		
10.	Silver revolution	Egg production/poultry		
11.	Silver Fibre Revolution	Cotton		
12.	Pink revolution	Prawn production		
13.	Golden revolution	Fruit production/overall horticulture development	Dr. K.L. Chadha	
14.	Brown revolution	Non-conventional fuel, cocoa, leather		
15.	Rainbow revolution	Agriculture	Dr. Nitish Kumar	
16.	Black revolution	Bio-fuel (Jatropha) production		
17.	Food chain revolution fruits and vegetables	Reduction in wastage of food grains,		
18.	Prabhani revolution	Okra		

Appendix-XIII: Popular Name of Horticultural Crops

S. No.	Popular Name	Crop
1.	King of fruits	Mango
2.	King of arid and semi arid fruits	Ber
3.	King of temperate fruits	Apple
4.	Queen of fruits	Litchi
5.	King of spices	Black Pepper
6.	Queen of spices	Cardamom
7.	King of vegetables	Potato
8.	Queen of vegetables	Asparagus
9.	Poor man's fruit	Ber
10.	Poor man's friend	Potato
11.	National fruit of India	Mango
12.	Glory of east	Chrysanthemum
13.	Autumn queen	Chrysanthemum
14.	Queen of night	*Cestrum nocturnum*
15.	King of flowers	Rose
16.	Queen of flowers	Jasmine
17.	Egg plant	Brinjal
18.	Butter fruit	Avocado
19.	Queen of beverages	Tea
20.	China's miracle fruit	Kiwi fruit
21.	Food of Gods	Cocoa
22.	Small holder's irrigated crop	Oil palm
23.	Oldest cultivated tropical fruit	Banana
24.	Tree of heaven	Coconut

Appendix XIV: List of People Considered Father of a Scientific Field

S. No.	Field	Person/s considered "Father"
1.	Agricultural Chemistry	Justus von Liebig
2.	Agro Climatology	Koppen
3.	Agro Meteorology	D.N.Walia
4.	Agronomy	Pietro Decrescenzi
5.	Antibiotics	Alexander Fleming
6.	ATP cycle	Lipmann
7.	Bacteriology	Robert Koch/ Louis Pasteur
8.	Biochemistry	Justus von Liebig
9.	Biogeography	Alfred Russel Wallace
10.	Biology	Aristotle
11.	Botany	Theophrastus
12.	Canning	Nicolus Appart
13.	Cooperative movement in India	F. Nicholson
14.	Crop rotation	Norfork (1730)
15.	Cytology	Robert Hooke
16.	Cytoplasmic Inheritance	Carl Correns
17.	DNA Finger Print Technique	Alec Jeffrey
18.	Ecology	Reiter
19.	Economic Ecology	M.S. Swaminathan
20.	Entomology	Jan Swammerdam
21.	Evolution Natural Selection	Charles Darwin
21.	Experimental Genetics	Thomas Hunt Morgan
22.	Extension	J. Paul Leagons
23.	Extension Education	A.Seaman and Knapp (in USA) and Leagans (in India)
24.	Fermentation	Louis Pasteur
25.	Field Plot Experiment	J.B. Boussingault
26.	Forest Pathology	Robert Haring
27.	Fruit and Vegetable Preservation	M. Nicholas Apart
28.	Genetic Engineering	Paul Berg
29.	Genetics	Gregor Johann Mendal
30.	Golden Revolution in India	Dr. K.L.Chadha
31.	Golden Rice	Dr. Ingo Potrykus
32.	Green Revolution	Dr. N.E. Borlaug

Contd.

S. No.	Field	Person/s considered "father"
33.	Green Revolution in India	M.S. Swaminathan
34.	Hybrid Cotton	C.T. Patel
35.	Hybrid Rice	Yuan Long Ping
36.	Hybrid Rice in Indian	E. A. Siddiqe
37.	Ichthyology	Peter Artedi
38.	Immunology	Edward Jenner
39.	Indian Ecology	R. Mishra
40.	Indian Mycology	E.J. Butler
41.	Indian Palaeobotany	Birbal Sahani
42.	Indian Phytopathology	E.J. Butler
43.	Indian Plant Pathology	E.J. Butler
44.	Indian Remote Sensing	Dr.P.R. Pishroty
45.	Indian Rust	Dr. K.C. Mehta
46.	Lichenology	Erik Acharius
47.	Medical Bacteriology	Robert Koch
48.	Microbiology	Antonie van Leeuwenhoek
49.	Modern Botany	Linnaeus/Bauchin
50.	Modern Cytology	Swanson
51.	Modern Genetics	T.H. Morgan
52.	Molecular Biology	Linus Pauling
53.	Molecular Biophysics	Gopalasamudram Narayana Iyer Ramachandran
54.	Mycology	Pler A. Micheli
55.	Nematology	N.A. Cobb
56.	Nitrogen Fixation	Winogradsky
57.	Ornamental Gardening	M.S. Randhawa
58.	Paleontology	George Cuvier
59.	Parasitology	F. Platter
60.	Pedology	V.V.Dokuchalev
61.	Plant Anatomy	Grew
62.	Plant Pathology	Anton De Bary
63.	Plant Physiology	Stephen Hales
64.	Plant Tissue Culture	G. Haberlandt
65.	Polygenic Inheritance	Kolreuter
66.	Protozoology	Antonie van Leeuwenhoek
67.	Pure Culture Technique	Oscar Brefeld
68.	Sociology	Auguste Compte

Contd.

S. No.	Field	Person/s considered "father"
69.	Soil Conservation	H.H. Bennett
70.	Soil Microbiology	S. N. Winogradsky
71.	Soil Science	Dokuchalev
72.	Soil Testing Technique	M. L. Troug
73.	Statistics	R. A. Fisher
74.	Super Rice	Dr. G.H. Khush
75.	Systematic Pomology	De Candole
76.	Taxonomy	Carolus Linnaeus
77.	Tillage and Weeds	Jethro Tull
78.	Toxicology	Paracelsus
79.	Virology	Martinus Beijerinck
80.	White Revolution	Dr. Varghese Kurien
81.	Zoology	Aristotle

Appendix-XV:List of Pesticides / Pesticides Formulations Banned in India

A. Pesticides Banned for manufacture, import and use (28 Nos.)

1. Aldrin
2. Benzene Hexachloride
3. Calcium Cyanide
4. Chlordane
5. Copper Acetoarsenite
6. CIbromochloropropane
7. Endrin
8. Ethyl Mercury Chloride
9. Ethyl Parathion
10. Heptachlor
11. Menazone
12. Nitrofen
13. Paraquat Dimethyl Sulphate
14. Pentachloro Nitrobenzene
15. Pentachlorophenol
16. Phenyl Mercury Acetate
17. Sodium Methane Arsonate
18. Tetradifon
19. Toxafen
20. Aldicarb
21. Chlorobenzilate
22. Dieldrine
23. Maleic Hydrazide
24. Ethylene Dibromide
25. TCA (Trichloro acetic acid)
26. Metoxuron
27. Chlorofenvinphos
28. Lindane (Banned vide Gazette Notification No S.O. 637(E) Dated 25/03/2011)-Banned for Manufecture,Import or Formulate w.e.f. 25th March,2011 and banned for use w.e.f. 25th March,2013.

B. Pesticide / Pesticide formulations banned for use but their manufacture is allowed for export (2 Nos.)

1. Nicotin Sulfate
2. Captafol 80% Powder

C. Pesticide formulations banned for import, manufacture and use (4 Nos)

1. Methomyl 24% L
2. Methomyl 12.5% L
3. Phosphamidon 85% SL
4. Carbofuron 50% SP

D. Pesticide Withdrawn(7 Nos)

1. Dalapon
2. Ferbam
3. Formothion
4. Nickel Chloride
5. Paradichlorobenzene (PDCB)
6. Simazine
7. Warfarin

List of Pesticides Refused Registration

S. No.	Name of Pesticides
1.	Calcium Arsonate
2.	EPM
3.	Azinphos Methyl
4.	Lead Arsonate
5.	Mevinphos (Phosdrin)
6.	2,4, 5-T
7.	Carbophenothion
8.	Vamidothion
9.	Mephosfolan
10.	Azinphos Ethyl
11.	Binapacryl
12.	Dicrotophos
13.	Thiodemeton / Disulfoton
14.	Fentin Acetate
15.	Fentin Hydroxide
16.	ChinomEthionate (Morestan)
17.	Ammonium Sulphamate
18.	Leptophos (Phosvel)

Pesticides Restricted for use in India

S. No.	Name of Pesticides
1.	Aluminium Phosphide
2.	DDT
3.	Lindane
4.	Methyl Bromide
5.	Methyl Parathion
6.	Sodium Cyanide
7.	Methoxy Ethyl Mercuric Chloride (MEMC)
8.	Monocrotophos
9.	Endosulfan*
10.	Fenitrothion
11.	Diazinon
12.	Fenthion
13.	Dazomet

Endosulfan*:- Endosulfan has been banned by the Supreme Court of India w.e.f. 13-05-2011 for production, use & sale, all over India, till further orders vide ad-Interim order in the Writ Petition (Civil) No. 213 of 2011.

Source: Directorate of Plant Protection, Quarantine & Storage Central Insecticides Board & Registration Committee, Faridabad.

Appendix XVI: Conversion Chart

Conversion Chart for Length, Volume, Mass and Temperature

From	To	Multiply By
Conversion chart for length		
Kilometers (Km)	Miles (Mi)	0.62
Kilometers (Km)	Feet (Ft)	3280.8
Meters (M)	Feet (Ft)	3.28
Centimeters (Cm)	Inches (In)	0.39
Millimeters (Mm)	Inches (In)	0.039
Inches (In)	Meters (M)	0.0254
Inches (In)	Centimeters (Cm)	2.54
Inches (In)	Millimeters (Mm)	25.40
Feet (Ft)	Meters (M)	0.30
Yards (Yd)	Meters (M)	0.91
Yards (Yd)	Kilometers (Km)	0.00091
Miles (Mi)	Kilometers (Km)	1.61
Volume		
Litres (L)	Quarts (Qt)	1.057
Litres (L)	Gallons (Gal)	0.264
Millilitres (Ml)	Cups (C)	0.0042
Millilitres (Ml)	Ounces (Oz)	0.0338
Ounces (Oz)	Millilitres (Ml)	29.57
Cups (C)	Millilitres (Ml)	236.6
Quarts (Qt)	Litres (L)	0.95
Gallons (Gal)	Litres (L)	3.785
Mass		
Kilogram (Kg)	Tons (Ton)	0.0011
Kilogram (Kg)	Pounds (Lb)	2.2046
Grams (G)	Ounces (Oz)	0.035
Grams (G)	Pounds (Lb)	0.002205
Milligrams (Mg)	Ounces (Oz)	0.000035
Ounces (Oz)	Milligrams (Mg)	28350
Ounces	Grams (G)	28.35
Pounds (Lb)	Kilograms (Kg)	0.454
Tons (Ton)	Kilograms (Kg)	907.18
Temperature		
Fahrenheit (F)	Celsius (C)	(Temperature (F) – 32)*5/9
Celsius (C)	Fahrenheit (F)	(Temperature (C)*9/5) + 32

References

1. Agricultural and Processed Food Products Export Development Authority (2014) *About APEDA*. Available at http://www.apeda.gov.in/apedawebsite/about_apeda/About_apeda.htm [Accessed 1st December 2014].
2. Central Insecticides Board (2014) *List of Pesticides / Pesticides Formulations Banned in India.* Available at http://cibrc.nic.in/list_pest_bann.htm [Accessed 1st December 2014].
3. Chadha, K.L. (2001) *Handbook of Horticulture,* New Delhi, Director of Information and Publication of Agriculture, ICAR.
4. Coconut Development Board (2014) *PACKAGE OF PRACTICES OF COCONUT.* Available at http://coconutboard.nic.in/package.htm [Accessed 1st December 2014].
5. Coffee Board (2014) *Statistics on Coffee.* Available at http://www.indiacoffee.org/coffee-statistics.html?page=CoffeeData#pro [Accessed 1st December 2014].
6. Department of Agriculture and Cooperation (2013) *Check list of Commercial Verities of Plantation Crops*. Available at http://agricoop.nic.in/ [Accessed 1st December 2013].
7. Directorate of Arecanut and Spices Development (2014) *About DASD.* Available at http://dasd.gov.in/index.php/home.html [Accessed 1st December 2014].
8. Directorate of Cashew and Cocoa Development (2014) *About DCCD.* Available at http://dccd.gov.in/dccd.htm [Accessed 1st December 2014].
9. Directorate of Oil Palm Research (2014) *Institute and History.* Available at http://www.dopr.gov.in/AboutUs.htm [Accessed 1st December 2014].
10. ICAR-Directorate of Medicinal Aromatic Plants Research (2014) *ICAR-DMAPR Annual Reports.* Available at http://www.dmapr.org.in/Publications/AnualReport.html [Accessed 1st December 2014].
11. Indian Council of Agricultural Research (2014) *DARE/ICAR Annual Reports.* Available at http://www.icar.org.in/en/taxonomy/term/52 [Accessed 1st December 2014].
12. National Horticulture Board (2014) *Aroma Plant.* Available at http://nhb.gov.in/bulletin-aromatic.html [Accessed 1st December 2014].

13. National Horticulture Board (2014) *Indian Horticulture Database (2013).* Available at http://nhb.gov.in/area-pro/Indian%20Horticulture%202013.pdf [Accessed 1st December 2014].
14. National Medicinal Plants Board (2014) *NMPB Publications (2012.* Available at http://nmpb.nic.in/index1.php?level=1&sublinkid=1254&lid=769 [Accessed 1st December 2014].
15. Spices Board of India(2014) *Cultivation Practices.* Available at http://www.indianspices.com/ [Accessed 1st December 2014].
16. Tea Board of India (2014) *About Tea Board.* Available at http://www.teaboard.gov.in/inner1.asp?param_link_id=110&mem_link_name=About%20Tea%20Board [Accessed 1st December 2014].
17. TNAU Agritech portal (2014) *Horticulture Crops.* Available at http://agritech.tnau.ac.in/horticulture/horti_index.html [Accessed 1st December 2014].

□□□□

List of Horticulture Books

S.No.	Title	Author	ISBN	Year
1	A Handbook of Soil-Plant-Water-Fertilizer and Manure Analysis	Durai, M.V.	9789381450185	2014
2	A Text Book of Seed Science and Technology	Padmavathi, S.	9789381450437	2012
3	Abiotic Stress Tolerance in Crop Plants: Breeding and Biotechnology	Roy, Bidhan	9788189422943	2009
4	Acid Soils: Their Chemistry and Management	Sarkar,A.K.	9789381450383	2013
5	Advances and Challenges in Agricultural Extension and Rural Development	Rathakrishnan,R.T.	9789380235035	2009
6	Advances in Preservation and Processing Technologies of Fruits and Vegetables	Rajarathnam, S.	9789380235523	2011
7	Advances in Protected Cultivation	Singh,Brahma	9789383305179	2014
8	Advances in Soil Borne Plant Diseases	Naik, Manjunath	9788189422813	2008
9	Agricultural Plant Biochemistry	G.Nagaraj	9789383305551	2015
10	Agri-Horticultural Biodiverstiy of Temperate and Cold Arid Regions	Nazeer, A.Zeerak	9789381450086	2012
11	Agro-enterprises for Rural Development and Livelihood Security	Sharma J.P.	9789380235851	2011
12	Agroforestry Systems for Resource Conservation and Livelihood Security in Lower Himalays	Panwar, P	9789381450215	2012
13	Agroforestry: Principles and Practices	Patra, Alok Kumar	9789381450765	2013
14	Agroforestry: Systems and Practices	Puri, Sunil &	9788189422622	2007
15	Agroforestry: Systems and Prospects	C.B.Pandey	9789381450970	2014
16	Agro-Informatics	Vanitha, G	9789380235714	2011
17	Agrotechnology Manual: Including Nursery Management and Practices	Etomes, M.N.	9789383305018	2014
18	Air Pollution and It's Impacts on Plant Growth	Shyam, S.	9788189422103	2006
19	An Introduction to Intellectual Property Rights	Pathak, Manju	9789383305124	2014
20	An Introduction to Nanotechnology	Rathinasamy, A.	9789381450413	2012
21	Aromatic Plants: Vol.01. Horticulture Science	Skaria, B.P. et.al.	9788189422455	2007
22	Auddaniki ke Adharbhoot Sidhant Tatha Nashijeev Prabandhan (PB)	Tiwari, A.K.	9789380235998	2012
23	Bananas and Plantains	C.K.Narayana	9789383305452	2015
24	Basics of Horticulture: 2nd Revised Edition (As Per Revised ICAR Syalbus) (PB)	K.V.Peter	9789383305735	2015
25	Basics of Mutation Breeding	Thirugnanakumar	9789383305193	2014
26	Basics of Research Methodology	Ekwal Imam	9789383305438	2015
27	Biochemistry, Molecular Biology and Biotechnology: Instant Notes	H.P.Gazera	9789383305520	2015
28	Bioinformatics in Agriculture: Tools and Applications	M.Balakrishnan	9789381450925	2014
29	Biostatistics: Basic Concepts and Methodology	K.S.Kushwaha	9789383305469	2014
30	Biotechnology in Horticulture: Methods and Applications	Peter, K.V.	9789381450918	2014
31	Botanicals as Ecofriendly Pesticides	Mahulikar, P.P.	9788189422578	2007

S.No.	Title	Author	ISBN	Year
32	Breeding and Biotechnology of Flowers (Set of 2 Vols.) Set Price	A.K.Singh	9789383305612	2014
33	Breeding and Biotechnology of Flowers: Vol.01: Commercial Flowers	A.K.Singh	9789383305353	2014
34	Breeding and Biotechnology of Flowers: Vol.02: Garden Flowers	A.K.Singh	9789383305407	2014
35	Breeding and Protection of Vegetables	Rana, M.K.	9789380235493	2011
36	Breeding of Horticultural Crops: Principles and Practices: 2nd Revised and Expanded Edition	N.Kumar	9789383305773	2015
37	Breeding,Biotechnology and Seed Production of Field Crops	Roy, Bidhan	9789381450680	2013
38	Climate Change and Chemicals: Environmental & Biological Aspects	Kibria, Golam.	9789380235301	2010
39	Climate Change and Food Security	Datta, M.	9788189422387	2008
40	Climate Change and Natural Resources Mgt.	Lenka, S.	9789381450673	2013
41	Climate Mitigation and Carbon Finance: Global Initiavities & Challenges	Sahoo, A.K.	9789381450024	2012
42	Climate Resilient Crops for the Future	K.V.Peter	9789383305599	2015
43	Climatic Variability: Impacts on Agriculture and Allied Sectors	Datta, M.:Ed.	9789381450949	2014
44	Clinical Veterinary Medicine: Practical Manual Series Vol 03	Sharma, Neelash	9788190851251	2009
45	Commercial Crops Technology: Vol.08. Horticulture Science Series	Kurian, A.	9788189422523	2007
46	Conducting A Successful Training Programme	Anandaraja	9789383305223	2015
47	Conservation Agriculture for Carbon Sequestration and Sustaining Soil Health	J.Somasundaram	9789383305322	2014
48	Current Trends in Global Environment	Bhatia, A.L.	9788189422004	2005
49	Cyanobacteria: Antibacterial Activity	K., Purshotam	9788190723770	2009
50	Detritus and Decomposition in Ecosystems	Reshi, Zafar	9788189422158	2007
51	Developments in Physiology,Biochemistry and Molecular Biology of Plants Vol 01	Bose, Bandana A. Hemantaranjan	9788189422028	2005
52	Developments in Physiology,Biochemistry and Molecular Biology of Plants Vol 02	Bose, Bandana & A.Hemantaranjan	9788189422929	2008
53	Dimensions of Extension Education	Mohapatra, B.P.	9789381450987	2014
54	Disasters: Strengthening Community Mitigation and Preparedness	Khanna, B.K.	9789380235455	2011
55	Diseases of Horticultural Crops Identification and Management	Sanjeev Kumar	9789383305643	2015
56	Drug Discovery and Development: Traditional Medicine and Ethnopharmacology	Patwardhan Bhushan	9788189422295	2007
57	Ecologically Based Integrated Pest Management (Set of 2 Vols.) (Set Price)	Abrol,Y.P.	9789380235950	2012
58	Economics,Marketing and Sales of Agricultural Products	Thakur, S.N.	9789381450659	2013
59	Elements of Entomology	H.Lewin Devasahayam	9789381450635	2014
60	Elements of Food Science	Jan, Safia	9789381450246	2013
61	Emerging Technologies of the 21st Century	A.K.Roy	9789383305339	2015
62	Environmental Changes and Natural Disasters	Babar, Md.	9788189422752	2007

S.No.	Title	Author	ISBN	Year
63	Essential Oils and their Applications	Kuntal Das	9789381450741	2013
64	Ethnomedicinal Plants Resource of Orissa: Vol.01	Sahoo, A.K.	9789380235752	2011
65	Experimental Phytochemical Techniques	Raaman, N.	9789380235943	2012
66	Extension of Technologies: From Labs to Farms	Anandaraja, N.	9788189422837	2008
67	Flower Crops: Cultivation and Management	Singh, A.K.	9788189422356	2006
68	Flowering Trees: Vol.12. Horticulture Science Series	Valsalakumari, P.K.	9788189422509	2008
69	Flowers for Trade: Vol.10. Horticulture Science Series	Sheela, V.L.	9788189422516	2008
70	Food and Nutritonal Security by Sustainable Agriculture	Mishra, K. Bijesh	9789383305049	2014
71	Food Science	Bawa, A.S.	9789381450147	2013
72	Fruit Crops: Vol.03. Horticulture Science Series	Radha, T.	9788189422462	2007
73	Fundamentals of Garden Designing: A Colour Encyclopedia	Roy, Rup Kumar	9789381450307	2013
74	Fundamentals of Ornamental Horticulture and Landscape Gardening	Tiwari, A.K.	9789381450079	2012
75	Fundamentals of Vegetable Production	Rana, M.K.	9789380235707	2011
76	Genetic Resources and Seed Enterprises: Management and Policies: In 2 Parts (Set Price)	Ram, Hari Har & R.Yadava	9788189422653	2007
77	Geographic Information System	Gurugnanam, B.	9788190851282	2009
78	Geospatial Technologies for Natural Resources Management	Soam, S.K.	9789381450802	2013
79	GIS: Fundamentals,Applications and Implementations	Elangovan, K.	9788189422165	2006
80	Green Agriculture: Newer Technologies	Behera, K.K.	9789381450277	2012
81	Handbook of Genetics and Biotechnology: 2nd Revised and Expanded Edition (PB)	R.S. Tomar	9789383305445	2014
82	Heterosis Breeding in Vegetable Crops	Rai, Nagendra	9788189422035	2006
83	Hill Agriculture: Economics and Sustainability	Sharma, Pawan	9789381450871	2014
84	Horticulture Science Series: Vol.01-12: Set of 12 Vols. in 12 Parts(Set Price)	Peter, K.V.	9788189422479	2008
85	Human Resource Development	Sannigrahi, A. K.	9789380235769	2011
86	ICTs for Agricultural Extension: Global Experiments,Innovations and Experiences	Saravanan, R. ed.	9789380235240	2010
87	Identification and Management of Horticultural Pests	Ranjith, A.M.	9789381450567	2013
88	Illustrated Plant Pathology: Basic Concepts	Darwin, Henry	9789380235080	2009
89	Indigenous Medicinal Plants and their Practical Utility	Lakshman, H.C.	9789381450116	2012
90	Information and Communication Technology for Agriculture and Rural Development	Saravanan, R.	9789380235882	2011
91	Information Technology,Plant Pathology and Biodiversity: Indian Print	Bridge, P.D.	9788189422080	2005
92	Innovations in Agribusiness Management	Lokanadan, K.	9789380235097	2009
93	Innovative Horticulture	Arunkumar, K.	9788189422738	2008
94	Intellectual Property Rights Demystified (PB)	Ramkumar, Mu.	9788189422875	2008
95	IPR: Drafting,Interpretation of Patent Specifications and Claims	Rathore,N.S.: Ed.	9789381450819	2013
96	Irrigation and Agricultural Drainage Engineering	Ranajit K. Biswas	9789383305247	2015
97	Management of Horticultural Crops: Vol.11 Horticulture Science Series	Pradeepkumar,T.	9788189422493	2008

S.No.	Title	Author	ISBN	Year
98	Mechanization of Cultivated Crops	Surendra Singh	9789383305759	2015
99	Medicinal Plants: Holistic Approaches	Zawar, Nudrat Sayed	9789380235585	2011
100	Medicinal Plants: Vol.02. Horticulture Science Series	Kurian, A. & A.Sankar	9788189422424	2007
101	Merging Plant Breeding with Crop Biotechnology	Yasin, J.K.:Ed.	9789381450598	2012
102	Methods and Techniques in Plant Physiology	A.Bhattacharya	9789383305506	2015
103	Microbial Biotechnology for Sustainable Agriculture, Horticulture and Forestry	Bagyaraj, D.J.	9789380235820	2011
104	Microbial Diversity and Functions	Bagyaraj, D.J.	9789381450109	2012
105	Microbial Diversity and Its Applications	Barbudde, S.B.	9789381450666	2013
106	Mobile Phones for Agricultural Extension	Saravanan, R	9789383305230	2014
107	Modern Biotechnology and Its Applications (Set of 2Vols.) Set Price	Behera, K.K.	9789381450833	2013
108	Modern Technologies for Sustainable Agriculture	Kumar, Sunil	9789381450611	2013
109	Modern Technology in Vegetable Production	Hazra, Pranab	9789380235325	2011
110	Nanotechnology in Soil Science and Plant Nutrition	Adhikari, Tapan	9789381450789	2013
111	Nutritional Disorders in Fruit Crops: Diagnosis and Mngt.	Prakash, M.	9789381450956	2013
112	Objective Agricultural Science (PB)	Brajendra	9789381450222	2012
113	Objective Agriculture Statistics (PB)	Kushwaha, K.S.	9789380235615	2011
114	Objective Agronomy (PB)	Vishwakarma,A.	9789380235127	2010
115	Objective Biochemistry (PB)	Sengar,R.S.	9789381450376	2013
116	Objective Horticulture (PB)	Verma, Anil	9789381450239	2012
117	Oilseeds: Properties,Products,Processing and Procedures	Nagaraj, G.	9788190723756	2009
118	Orchids: Cultivation and Management	Tiwari, A.K.	9789383305001	2014
119	Organic Spices	Parthasarathy, V.A	9788189422844	2008
120	Ornamental Plants	Sabina, George	9788190851268	2009
121	Pest Management and Residual Analysis in Horticultural Crops	Gulati Rachna	9789381450710	2013
122	Phal Evam Sabji Parirakshan Digdarshika (PB)	Kumari, Arunima	9789380235561	2011
123	Physio-Biochemistry and Biotechnology of Vegetable Crops	Rana, M.K.	9789380235318	2011
124	Physiological Disorders of Fruit Crops	Savreet, Sandhu	9789381450581	2013
125	Phytochemical Techniques	Raaman, N.	9788189422301	2006
126	Phyto-nematodes in Crops: Their Identification, Treatment and Management	Sharma, G.L.	9789380235837	2011
127	Plant Diseases: Identification and Management (With Illustrations)	J.P.Rai	9789383305315	2014
128	Plant Taxonomy and Biosystematics: Classical and Modern Methods	T.S.Rana	9789383305414	2014
129	Postharvest Management and Processing of Fruits and Vegetables: Instant Notes (PB)	Sharma, Satish	9789380235202	2010
130	Postharvest Techniques and Management of Dry Flowers	V.Ponnuswami	9789380325868	2011
131	Postharvest Technologies for Commerical Floriculture	Verma, Anil	9789381450048	2012
132	Postharvest Technology and Engineering: An Illustrated Guide	Dev Raj	9789381450451	2012
133	Postharvest Technology and Processing of Horticultural Crops	P.S.Pandit	9789383305278	2014

S.No.	Title	Author	ISBN	Year
134	Postharvest Technology of Horticultural Crops: Practical Manual Series Vol 02	Sharma, Satish	9788190851206	2009
135	Postharvest Technology of Horticultural Crops: Vol.07. Horticulture Science Series	Sudheer, K.P.	9788189422431	2007
136	Practical Manual of Horticultural Crops (Set of 2 Vols.)	Verma Anil	9789383305711	2015
137	Practical Manual of Horticultural Crops: Vol.01: Production Technology	Verma Anil	9789383305704	2015
138	Practical Manual of Horticultural Crops: Vol.02: Processing and Post Harvest Technology	Verma Anil	9789383305698	2015
139	Precision Farming in Horticulture	Singh, Jitendar	9789381450475	2013
140	Propagation of Horticultural Crops: Vol.06. Horticulture Science Series	Rajan, S.	9788189422486	2007
141	Propagation of Horticultural Plants: Arid and Semi-Arid Regions	R.S.Singh & R.Bhargava	9789383305254	2014
142	Protected Cultivation of Horticultural Crops	Singh, D.K	9789383305155	2014
143	Quantitative Genetics and Crop Breeding	Thirugnanakumar	9789380235981	2012
144	Remember Your Humanity: Pathway to Sustainable Food Security	Swaminathan,M.S.	9789381450178	2012
145	Samekit Krishi Pranali (Hindi)	Kumar, Sanjeev	9789381450154	2012
146	Seed Production of Field Crops	Mondal, S.S.	9788190723763	2009
147	Seed Science and Technology	Vanangamudi,K.	9789383305117	2014
148	Soil Sampling and Methods of Analysis	Pal, Sushant	9789381450574	2013
149	Soil Testing and Analysis: Plant, Water and Pesticide Residues	Brajendra, Patiram	9788189422707	2007
150	Spices: Vol.05. Horticulture Science Series	Nybe, E.V.	9788189422448	2007
151	Statistical Designs and Analysis for Agricultural Field Experiments	Katyal, Vijay	9789381450840	2013
152	Sustainable Environmental Science	Sahu, D.D.	9789381450208	2012
153	Systematics of Fruit Crops	Sharma, Girish	9789380235066	2009
154	Technologies for Sustainable Green Environment	Davamani, V.	9789381450420	2012
155	Temperate Horticulture: Current Scenario	Kishore, D.K.	9788189422363	2006
156	The Basics of Human Civilization: Food, Agriculture and Humanity (Vol. I)	Prem Nath	9789381450734	2013
157	The Basics of Human Civilization: Food, Agriculture and Humanity (Vol. II) Food	Prem Nath	9789383305377	2014
158	The Pomegranate	Hiwale, S.S.	9789380235158	2009
159	The Science of Horticulture Vol 01	Peter, K.V.: eds.	9789380235479	2011
160	The Science of Horticulture Vol 02	Peter, K.V.: eds.	9789380235486	2011
161	Traditional Agricultural Practices: Applications and Technical Implementations	Rathakrishnan,R.T.	9789380235028	2009
162	Tuber and Root Crops: Vol.09. Horticulture Science Series	Palaniswami, M.S.	9788189422530	2008
163	Turning Plants Into Medicines: Novel Approaches	T.Parimelazhagan	9789381450468	2013
164	Underutilized and Underexploited Horticultural Crops: Vol 01	Peter, K.V.	9788189422608	2007
165	Underutilized and Underexploited Horticultural Crops: Vol 02	Peter, K.V.	9788189422691	2007
166	Underutilized and Underexploited Horticultural Crops: Vol 03	Peter, K.V.	9788189422851	2008

S.No.	Title	Author	ISBN	Year
167	Underutilized and Underexploited Horticultural Crops: Vol 04	Peter, K.V.	9788189422905	2008
168	Underutilized and Underexploited Horticultural Crops: Vol 05	Peter, K.V.	9789380235288	2010
169	Value Additions in Flowers and Orchids	De, L.C.	9789381450123	2011
170	Vegetable Crops: Genetics Resources and Improvements	D.K. Singh	9789380235509	2012
171	Vegetable Crops: Vol.04. Horticulture Science Series	Gopalakrishnan	9788189422417	2007
172	Women in Agriculture and Rural Development	S. Shakuntala	9788189422998	2009
173	Youth in Agriculture and Rural Development	K.Narayana Gowda:Ed.	9789381450758	2013